Soil Erosion: Dust Control and Sand Stabilization, Volume II

Soil Erosion: Dust Control and Sand Stabilization, Volume II

Editor

Itzhak Katra

MDPI • Basel • Beijing • Wuhan • Barcelona • Belgrade • Manchester • Tokyo • Cluj • Tianjin

Editor
Itzhak Katra
Ben-Gurion University of the Negev
Israel

Editorial Office
MDPI
St. Alban-Anlage 66
4052 Basel, Switzerland

This is a reprint of articles from the Special Issue published online in the open access journal *Applied Sciences* (ISSN 2076-3417) (available at: https://www.mdpi.com/journal/applsci/special_issues/dust_control_sand_stabilization).

For citation purposes, cite each article independently as indicated on the article page online and as indicated below:

LastName, A.A.; LastName, B.B.; LastName, C.C. Article Title. *Journal Name* **Year**, *Volume Number*, Page Range.

ISBN 978-3-0365-6764-8 (Hbk)
ISBN 978-3-0365-6765-5 (PDF)

Contents

About the Editor . . . . **vii**

Preface to "Soil Erosion: Dust Control and Sand Stabilization, Volume II" . . . . **ix**

Itzhak Katra
Special Issue on Soil Erosion: Dust Control and Sand Stabilization (Volume II)
Reprinted from: *Appl. Sci.* **2023**, *13*, 1727, doi:10.3390/app13031727 . . . . **1**

Rattan Lal
Fate of Soil Carbon Transported by Erosional Processes
Reprinted from: *Appl. Sci.* **2022**, *12*, 48, doi:10.3390/app12010048 . . . . **3**

Rattan Lal
Biophysical Controls That Make Erosion-Transported Soil Carbon a Source of Greenhouse Gases
Reprinted from: *Appl. Sci.* **2022**, *12*, 8372, doi:10.3390/app12168372 . . . . **17**

Pua Bar Kutiel and Michael Dorman
The Importance of Annual Plants and Multi-Scalar Analysis for Understanding Coastal Dune Stabilization Process in the Mediterranean
Reprinted from: *Appl. Sci.* **2021**, *11*, 2821, doi:10.3390/app11062821 . . . . **31**

Tania L.F. Bird, Pua Bar (Kutiel), Elli Groner and Amos Bouskila
Asynchrony Drives Plant and Animal Community Stability in Mediterranean Coastal Dunes
Reprinted from: *Appl. Sci.* **2021**, *11*, 6214, doi:10.3390/app11136214 . . . . **45**

Hongchao Dun, Guowei Xin, Ning Huang, Guangtian Shi and Jie Zhang
Wind-Tunnel Studies on Sand Sedimentation around Wind-Break Walls of Lanxin High-Speed Railway II and Its Prevention
Reprinted from: *Appl. Sci.* **2021**, *11*, 5989, doi:10.3390/app11135989 . . . . **59**

Hadas Raveh-Amit, Avi Sharon, Itzhak Katra, Terry Stilman, Shannon Serre, John Archer and Matthew Magnuson
Limiting Wind-Induced Resuspension of Radioactively Contaminated Particles to Enhance First Responder, Early Phase Worker and Public Safety—Part 1
Reprinted from: *Appl. Sci.* **2022**, *12*, 2463, doi:10.3390/app12052463 . . . . **73**

Matthew Magnuson, Terry Stilman, Shannon Serre, John Archer, Ryan James, Xiaoyan Xia, et al.
Part 2: Stabilization/Containment of Radiological Particle Contamination to Enhance First Responder, Early Phase Worker, and Public Safety
Reprinted from: *Appl. Sci.* **2022**, *12*, 3861, doi:10.3390/app12083861 . . . . **87**

Justus Freer, Maximilian Lübeck, Johannes L. Sieger, Bernd G. Lottermoser and Marius Braun
Effectiveness of Food Processing By-Products as Dust Suppressants for Exposed Mine Soils: Results from Laboratory Experiments and Field Trials
Reprinted from: *Appl. Sci.* **2022**, *12*, 11551, doi:10.3390/app122211551 . . . . **111**

Johannes L. Sieger, Bernd G. Lottermoser and Justus Freer
Evaluation of Protein and Polysaccharide Biopolymers as Dust Suppressants on Mine Soils: Laboratory Experiments
Reprinted from: *Appl. Sci.* **2023**, *13*, 1010, doi:10.3390/app13021010 . . . . **131**

Meni Ben-Hur, Reut Cohen, Michael Danon, Uri Nachshon and Itzhak Katra
Evaluation of Groundwater Salinization Risk Following Application of Anti-Dust Emission Solutions on Unpaved Roads in Arid and Semiarid Regions
Reprinted from: *Appl. Sci.* **2021**, *11*, 1771, doi:10.3390/app11041771 . . . . **169**

Aviv Rubinstein, Meni Ben-Hur and Itzhak Katra
Dust Emission Thresholds in Loess Soil Under Different Saltation Fluxes
Reprinted from: *Appl. Sci.* **2020**, *10*, 5949, doi:10.3390/app10175949 . . . . **181**

About the Editor

Itzhak Katra

Itzhak Katra received his PhD degree at the Department of Geography, Bar Ilan University, Israel. He was a postdoctoral fellow at the Division of Earth Sciences and Ecosystem Sciences, Desert Research Institute (DRI), Nevada, USA. Katra is a Professor at the Department of Geography and Environmental Development, Ben-Gurion University of the Negev, Israel, and is the Head of the Aeolian Simulation Laboratory. The overriding aim of his research is to understand the dynamic processes of soil erosion in arid areas, sand transport, dust emission, and related air pollution. Katra has published papers in journals of earth sciences, environmental sciences, multidisciplinary sciences, and applied sciences.

Preface to "Soil Erosion: Dust Control and Sand Stabilization, Volume II"

Soil erosion caused by wind is significant to Earth systems and human health. Soil-derived dust particles with origins in various source areas constitute one of the major components of global aerosols. There is a strong interest in understanding the factors and processes of soil erosion caused by wind as well as in developing and applying methods to control dust emission from soils and to stabilize active sands. This Special Issue contains information on applications of natural and synthetic materials to reduce soil erosion, development of materials and methods, experimental methods and modeling, impacts on soil quality and environments, and quantification of the efficiency in dust control and sand stabilization applications. Eleven papers were accepted for publication: two review papers and nine research papers, one of which was selected as a featured paper.

Itzhak Katra
Editor

applied sciences

MDPI

Editorial

Special Issue on Soil Erosion: Dust Control and Sand Stabilization (Volume II)

Itzhak Katra

Department of Geography and Environmental Development, Ben Gurion University of the Negev, Beersheba 8410501, Israel; katra@bgu.ac.il

1. Introduction

This is the second volume of the Special Issue on Soil Erosion: Dust Control and Sand Stabilization, following the first volume published in 2020 [1]. Many soils throughout the world are subjected to the impacts of extensive land use on soils, including agricultural areas, unpaved roads, mines and quarries, waste soils, and active sand dunes and sand sheets. There is a strong interest in understanding the factors and processes of soil erosion by wind, as well as in developing and applying methods to control dust emissions from soils and to stabilize active sands.

Eleven papers were accepted for publication: two review papers and nine research papers, one of which was selected as a featured paper. The review papers by Lal provide us with a complete picture of soil carbon transported by erosional processes. One review paper [2] deliberates the fate of soil carbon due to soil erosion by wind and water being transported over the landscape. The second review paper [3] concerns the biophysical controls that make erosion-transported soil carbon a source of greenhouse gases. Conversely, three of the research papers focus on sand stabilization. The featured paper, by Bar-Kutiel and Dorman [4], demonstrates the importance of studying vegetation at multiple entity-defined scales through analyses of the annual vegetation on coastal Mediterranean sand dunes. The paper by Bird et al. [5] concerns the effects of environmental degradation on ecosystem stability and productivity, which have destabilizing consequences beyond biodiversity loss in Mediterranean coastal dunes. Dun et al. [6] explored sand transport around a high-speed railway using three models of wind-break walls in a wind tunnel experiment with the reproduced movement process of windblown sand. The other papers in this Special Issue focus on dust emissions from soils and control methods. The papers of Raveh-Amit et al. [7] and Magnuson et al. [8] are two parts of a project on limiting dust re-suspension by wind and soil stabilization of radiologically contaminated surfaces. Another research project is represented here by the two papers of Freer et al. [9] and Sieger et al. [10] on the effectiveness of biopolymers from food processing by-products as dust suppressants in mine soils through laboratory and field experiments. The environmental risk of applying dust suppressants in soils was investigated by Ben-Hur et al. [11], providing an evaluation of groundwater salinization risk following application of brine on calcareous soil in an arid region. The last paper on dust, by Rubinstein et al. [12], is a study on the role of soil particle size distribution in dust emission rates in Loess soils through laboratory wind tunnel experiments.

A wide range of dust control products have been tested for soil stabilization and dust control. However, there is a need to investigate the possible environmental impacts of diverse dust suppression substances, including the toxicity of atmospheric particulate matter, when dust is emitted from the treated soils.

Citation: Katra, I. Special Issue on Soil Erosion: Dust Control and Sand Stabilization (Volume II). *Appl. Sci.* **2023**, *13*, 1727. https://doi.org/10.3390/app13031727

Received: 18 January 2023
Accepted: 21 January 2023
Published: 29 January 2023

Funding: This research received no external funding.

Acknowledgments: This Special Issue is the result of the long-term efforts of the authors, the reviewers, and the editorial team. Special thanks to the Section Managing Editor, Applied Sciences, MDPI.

Conflicts of Interest: The authors declare no conflict of interest.

References

1. Katra, I. Soil Erosion: Dust Control and Sand Stabilization. *Appl. Sci.* **2020**, *10*, 8044. [CrossRef]
2. Lal, R. Fate of Soil Carbon Transported by Erosional Processes. *Appl. Sci.* **2022**, *12*, 48. [CrossRef]
3. Lal, R. Biophysical Controls That Make Erosion-Transported Soil Carbon a Source of Greenhouse Gases. *Appl. Sci.* **2022**, *12*, 8372. [CrossRef]
4. Bar Kutiel, P.; Dorman, M. The Importance of Annual Plants and Multi-Scalar Analysis for Understanding Coastal Dune Stabilization Process in the Mediterranean. *Appl. Sci.* **2021**, *11*, 2821. [CrossRef]
5. Bird, T.L.F.; Bar, P.; Groner, E.; Bouskila, A. Asynchrony Drives Plant and Animal Community Stability in Mediterranean Coastal Dunes. *Appl. Sci.* **2021**, *11*, 6214. [CrossRef]
6. Dun, H.; Xin, G.; Huang, N.; Shi, G.; Zhang, J. Wind-Tunnel Studies on Sand Sedimentation Around Wind-Break Walls of Lanxin High-Speed Railway II and Its Prevention. *Appl. Sci.* **2021**, *11*, 5989. [CrossRef]
7. Raveh-Amit, H.; Sharon, A.; Katra, I.; Stilman, T.; Serre, S.; Archer, J.; Magnuson, M. Limiting Wind-Induced Resuspension of Radioactively Contaminated Particles to Enhance First Responder, Early Phase Worker and Public Safety—Part 1. *Appl. Sci.* **2022**, *12*, 2463. [CrossRef]
8. Magnuson, M.; Stilman, T.; Serre, S.; Archer, J.; James, R.; Xia, X.; Lawrence, M.; Tamargo, E.; Raveh-Amit, H.; Sharon, A. Part 2: Stabilization/Containment of Radiological Particle Contamination to Enhance First Responder, Early Phase Worker, and Public Safety. *Appl. Sci.* **2022**, *12*, 3861. [CrossRef]
9. Freer, J.; Lübeck, M.; Sieger, J.L.; Lottermoser, B.G.; Braun, M. Effectiveness of Food Processing By-Products as Dust Suppressants for Exposed Mine Soils: Results from Laboratory Experiments and Field Trials. *Appl. Sci.* **2022**, *12*, 11551. [CrossRef]
10. Sieger, J.L.; Lottermoser, B.G.; Freer, J. Evaluation of Protein and Polysaccharide Biopolymers as Dust Suppressants on Mine Soils: Laboratory Experiments. *Appl. Sci.* **2023**, *13*, 1010. [CrossRef]
11. Ben-Hur, M.; Cohen, R.; Danon, M.; Nachshon, U.; Katra, I. Evaluation of Groundwater Salinization Risk Following Application of Anti-Dust Emission Solutions on Unpaved Roads in Arid and Semiarid Regions. *Appl. Sci.* **2021**, *11*, 1771. [CrossRef]
12. Rubinstein, A.; Ben-Hur, M.; Katra, I. Dust Emission Thresholds in Loess Soil Under Different Saltation Fluxes. *Appl. Sci.* **2020**, *10*, 5949. [CrossRef]

applied sciences

Review

Fate of Soil Carbon Transported by Erosional Processes

Rattan Lal

Rattan Lal Center for Carbon Management and Sequestration, School of Environment and Natural Resources, The Ohio State University, Columbus, OH 43210, USA; lal.1@osu.edu; Tel.: +1-614-292-9069

Abstract: The accelerated process of soil erosion by water and wind, responsible for transport and redistribution of a large amount of carbon-enriched sediments, has a strong impact on the global carbon budget. The breakdown of aggregates by erosivity of water (raindrop, runoff) and wind weakens the stability of soil C (organic and inorganic) and aggravates its vulnerability to degradation processes, which lead to the emission of greenhouse gases (GHGs) including CO_2, CH_4, and N_2O, depending on the hydrothermal regimes. Nonetheless, a part of the eroded soil C may be buried, reaggregated and protected against decomposition. In coastal steep lands, (e.g., Taiwan, New Zealand) with a short distance to burial of sediments in the ocean, erosion may be a sink of C. In large watersheds (i.e., Amazon, Mississippi, Nile, Ganges, Indus, etc.) with a long distance to the ocean, however, most of the C being transported is prone to mineralization/decomposition during the transit period and is a source of GHGs (CO_2, CH_4, N_2O). Land use, soil management and cropping systems must be prudently chosen to prevent erosion by both hydric and aeolian processes. The so-called plague of the soil, accelerated erosion by water and wind, must be effectively curtailed.

Keywords: global warming; soil erosion; carbon erosion; gaseous emissions; methanogenesis; nitrification/denitrification; dust; enrichment ratio; burial of carbon; aggregate disruption

Citation: Lal, R. Fate of Soil Carbon Transported by Erosional Processes. *Appl. Sci.* **2022**, *12*, 48. https://doi.org/10.3390/app12010048

Academic Editor: Itzhak Katra

Received: 27 October 2021
Accepted: 15 December 2021
Published: 21 December 2021

1. Introduction

Accelerated soil erosion by water and wind is a predominant process that impacts the soil carbon (C) budget [1] through a range of interacting processes [2]. The severe problem of accelerated soil erosion is widely regarded as a global menace, which is also threatening the agricultural resource base [3] that sustains planetary processes [4] and generates numerous ecosystem services. Erosional processes include breakdown of structural aggregates, and selective removal and redistribution of the sediment and displaced C on the landscape. Displaced C may be mineralized or buried [5] depending on the site-specific conditions. Whereas reaggregation and deep burial may stabilize the displaced C against microbial processes [6], decomposition induced by aggregation breakdown and alterations in soil moisture and temperature regimes over the landscape may affect the emission of carbon dioxide (CO_2) under aerobic environments [7] and those of methane (CH_4) and nitrous oxide (N_2O) under anaerobic conditions [8]. Therefore, erosion-induced transport of soil C followed by its redistribution en route to depositional/depression site(s) may lead to emission of all three greenhouse gases (GHGs). With a large amount of sediment and sediment-laden C being transported over the landscape, and the magnitude of sediment being increased because of anthropogenic activities, it is critically important to understand the complex and interactive processes affecting the fate of C being displaced and transported over the landscape (e.g., sediment) and emission of GHGs to the atmosphere (i.e., windborne dust). Emissions of all three GHGs by erosional processes, at all landscape positions from summit to foot slope and into the atmosphere, must be understood, quantified, and accounted for the purpose of compiling the soil C budget under diverse land use and soil management systems.

Novara et al. [9] reported that transport of soil sediments by erosional processes increased SOC mineralization by 43% under Mediterranean conditions. The global menace of soil erosion (hydric and aeolian) is projected to be aggravated by climate change because of the increase in climatic erosivity [10] and the frequency of extreme events [11,12]. Therefore, the objective of this review article is to deliberate the fate of soil C being transported over the landscape, identify ecosystem conditions that lead to either mineralization and emission of GHGs or to sequestration of the transported C and its stabilization, understand the net effect of erosion/deposition processes as a source or sink of GHGs, explain the effect of global warming on erosional processes, and outline some researchable priorities. Rather than being a comprehensive and inclusive review of the available literature, this article merely provides some pertinent examples of the processes, factors and causes which control the fate of carbon being transported by erosional processes.

2. Selectivity of the Soil Erosion Process

Global soil erosion by water and wind, exacerbated by agricultural expansion and intensification and other anthropogenic activities, transports, and redistributes large amounts of soil organic C (SOC) and soil inorganic carbon (SIC) with significant impacts on global C budget, atmospheric chemistry, and water quality. However, hydric erosion may have a more significant impact on transport of SOC and emission of GHGs. Indeed, soil erosion by water is a multistage process encompassing a range of pedological processes. Some of these processes are described below:

i. **Breakdown of aggregates** and leading to the exposure of the hitherto protected SOC against microbial processes and environmental conditions, is the first process. Macroaggregates, containing labile fractions, are broken apart by the kinetic energy and momentum of the impacting raindrops and velocity of the runoff flow. The SOC thus exposed is accessible to microbial processes.

ii. **Removal of the colloidal and light fractions** (clay, fine silt, SOC, SIC, sesquioxides) is aggravated by the breakdown of aggregates. Thus, the sediments have a high C-enrichment ratio (CER). The latter is defined as the ratio of C in sediment to that in the original soil from which the sediment are derived [13]. A high CER is observed in both hydric and aeolian sediments.

iii. **Redistribution of the sediment and associated soil C** is a major pedological process on actively eroding landscapes. Furthermore, redistribution is accompanied by possible mineralization of biomass-C (into CO_2, CH_4, and N_2O), depending on the hydrothermal regimes, Whereas CO_2 is the primary GHG evolved and emitted under aerobic conditions, CH_4 and N_2O may be evolved under anaerobic environments.

iv. **Deposition of sediments** SOC-laden sediments are deposited at depressional and other sites following Stokes Law. Accordingly, heavier fractions (e.g., gravels, sand and coarse silts are deposited in vicinity of the eroding sites and SOC and clay fractions are deposited either latter or carried farther away. Windblown sediments originating from the Sahara have been observed in the Caribbean and northern Europe. This process is the principal cause of eutrophication of water, pollution of air and responsible for the off-site adverse effects of the erosional process.

v. **Burial of SOC:** and SIC deposition following Stokes Law leads to burial and stratification, with coarser heavier sediments deposited first and the lighter fractions (SOC, SIC, clay particles) later. With multiple events over time, the layering can be observed in active depositional sites which are not disturbed by farm operations or natural perturbations.

vi. **Reaggregation** Some of the dispersed clay, released by the breakdown of aggregates (step 1) and other pedological processes aggravating the slaking of structural units, may interact with SOC and cations and the reformation of aggregates. Therefore, some of the buried SOC may be stabilized through encapsulation within reformatted aggregates.

vii. **Gaseous Emission from Eroding and Depositional Sites**: Erosional processes may aggravate emissions of GHGs from both eroding and depositional sites because of the drastic perturbations of soil structure. The magnitude and type of gaseous emission (CO_2, CH_4, and N_2O), depend on site-specific conditions and their spatial and temporal variations because of natural and anthropogenic processes.

However, pedological processes involved in the above listed erosional stages vary widely, depending on the stage of the erosion/deposition process and prevalence of the specific hydrothermal regime of the specific landscape position. The third stage, involving redistribution of soil C, is not uniform but affected by preferential transport and deposition. The site of SOC deposition depends on topography (slope gradient, length, shape and aspect), in-transit distance of sediment containing SOC [14] and the time taken for the sediments to reach the deposition site. Because of the selective removal of the light fraction, CER of the sediments may range from 1.3 to 40.0 [6]. Furthermore, the average SOC fluxes as particles (4.7 g $C/m^2 \cdot yr$) was found to be 18 times larger than that of the dissolved organic C (DOC). Wang and colleagues also observed that the cumulative emission of soil CO_2 slightly decreased at the erosion site but increased by 56% and 27% at the transport and depositional zone, respectively, in comparison to non-eroded soils. Site-factors (i.e., slope gradient, slope aspect, slope shape, and the landscape position) are the control of specific biogeochemical and biogeophysical processes that determine transformation of soil and emission of specific GHGs such as CO_2 by aerobic mineralization, CH_4 by methanogenesis catalyzed by anaerobiosis (poor drainage and high degree of saturation), and N_2O by nitrification/denitrification processes under variable hydrothermal regimes.

3. Soil Erosion as a Source or Sink of Carbon

There are numerous interacting processes affecting the fate of SOC and SIC transported by erosional processes. The net effect as a source or sink depends on the balance of emission vs. sequestration, as outlined in Figure 1. Factors which increase the source from erosion-induced transport include disruption/breakdown of aggregates, a selective removal of the low-density/lighter fractions, fast-transport landscape vs. slow-transport landscape, and severe degradation of the soil at the eroded site so that its agronomic productivity is curtailed (Figure 1). There could be deposition in short-distance anaerobic lakes that would be a source of carbon, and in long-distance aerobic environment with reaggregation at depositional sites leading to sequestration of C. Some conditions which may lead erosion-induced transport of soil C as a sink may include a fast transport landscape with short transit time to the deposition site, deep burial of C, rapid reaggregation of buried C so that it is protected against microbial processes, and slight or moderate degradation of soil at the eroded site so that its agronomic productivity is not jeopardized (Figure 1).

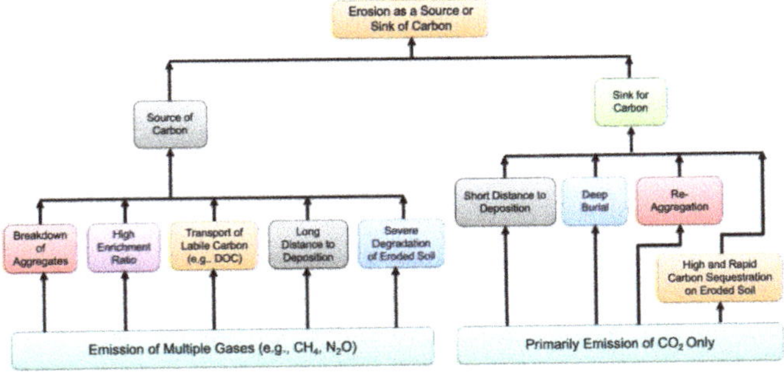

Figure 1. Pedological processes which impact emission of greenhouse gases from erosion-induced transport of soil carbon. Available literature highlights the importance of soil, climate and land attributes which aggravate the breakdown of aggregates and severely degrade the quality and productivity of eroded soil which has a long recovery period. Some steep land coastal ecoregions with a short transit time can be a sink of eroded soil carbon. Globally, however, accelerated erosion (hydric, aeolian, tillage) is a source of greenhouse gases especially those with high global warming potential (i.e., CH_4, N_2O).

Some critical factors which impact the gaseous processes during the transit phase and at the depositional site are briefly described below.

3.1. Slope

Attributes of land slope that determine specific pedotransformation include gradient (steepness), aspect (north vs. south facing), shape (convex vs. concave), position (summit, side slope, foot slope), and length (short vs. long). The magnitude of sediment generated, and the amount of soil C transported laterally also depend on the soil surface as altered by agronomic management and its interaction with the slope attributes. While assessing the scale-dependency of erosion-induced variation in CO_2 emission from terraced slopes, Hu et al. [15] observed that emissions tend to decline with increase in slope gradient. The declining trend is attributed to increased runoff and more soil erosion on steeper slopes and making eroded soil less habitable for soil biota in terms of their activity and species diversity.

Slope attributes also affect the deposition of sediments and the associated soil C. Lense et al. [16] evaluated losses of soil C by water erosion in a tropical watershed characterized by a wide land-use diversity. Lense and colleagues observed that of the 126.5 Mg/yr. of the total eroded SOC from the watershed, 111.6 Mg/yr. were deposited in relief depressions and only 14.9 Mg/yr. reached the water body system. Wei et al. [17] observed that bulk soil samples obtained from the summit landscape position emitted the greatest cumulative CO_2-C (0.49 ± 0.04 g C/kg) among all the landscape positions. Furthermore, the CO_2 emission rate from coarse-sized aggregate fraction at depositional slope positions (toe-slope and foot-slope) was significantly greater than those from the eroded slope positions (summit, shoulder slope and back slope).

Erosion-induced losses of soil C and the attendant CO_2 emissions from a loess and black soil in China were studied by Gao et al. [18] for three slope gradients (5°, 15°and 25°) using a rainfall simulator. On average, SOC loss from loess soil was about 1.8 times that from black soil although the SOC concentration in original black soil was 56% higher than that in the loess soil. The cumulative CO_2 emissions from the eroding slopes of the loess ranged from 15.4 to 19.7 g C/m^2 compared with 28.1 to 59.6 g C/m^2 for the black soils. However, when the slope gradient increased from 5° to 25°, the cumulative soil CO_2 emission decreased by 23.8% on black soil and 12.6% on loess soil. These observations documenting decrease in on-site emissions from steep slopes are similar to those by Hu et al. [15] discussed above.

3.2. Soil Attributes

As is documented by Gao et al. [18] regarding the difference in SOC loss from black soil vs. a loess soil, erosion-induced emissions of GHGs are also strongly affected by soil physical, chemical, and biological properties, which are indeed strong moderators of the rate and total magnitude of GHGs emissions. Soil hydrological properties, as affected by water repellency, also impact emissions of GHGs. In the context of the western U.S. wildfires, which have increased in intensity and scale because of the climate change, Samburova et al. [19] assessed the impact of the fire-induced soil water repellency on GHG emissions. The water repellency was characterized by: water drop penetration time (WDPT), effective contact angle and relative sorptivity of exposed silica sand (used as soil surrogate). All sand samples, exposed to either fire emissions or fulvic acid were characterized by WDPT > 81 s, effective contact angle of 78–87°, and relative sorptivity between 0.31 and 0.49 compared with untreated sand samples with WDPT < 0.5 s, effective contact angle of 48°, and relative sorptivity of 1.

Soil attributes affect emissions of GHGs from eroding and depositional sites through their effects on SOC stability. Based on an experiment on SOC stability in sub-tropical China, Nie et al. [20] observed that eroding and depositional sites had different SOC contents but had similar organic functional groups. However, SOC in eroded soils was more stable than that in depositional soils in the surface 0–5 cm and 5–10 cm layers only. Differences in SOC stability were attributed to differences in soil properties such as texture, bulk density, pH, SOC content, DOC, Fe content, soil aggregates, depth, erosion, and deposition.

A study on the effects of soil properties on water erosion and emissions of GHGs on soil under diverse land uses in Spain by Gispert et al. [21] showed that soils with a higher SOC content showed proportionally lesser CO_2 emissions and were a C sink. In comparison, the shallowness of the soil profile, overgrazing, and frequent fire occurrence were elements that threatened soil ecosystem integrity and must be addressed.

3.3. Hydro-Thermal Regime and Aggregation

Emission of GHGs depends on soil moisture and temperature regimes, as moderated by aggregation, in relation to stability of aggregates along with size and continuity of pores, through their strong impacts on activity and species diversity of soil biota. For example, based on a 19-day simulated depositional experiment using three soils of contrasting texture (Ultisol, Mollisol, and Entisol) and two moisture regimes, Mao et al. [22] observed that deposition increased C emissions under both drying and wetting conditions for the Ultisol and Entisol, but the trend in Mollisol varied with the soil moisture regime. Gaseous emissions increased under drying but decreased under wetting conditions. Thus, the effect of deposition on GHG emissions depend on the soil moisture regime and its control such as texture, clay minerals, SOC content and aggregation.

Soil aggregation (structure) is a key attribute in stabilization of SOC [23]. In addition, degree and stability of soil aggregates is an important factor affecting emission of GHGs. An incubation study by Wei et al. [17] involving soil aggregates obtained from diverse landscape positions indicated that CO_2 emissions from coarse-size aggregate fractions (0.024 ± 0.009 g C/kg·d) was six times higher than that from small size aggregate fractions (0.0038 ± 0.0011 g C/kg·d) at the depositional toe slope position. Wei and colleagues concluded that the breakdown of aggregates (stage 1 of the erosional process) by accelerated erosion impacts both aggregate size distribution and CO_2 emissions from aggregates at different slope positions.

Soil aggregates, being transported by fluvial processes, may increase settling velocities of mineral particles contained within the aggregate and skew the distribution of SOC along the slope. Hu et al. [14] observed that redeposition of eroded SOC within terrestrial ecosystems increased by 64% considering the actual aggregate size, and the CO_2 emission rates also differed across settling fractions depending on soil type. Hu and colleagues also observed that over a 50-day incubation, CO_2 emissions from sediment was 114% greater than that from the non-eroded soil, probably because of enrichment with the labile fractions.

Chaplot and Cooper [24], based on their study comprising 24 locations of a typical hillslope of the South African Highveld with a wide range of soil texture, concluded that the increase in aggregate stability resulted in a significant increase in particulate and DOC concentrations in the eroded sediments and in losses of gaseous organic C. In contrast, however, high aggregate stability induced low total loss of particulate and DOC because of increase in water infiltration rate and reduced transport in runoff. In the context of wind erosion risks of Histosols in the U.S., Zobeck et al. [25] observed that variations in sediments and dust emissions is linked to soil properties. A study based on simulated rainfall in subtropical China by Huang et al. [26] indicated that soil bulk density, moisture content, and DOC were the major factors controlling erosion-induced SOC mineralization.

3.4. Other Factors Affecting Long-Term Biogeochemical and Biogeophysical Transformation

Stability of erosion-induced transport of SOC and its fate as a source or sink of GHGs is governed by factors which impact its long-term biogeochemical and biogeophysical transformation. Notable among these are elevation, vegetation, land use, frequency and intensity of natural or managed fires, and soil biodiversity in relation to activity and species diversity of macro, meso and micro fauna and flora. In addition to biochemical processes, biogeophysical transformation of aggregates (e.g., breakdown or slaking vs. reaggregation of the dispersed materials) is an important process governing the fate of SOC transported by erosional process as a source of or sink of GHGs (1,2,5,6,7,14,19). Apparently, scale is also an important variable affecting the net balance of SOC in relation to erosional processes (15) Factors elevating soil moisture regime may also aggravate C loss from mineral soils (26).

4. Soil Erosion and the Global Carbon Budget

The magnitude of sediment displacement globally by water erosion increased from 14.0 Gt/yr. during the prehuman era to 36.6 Gt/yr. at present in the absence of reservoir trapping [27]. There has also been a strong increase in windblown sediments due to the expansion of agriculture and other anthropogenic activities. Despite the severe environmental consequences of water erosion, its effects on the global carbon budget (GCB) have neither been adequately researched nor been accounted for.

4.1. Water Erosion

The GCB is strongly affected by accelerated erosion by water and reportedly emits ~1.1 Pg C/yr. [2]. However, the C emitted by erosion is not accounted for in the GCB. In India, Mandal et al. [28] reported that erosion transported about 114.4 Tg of C/yr. of which 34.6 Tg C is emitted into the atmosphere. Using 13C isotopic signature in abandoned Mediterranean agricultural land, Novara et al. [9] observed that previous estimates have not considered that erosion transported SOC can be released to the atmosphere as a range of GHGs (CO_2, CH_4, N_2O). The data by Worrell and colleagues from U.K. rivers showed that soil erosion is a net source of GHGs with median emission factors of 5.5, 4.4, and 0.3 Mg CO_2 eq/yr. for one Mg of fluvial C, gross C erosion, and gross soil erosion respectively. Worrell and colleagues concluded that gross soil erosion would only be a sink of both C and GHGs only if all the following criteria are met: (i) gross soil erosion rates are <91 Mg/km^2·yr., (ii) the eroded C is completely replaced by new SOM, and (iii) if less than half of the gross erosion makes it into the ocean network. Similar conclusions were arrived at by Lal et al. [5]. In this context, Müller-Nedebock and Chaplot [29] also highlighted the importance of sheet erosion and its impact on the lateral transport of SOC and the attendant impact on GCB. Based on a study involving data from 240 runoff plots over entire rainy seasons from different regions of the world, Müller-Nedebock and Chaplot reported that the media in particulate organic carbon loss (POCL) was 9.9 g C/m^2·yr. with the highest value of 10.8 g C/m^2·yr. for semi-arid soils followed by 6.4 g C/m^2·yr. for tropical soils and 1.7 g C/m^2·yr. for temperate soils. Assuming the mean POCL of 27.2 g C/m^2·yr., the total amount of SOC displaced by sheet erosion from its source was estimated at 1.32 $\pm$ 0.20 Gt C/yr. or about 13.2% of the 10 Gt of C emitted from the fossil fuel combustion. Müller-Nedebock and Chaplot also observed that erosion-induced CO_2 emissions may be limited in clayey soils but severe in sandy soils, and that sheet erosion is an important and efficient mechanism of detachment and transport of surficial material (e.g., silt, clay, SOC).

It is also recognized that soil erosion can be a sink of C in coastal ecosystems and steep terrains. For example, a study in New Zealand documented that transport of sediments and POC from coastal steep lands is a sink. Dymond [30] observed that the North Island of New Zealand exports. 1.9 Tg of POC/yr. to the sea and sequesters 1.25 $\pm$ 0.3 to 0.6 Tg POC·yr. through regenerating soils. In contrast, the South Island of New Zealand exports 2.9 $\pm$ 0.7–1.5 Tg POC/yr. and sequesters about the same amount. With 80% efficiency of burial at seas, New Zealand has a net sink of 3.1 $\pm$ 2–2.5 Tg C/yr.

4.2. Emissions from Tillage-Induced Soil Erosion

Tillage-induced soil displacement downslope is a significant process on cultivated steep lands. Soil disturbance by plow tillage (PT) or its lack under notill (NT) can have variable effects on the magnitude of SOC removal by erosional processes and pedological processes involved in gaseous emissions. In the Kwa Zulu-Natal Province of South Africa, Chaplot et al. [31] evaluated the effects of NT vs. PT on losses of SOC from soil. Chaplot and colleagues observed that soil under NT had greater SOC density than that under PT

(17.70 vs. 13.19 kg/m^3), lower gaseous emissions by 4.4% (10.40 vs. 10.88 g CO_2-C/m^2) but reduced the release of CO_2 from eroded sediments (0.185 vs. 0.778 g CO_2C/m^2) representing 76.3% decline. In the case of PT, cumulative emissions over a 141-day period were 19% greater in sediments (0.048 g CO_2C/g C) compared to soils (0.04 g CO_2C/g C). In the case of NT, emissions were 33% lower in sediments (0.024 g CO_2C/g C) compared to soils (0.032 g CO_2C/g C). The authors' hypothesized these trends to a high aggregate stability and better physical protection of SOC within stable aggregates. Conservation agriculture (CA), based on NT with residue mulching and cover cropping, is widely reported to sequester SOC [32]. A modeling study by Gaiser et al. [33] showed higher losses of SOC under PT than NT through enhanced CO_2 emissions and increased losses through intensified erosion. Gaiser and colleagues suggested that tillage effects on SOC losses through soil erosion must also be accounted for in the GCB. The loss of SOC from eroded soil is also determined by the sediment size distribution and its MRT [34].

Plow tillage is also a source of dust. In some cases, tillage-induced emissions of dust are more than that by wind. Funk et al. [35] observed that tillage-induced emissions are determined by the soil moisture content at the time of tillage. Funk and colleagues found that the threshold level of soil water content for fine dust emissions of soil was 2% to 5% for sandy soil, 5% to 10% for silty soils, ~30% for clayey soils, and 25% to 45% for organic soils.

4.3. Grazing Systems and Gaseous Emissions by Hydric Erosion

Similar to mechanical tillage, overgrazing can also accelerate hydric erosion and influence gaseous emissions. In addition to reducing the plant cover, overgrazing may aggravate risks of hydric erosion by altering soil bulk density, water infiltration rate, surface runoff and transport of POC and DOC and the attendant emission of GHGs into the atmosphere. Based on an experiment involving some sandy loam Aerosols in South Africa, Mchunu and Chaplot [36] assessed the effects of three levels of plant covers (100%, 25–50% and 0–5%) on loss of SOC under simulated rainfall. Plant C input into the soil profile and SOC stocks (g $C/m^2 \cdot$yr.) in 0–0.02 m layer, respectively, were 1950 $\pm$ 180 and 300 $\pm$ 16.2 at 100% cover. In comparison, soil C input by plants decreased by 38.5 $\pm$ 3.5% at 25–50% and by 75.4 $\pm$ 6.9% at 0–5% cover. The losses of SOC by water erosion were 0.75 g C/m^2 at 100% cover and increased by 66% at 25–50% cover (3.76 $\pm$ 1.8 g C/m^2) to 213% at 0–5% cover (7.08 $\pm$ 2.9 g C/m^2). Furthermore, these losses of SOC were mostly in POC form, which being a labile fraction can be easily mineralized with the attendant emission of GHGs.

4.4. Wind Erosion and Gaseous Emissions

Similar to hydric erosion, wind erosion is also a selective process and involves preferential removal of light and colloidal fractions such as SOC and SIC constituents of the surface layer. The CER for windblown dust for rangeland in western Queensland, Australia is reported at 1 to 2 for sandy soil and 9 to 41 for clayey soil [37]. Dust may contain as much as 15–20% of SOC content. In addition to gaseous emissions and loss of soil productivity, the dust generated by wind erosion is a serious environmental hazard in arid and semi-arid climates. Among vegetative barriers (windbreaks and shelterbelts) and mulch farming systems (e.g., CA), enzyme-induced carbonate precipitation is also used to stabilize loose soil that creates dust. To stabilize loose soil and mitigate dust in Khuzestan, Iran, Baziar et al. [38] used a soybean enzyme, which is reportedly much easier and more economical to prepare than that from leek bean.

The problem of wind erosion is no longer confined to developing countries. Indeed, wind erosion is also a serious issue even in northern Germany, and on Histosols throughout the U.S. In northern Germany, Nerger et al. [13] reported that soils under maize monoculture are prone to severe wind erosion. Nerger and colleagues reported that SOC stock decreased by 49.4 and 2.44 kg/m^2 from 1999 to 2009. Total soil loss by wind erosion during 16 events was 48.9 kg/m^2, and suspended material had a CER of 2.96 (compared with CER of 0.98 for saltation). Soil erosion loss in a single event can be as much as 12.6 kg/m^2. Histosols, with soil organic matter (SOM) content of >20% in the upper 80 cm profile, cover 21 M ha in 42 states of the U.S. [25]. Intensively cultivated Histosols are prone to wind erosion with adverse impact on productivity and degradation of the environment (i.e., soil, water, air, vegetation). Particle density of Histosols can be i~1.6 Mg/m^3. Dust emissions are common on plowed/dry Histosols [25].

In Adam County, Washington, Feng and Sharrat [39] assessed the wind erosion hazard by using the Wind Erosion Prediction System (WEPS) at 14.25 Mg soil/ha·yr. (6.4 tn/ac·yr) and a severe PM10 (particulate matter at <10 micron m in diameter) loss. Wind erosion affects ~20 Mha of land in Iran, and also affects GHG emissions from soil. Based on experiments conducted at the Iran Research Institute of Forests and Ranges, Kamali et al. [40] reported that the highest rate of CO_2 emissions in July was 4.90 g CO_2/m^2·d in severely eroded lands and the lowest in January of 0.086 g CO_2/m^2·d in less eroded lands. Kamali and colleagues hypothesized that increase in erosion intensity aggravated CO_2 emission rates at severe erosion.

A study on soil losses from spring dust emissions in northern China by Song et al. [41] indicated that losses of SOC, total N and total P in spring were 0.985 ± 0.149, 0.094 ± 0.014 and 0.089 ± 0.013 Tg/yr., respectively. However, the fate of SOC, TN, etc., is not known. In Australia, Chappell et al. [42] estimated SOC dust emission at 5083 Tg CO_2 eq/yr. for the country and 0.4 Tg CO_2 eq/yr. for agricultural soils. Chappell and colleagues concluded that omission of SOC dust emission from C cycling and C accounting is a significant source of uncertainty in the GCB. Thus, quantification of the release of CO_2 from SOC dust to the atmosphere and contribution of SOC deposition to downward C sinks is essential.

5. Soil Erosion and the Global Warming

Soil degradation by erosion is a serious global issue and it may be changed by the current and projected global warming. In arid or semi-arid areas, which may experience more rains with climate change, soil erosion could be less due to vegetation growth and more ground cover. The soil erosion hazard may also depend on the amount and intensity of the rain and the nature and dynamics of the emergent vegetation. The climate-induced dynamics may also differ among ecoregions (e.g., polar vs. tropical, and fire-prone vs. fire-free). Enzymatic and decomposition processes may be aggravated by the increase in temperature. In general, soil erosion hazard is linked to the fate of vegetation/land cover in relation to the climate change and may decrease with an increase in vegetation cover and the vice versa.

In agroecosystems or the managed landscape, soil erosion hazard may be aggravated during the 21st century [43] because of the anthropogenic climate change. Soil erosion is affected by interaction among a range of factors, such as climate (erosivity), soil (erodibility), slope (gradient, length, aspect, shape), land use (cropland, grazing land, plantations), and management (soil, crop, inputs). However, most of these parameters are also affected by the current and projected global warming (Figure 2). In the tropics and subtropics, soil erosion and global warming are mutually reinforcing processes. An increase in global warming in the tropics may increase erosion risks by both hydric and aeolian processes through an increase in climatic erosivity and soil erodibility, and both of these factors are aggravated by an increase in frequency and intensity of extreme events and the attendant decline in vegetation cover (Figure 2). However, the soil erosion hazard (hydric and aeolian) may decrease in temperate regions depending on the manner in which global warming affects the key parameters or factors of soil erosion. Based on a study in three catchments in Great Britain, Ciampalini et al. [44] observed that climactic parameters respond differently depending on their land use and management. Ciampalini and colleagues reported that an increase in rainfall increases soil erosion, but warmer temperatures in the U.K. may also lower erosion risks because of better vegetation growth. Warmer temperature can limit soil erosion risks by increasing primary productivity, improving rainfall interception, enhancing water infiltrability, and reducing soil erodibility. Ciampalini et al. observed that an increase in temperature in the U.K. may increase the rainfall thresholds to generate soil loss, and, thus, soil erosion rates could decrease by 33% from 2070 to 2099 because of the negative-feedback mechanisms limiting soil loss by runoff. Modeling studies on soil erosion under sugar beet in Central Europe by Scholz et al. [45] also reported a decline in soil erosion risks for the period of 2070–2099. Scholz and colleagues observed that the intra-annual precipitation change resulted in a net decrease of rainfall amounts in erosion-sensitive months and an overall increase of rainfall in periods when the region is less prone to erosion. Consequently, the predicted average soil erosion losses under climate change declined in all tillage systems by 11% to 24%.

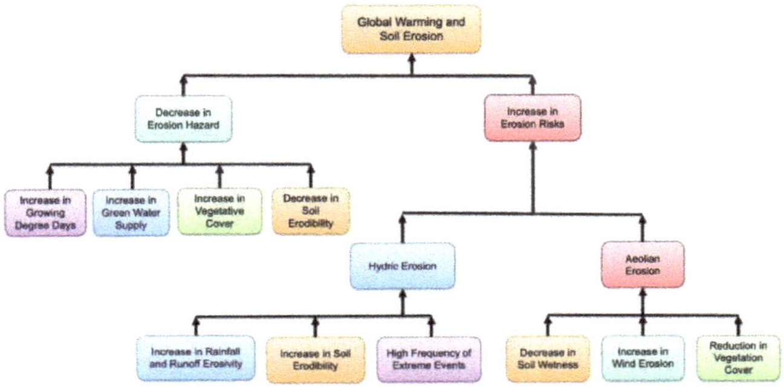

Figure 2. Effects of global warming on soil erosion risks which may decrease in high latitudes (boreal and temperate regions such as U.K.) but increase in the tropics (humid, subhumid, semi-arid and arid regions) and where the rainfall erosivity may increase (i.e., northeastern and northwestern U.S.) and wherever the frequency of extreme events may increase.

In contrast to the observation from the U.K., Segura et al. [10] reported an increase in vulnerability to erosion in some states of the U.S. (e.g., Ohio, Maryland, Indiana, Vermont, Illinois) with future increase in rainfall erosivity due to the projected climate change. Segura and colleagues based their conclusions on evaluation of changes in rainfall erosivity (R) from 1970 to 2090 across the U.S. under nine climate conditions. They observed that trends in R vary widely spatially with strong trends of increasing R in the northeastern and northwestern U.S., but weaker or inconsistent trends in the midwestern and southeastern U.S. An increase in soil erosion hazard in the U.S. with change in rainfall erosivity by 16–58% has also been reported by Nearing [12]. The response of the increase in risks of water and wind erosion by climate change may also occur in tropical and subtropical climates. Borrelli et al. [46] predicted a more vigorous hydrological cycle, which could increase global water erosion +30 to +60%. Thus, adoption of low erosion-producing practices can drastically reduce the risks of accelerated erosion [47]. Indeed, the adoption of conservation-effective measures can trump adverse impacts of climate change on soil erosion [48].

It is the breakdown of aggregates by high climate erosivity and weak aggregate strength (high erodibility) that exposes SOC/SIC to microbial processes and climactic parameters and aggravates the emission of GHGs. It is thus important to assess the risks of their mineralization by biotic and abiotic mechanisms. On the other hand, restoration of eroded soils/landscapes, via afforestation or establishment of any perennial vegetation cover, can create a positive soil/ecosystem C budget and set-in-motion recarbonization of the terrestrial biosphere through sequestration of atmospheric CO_2 as soil humus and secondary carbonates, etc. Sequestration of C in soil and biomass has a tremendous potential to create a substantial drawdown of atmospheric CO_2 through nature-based solutions [49].

6. Some Researchable Priorities

All other factors remaining equal, soil erosion is affected by land use and vegetation cover. Thus, it is a more serious problem in agricultural than in natural landscapes. However, land area under agroecosystems may increase during the 21st century and risks of soil erosion may be aggravated by an intensive land use. Intensification of the agroecosystems may be necessitated by the rising demands of the growing and increasingly affluent human population. Thus, adoption of conservation-effective measures and understanding of the fate of C being transported by erosional processes is important to promoting nature-positive agroecosystems. Therefore, there is a strong need to study the fate of erosion-induced transport of SOC in both agricultural and natural landscapes.

Key questions about future research priorities on soil erosion in agricultural landscapes include the following [50]: (i) understanding of the nexus between on-site and off-site effects of erosion, especially with regard to soil C budget and the emission of GHGs, (ii) evaluating site-specific adaptation of conservation-effective measures to reduce sediment connectivity and transport of C-laden sediments from hillslopes to eroded channels and eventually to depositional sites, (iii) identifying early signs of the on-set of a severe erosional process with a drastic impact on GCB and emission of erosion-induced GHGs (CO_2, CH_4, and N_2O) into the atmosphere, (iv) quantifying the regional and global impacts of accelerated erosion on GCB, (v) assessing in-depth the processes affecting erosion-induced emissions of GHGs and their impacts on GCB, and (vi) evaluating the ramifications and consequences of climate change and the magnitude of feedback related to erosion-induced emissions of GHGs.

Additional research is also needed on transport of SIC by aeolian processes and the fate of both SOC and SIC transported by wind erosion. There is a strong lack of scientific data on the impact of erosional processes (water and wind) on the transport of SIC and its fate during the redistribution and depositional phases, and its relationship with SOC. Based on a study of 62 erosional sites and 35 depositional sites of the Chinese Loess Plateau, Tong et al. [51] observed that the mean SOC and SIC contents in a depositional site (0–25 cm) increased by 24.4% and 15.4%, respectively. Further, SIC was significantly negatively correlated with SOC across all erosional sites. However, SIC was significantly positively correlated with SOC at all depositional sites.

The effects of complex and interacting processes must be assessed in order to understand and determine their impacts on the GCC. In this context, Kuhn et al. [52] argued that the eco-geomorphological perspective on soil C movement through the landscape can address any controversy with regard to erosion of C being a source or a sink. Soil erosion induced by water runoff [53] and by windblown dust [42] are important pathways of lateral transport of soil C in terrestrial landscapes. The 137Cs technique has been used as a tracer to assess removal or deposition of soil along the toposequence [54,55]. Alewell et al. [56] also recommended the 137Cs technique to assess emissions of gases from eroded SOC. However, the 137Cs technique is based on radioactive fallouts, mainly during the 1950s and 1960s, when nuclear tests were being conducted. With half-life of around 30 years, there are only a few sites where the isotopes can still be detected. Thus, there is a strong need to develop/identify new techniques.

However, how these alluvial and aeolian processes affect the soil C budget at diverse erosional, redistributional, and depositional landform positions is critical to assessing the fate of soil C being transported. Understanding and quantifying the C gains and losses at different landscape positions in soils of fragile agroecosystems for both SOC and SIC is essential to obtaining reliable estimates of the GCB. Estimates of the effects of hydric erosion on the GCB range from a sink of 0.06–1.0 vs. the source of 0.27–1.14 Pg C/yr. [57]. Such a large range necessitates more watershed-based studies using some innovative techniques.

Soil structure is an important determinant of the stability of SOM and its MRT in terrestrial ecosystems [24]. It is the breakdown of structural units and reaggregation of the dispersed clay that determines the emission of GHGs. Thus, additional research is needed on strengthening the understanding of the dynamics of aggregation during and after the erosional processes at diverse landscape positions.

Soil erosion response to climate change is a high researchable priority. In addition to modelling, past climate change and its effects on soil processes (soil C dynamics) can also be studied from soil profile properties, especially those of the buried soil horizon.

7. Summary

Accelerated erosion, both hydric and aeolian, involves selective removal of soil C (both SOC and SIC) along with clay and fine silt fractions. On-site, the eroded soil becomes progressively depleted of the light and colloidal fractions, leaving coarse material (gravels, sand, course silt) behind. Because CER of the eroded material is high, erosion has severe adverse effects on the functionality of eroding landscapes. Globally, soil erosion by water and wind transports a large amount of C, but the fate of C being transported is governed by complex pedotransformative processes, and its stability against decomposition is governed by site-specific biophysical conditions and hydrothermal regimes. Therefore, soil erosion not only affects soil properties and processes on-site, but also along the landscape over which the water or windborne sediments are being redistributed and the depressional sites or barriers where the sediments are being deposited. Whereas a proportion of soil C being buried at depositional sites or into the water bodies may be protected against mineralization and may even be encapsulated within reformed aggregates, most of the C redistributed over the landscape and part of those carried into depositional sites is subject to decomposition by both biotic and abiotic processes. Off-site, a large proportion of transported C is released into the atmosphere as CO_2, CH_4, and N_2O, depending on the biophysical and hydrothermal regimes. In total, the accelerated soil erosion has a drastic impact on the GCB. How much of soil C (SOC and SIC) is being removed by erosional processes and how it is being mineralized en route to depositional sites has significant environmental consequences, especially as a source or sink of GHGs, depending on the site-specific conditions. Nonetheless, the C displaced by erosional processes is not accounted for in the current processes of compiling the GCB. Not only should the impact of erosional process on the global C cycle be accounted for, but its impact as a source of GHGs should be considered in evaluating the economic and environmental impacts of erosion. The net effects of the erosional processes as a source or sink depend on a multitude of site-specific and highly interactive factors. Despite the on-site and off-site effects on productivity and off-site effects on the global C cycle and emission of GHGs, a prudent strategy is to minimize the risks of accelerated soil erosion by adoption of the conservation-effective measures.

The global menace of soil erosion by water and wind may be aggravated by global warming, especially in the tropics and subtropics, and with agricultural intensification of agroecosystems. Thus, there is a strong need to:

i. quantify the global magnitude of soil C (SOC, SIC) being transported by the erosional processes in relation to land use, farming/cropping systems, landscape characteristics, soil physical/chemical properties, and the dynamics of hydrothermal regimes along the landscape and at the depositional sites;
ii. account for the global amount of C transported by hydric and aeolian processes in the GCB;
iii. evaluate the site-specific conditions which make the eroded soil C a source (emission of CO_2, CH_4, and N_2O) of GHGs or a sink if some C is buried, reaggregated, and taken out of circulation;
iv. identify and implement site-specific conservation-effective management practices, which minimize the risks of soil erosion by water, wind, and other anthropogenic activities.

Funding: This research received no external funding.

Institutional Review Board Statement: Not Applicable.

Informed Consent Statement: Not Applicable.

Data Availability Statement: The data quoted in the manuscript are available in the specific reference number listed at the end of each statement or the paragraph.

Conflicts of Interest: The author declares no conflict of interest.

References

1. Chappell, A.; Webb, N.P.; Viscarra Rossel, R.A.; Bui, E. Australian net (1950–1990) soil organic carbon erosion: Implications for CO 2 emission and land-atmosphere modelling. *Biogeosciences* **2014**, *11*, 5235–5244. [CrossRef]
2. Lal, R. Soil Erosion and Gaseous Emissions. *Appl. Sci.* **2020**, *10*, 2784. [CrossRef]
3. Flanagan, D.C.; Ascough, J.C.; Nieber, J.L.; Misra, D.; Douglas-Mankin, K.R. Advances in soil erosion research: Processes, measurement, and modeling. *Trans. ASABE* **2013**, *56*, 455–463. [CrossRef]
4. Lal, R. Soil Degradation and Food Security in South Asia. In *Climate Change and Food Security in South Asia*; Springer: Cham, The Netherlands, 2010; pp. 137–152.
5. Lal, R.; Griffin, M.; Apt, J.; Lave, L.; Morgan, G. Response to Comments on "Managing Soil Carbon". *Science* **2004**, *305*, 1567. [CrossRef]
6. Wang, X.; Cammeraat, E.L.H.; Romeijn, P.; Kalbitz, K. Soil organic carbon redistribution by water erosion—The role of CO_2 emissions for the carbon budget. *PLoS ONE* **2014**, *9*, e0096299. [CrossRef]
7. Worrall, F.; Burt, T.P.; Howden, N.J.K. The fluvial flux of particulate organic matter from the UK: Quantifying instream losses and carbon sinks. *J. Hydrol.* **2014**, *519*, 611–625. [CrossRef]
8. Lal, R. Regenerative agriculture for food and climate. *J. Soil Water Conserv.* **2020**, *75*, 123A–124A. [CrossRef]
9. Novara, A.; Keesstra, S.; Cerda, A.; Pereira, P.; Gristina, L. Understanding the role of soil erosion on co(2)-c loss using C-13 isotopic signatures in abandoned Mediterranean agricultural land. *Sci. Total Environ.* **2016**, *550*, 330–336. [CrossRef] [PubMed]
10. Segura, C.; Sun, G.; Mcnulty, S.; Zhang, Y. Potential impacts of climate change on soil erosion vulnerability across the conterminous United States. *J. Soil Water Conserv.* **2014**, *69*, 171–181. [CrossRef]
11. Nearing, M.A.; Pruski, F.F.; O'Neal, M.R. Expected climate change impacts on soil erosion rates: A review. *J. Soil Water Conserv.* **2004**, *59*, 43–50.
12. Nearing, M. Potential changes in rainfall erosivity in the US with climate change during the 21st century. *J. Soil Water Conserv.* **2001**, *56*, 229–232.
13. Nerger, R.; Funk, R.; Cordsen, E.; Fohrer, N. Application of a modeling approach to designate soil and soil organic carbon loss to wind erosion on long-term monitoring sites (BDF) in Northern Germany. *Aeolian Res.* **2017**, *25*, 135–147. [CrossRef]
14. Hu, Y.; Kuhn, J. Erosion-induced exposure of SOC to mineralization in aggregated sediment. *Catena* **2016**, *137*, 517–525. [CrossRef]
15. Hu, Y.X.; Schneider, V.; Kuhn, B.; Guo, S.L.; Kuhn, N.J. Capturing the Scale Dependency of Erosion-Induced Variation in CO_2 Emissions on Terraced Slopes. *Front. Environ. Sci.* **2021**, *9*, 211. [CrossRef]
16. Lense, G.H.E.; Moreira, R.S.; Parreiras, T.C.; Avanzi, C.; Mincato, R.L. Modeling of soil organic carbon loss by water erosion on a tropical watershed. *Rev. Cienc. Agron.* **2021**, *52*, e20207257. [CrossRef]
17. Wei, S.; Zhang, X.; McLaughlin, N.; Yang, X.; Jia, S.; Chen, X. Effect of breakdown and dispersion of soil aggregates by erosion on soil CO_2 emission. *Geoderma* **2016**, *264*, 238–243. [CrossRef]
18. Gao, X.; Hu, Y.; Sun, Q.; Du, L.; Duan, P.; Yao, L.; Guo, S. Erosion-induced carbon losses and CO_2 emissions from Loess and Black soil in China. *Catena* **2018**, *171*, 533–540. [CrossRef]
19. Samburova, V.; Shillito, R.M.; Berli, M.; Khlystov, A.Y.; Moosmuller, H. Effect of Biomass-Burning Emissions on Soil Water Repellency: A Pilot Laboratory Study. *FIRE* **2021**, *4*, 24. [CrossRef]
20. Nie, X.D.; Yuan, Z.J.; Huang, B.; Liao, Y.S.; Zhang, X.Q.; Li, Z.W.; Li, D.Q. Effects of water erosion on soil organic carbon stability in the subtropical China. *J. Soils Sediments* **2019**, *19*, 3564–3575. [CrossRef]
21. Gispert, M.; Pardini, G.; Colldecarrera, M.; Emran, M.; Doni, S. Water erosion and soil properties patterns along selected rainfall events in cultivated and abandoned terraced fields under renaturalisation. *Catena* **2017**, *155*, 114–126. [CrossRef]
22. Mao, N.; Li, W.R.; Wei, X.R.; Shao, M.A. Soil moisture- and texture-dependent effects of soil deposition on evaporation and carbon emission. *Soil Tillage Res.* **2020**, *204*, 104703. [CrossRef]
23. Six, J.; Feller, C.; Denef, K.; Ogle, S.M.; de Moraes Sa, J.C.; Albrecht, A. Soil organic matter, biota and aggregation in temperate and tropical soils–effects of no-tillage. *Agron. Sustain. Dev.* **2002**, *22*, 755–775. [CrossRef]
24. Chaplot, V.; Cooper, M. Soil aggregate stability to predict organic carbon outputs from soils. *Geoderma* **2015**, *243*, 205–213. [CrossRef]
25. Zobeck, T.M.; Baddock, M.; Scott Van Pelt, R.; Tatarko, J.; Acosta-Martinez, V. Soil property effects on wind erosion of organic soils. *Aeolian Res.* **2013**, *10*, 43–51. [CrossRef]
26. Huang, W.; Hall, S.J. Elevated moisture stimulates carbon loss from mineral soils by releasing protected organic matter. *Nat. Commun.* **2017**, *8*, 1774. [CrossRef] [PubMed]
27. Walling, D.E. The changing sediment loads of the world's rivers. *Ann. Warsaw Univ. Life Sci. SGGW L. Reclam.* **2008**, *39*, 3–20. [CrossRef]
28. Mandal, D.; Giri, N.; Srivastava, P. The magnitude of erosion-induced carbon (C) flux and C-sequestration potential of eroded lands in India. *Eur. J. Soil Sci.* **2020**, *71*, 151–168. [CrossRef]

29. Müller-Nedebock, D.; Chaplot, V. Soil carbon losses by sheet erosion: A potentially critical contribution to the global carbon cycle. *Earth Surf. Process. Landf.* **2015**, *40*, 1803–1813. [CrossRef]
30. Dymond, J.R. Soil erosion in New Zealand is a net sink of CO_2. *EARTH Surf. Process. Landf.* **2010**, *35*, 1763–1772. [CrossRef]
31. Chaplot, V.; Mchunu, C.; Manson, A.; Lorentz, S.; Jewitt, G. Water erosion-induced CO_2 emissions from tilled and notilled soils and sediments. *Agric. Ecosyst. Environ.* **2012**, *159*, 62–69. [CrossRef]
32. Lal, R. Sequestering carbon and increasing productivity by conservation agriculture. *J. Soil Water Conserv.* **2015**, *70*, 55A–62A. [CrossRef]
33. Gaiser, T.; Stahr, K.; Billen, N.; Mohammad, M.A.R. Modeling carbon sequestration under zero tillage at the regional scale. The effect of soil erosion. *Ecol. Modell.* **2008**, *218*, 110–120. [CrossRef]
34. Polyakov, V.O.; Lal, R. Soil organic matter and CO_2 emission as affected by water erosion on field runoff plots. *Geoderma* **2008**, *143*, 216–222. [CrossRef]
35. Funk, R.; Reuter, H.I.; Hoffmann, C.; Engel, W.; Otti, D. Effect of moisture on fine dust emission from tillage operations on agricultural soils. *Earth Surf. Process. Landf.* **2008**, *33*, 1851–1863. [CrossRef]
36. Mchunu, C.; Chaplot, V. Land degradation impact on soil carbon losses through water erosion and CO_2 emissions. *Geoderma* **2012**, *177–178*, 72–79. [CrossRef]
37. Webb, N.P.; Strong, C.L.; Chappell, A.; Marx, S.K.; Mctainsh, G.H. Soil organic carbon enrichment of dust emissions: Magnitude, mechanisms and its implications for the carbon cycle. *Earth Surf. Process. Landf.* **2013**, *38*, 1662–1671. [CrossRef]
38. Baziar, M.H.; Sanaie, M.; Amirabadi, O.E. Mitigation of Dust Emissions of Silty Sand Induced by Wind Erosion Using Natural Soybean Biomaterial. *Int. J. Civ. Eng.* **2021**, *19*, 595–606. [CrossRef]
39. Feng, G.; Sharratt, B. Scaling from Field to Region for Wind Erosion Prediction Using the Wind Erosion Prediction System and Geographical Information Systems. *J. Soil Water Conserv.* **2007**, *62*, 321.
40. Kamali, N.; Siroosi, H.; Sadeghipour, A. Impacts of wind erosion and seasonal changes on soil carbon dioxide emission in southwestern Iran. *J. Arid Land* **2020**, *12*, 690–700. [CrossRef]
41. Song, H.Q.; Zhang, K.S.; Piao, S.L.; Liu, L.L.; Wang, Y.P.; Chen, Y.M.; Yang, Z.L.; Zhu, L.L.; Wan, S.Q. Soil organic carbon and nutrient losses resulted from spring dust emissions in Northern China. *Atmos. Environ.* **2019**, *213*, 585–596. [CrossRef]
42. Chappell, A.; Webb, N.P.; Butler, H.J.; Strong, C.L.; McTainsh, G.H.; Leys, J.F.; Viscarra Rossel, R.A. Soil organic carbon dust emission: An omitted global source of atmospheric CO_2. *Glob. Chang. Biol.* **2013**, *19*, 3238–3244. [CrossRef] [PubMed]
43. Lal, R. Soil degradation by erosion. *L. Degrad. Dev.* **2001**, *12*, 519–539. [CrossRef]
44. Ciampalini, R.; Constantine, J.A.; Walker-Springett, K.J.; Hales, T.C.; Ormerod, S.J.; Hall, I.R. Modelling soil erosion responses to climate change in three catchments of Great Britain. *Sci. Total Environ.* **2020**, *749*, 141657. [CrossRef] [PubMed]
45. Scholz, G.; Quinton, J.N.; Strauss, P. Soil erosion from sugar beet in Central Europe in response to climate change induced seasonal precipitation variations. *Catena* **2008**, *72*, 91–105. [CrossRef]
46. Borrelli, P.; Robinson, D.A.; Panagos, P.; Lugato, E.; Yang, J.E.; Alewell, C.; Wuepper, D.; Montanarella, L.; Ballabio, C. Land use and climate change impacts on global soil erosion by water (2015–2070). *Proc. Natl. Acad. Sci. USA* **2020**, *117*, 21994–22001. [CrossRef]
47. Labrière, N.; Locatelli, B.; Laumonier, Y.; Freycon, V.; Bernoux, M. Soil erosion in the humid tropics: A systematic quantitative review. *Agric. Ecosyst. Environ.* **2015**, *203*, 127–139. [CrossRef]
48. Garbrecht, J.D.; Nearing, M.A.; Steiner, J.L.; Zhang, X.J.; Nichols, M.H. Can conservation trump impacts of climate change on soil erosion? An assessment from winter wheat cropland in the Southern Great Plains of the United States. *Weather Clim. Extrem.* **2015**, *10*, 32–39. [CrossRef]
49. Lal, R.; Smith, P.; Jungkunst, H.F.; Mitsch, W.J.; Lehmann, J.; Ramachandran Nair, P.K.; McBratney, A.B.; De Moraes Sá, J.C.; Schneider, J.; Zinn, Y.L.; et al. The carbon sequestration potential of terrestrial ecosystems. *J. Soil Water Conserv.* **2018**, *73*, 145A–152A. [CrossRef]
50. Garcia-Ruiz, J.M.; Begueria, S.; Lana-Renault, N.; Nadal-Romero, E.; Cerda, A. Ongoing and Emerging Questions in Water Erosion Studies. *L. Degrad. Dev.* **2017**, *28*, 5–21. [CrossRef]
51. Tong, L.S.; Fang, N.F.; Xiao, H.B.; Shi, Z.H. Sediment deposition changes the relationship between soil organic and inorganic carbon: Evidence from the Chinese Loess Plateau. *Agric. Ecosyst. Environ.* **2020**, *302*, 107076. [CrossRef]
52. Kuhn, N.J.; Hoffmann, T.; Schwanghart, W.; Dotterweich, M. Agricultural soil erosion and global carbon cycle: Controversy over? *Earth Surf. Process. Landf.* **2009**, *34*, 1033–1038. [CrossRef]
53. Gasper, D.; Shah, A.; Tankha, S. The Framing of Sustainable Consumption and Production in SDG 12. *Glob. Policy* **2019**, *10*, 83–95. [CrossRef]
54. Zhang, X.C. Spatial sensitivity of predicted soil erosion and runoff to climate change at regional scales. *J. Soil Water Conserv.* **2006**, *61*, 58–64.
55. Quine, T.A.; van Oost, K. Quantifying carbon sequestration as a result of soil erosion and deposition: Retrospective assessment using caesium-137 and carbon inventories. *Glob. Chang. Biol.* **2007**, *13*, 2610–2625. [CrossRef]
56. Kirkels, F.M.S.A.; Cammeraat, L.H.; Kuhn, N.J. The fate of soil organic carbon upon erosion, transport and deposition in agricultural landscapes—A review of different concepts. *Geomorphology* **2014**, *226*, 94–105. [CrossRef]
57. Alewell, C.; Schaub, M.; Conen, F. A method to detect soil carbon degradation during soil erosion. *Biogeosciences* **2009**, *6*, 2541–2547. [CrossRef]

applied sciences

MDPI

Review

Biophysical Controls That Make Erosion-Transported Soil Carbon a Source of Greenhouse Gases

Rattan Lal

CFAES Rattan Lal Center for Carbon Sequestration and Management, The Ohio State University, Columbus, OH 43210, USA; lal.1@osu.edu

Abstract: Soil erosion is a selective process which removes the light fraction comprised of soil organic carbon (SOC) and colloidal particles of clay and fine silt. Thus, a large amount of carbon (C) is transported by erosional processes, and its fate (i.e., emission, redistribution, burial, and translocation into aquatic ecosystems) has a strong impact on the global carbon cycle. The processes affecting the dynamics of soil C emission as greenhouse gases (i.e., CO_2, CH_4, N_2O), or its deposition and burial, vary among different stages of soil erosion: detachment, transport, redistribution, deposition or burial, and aquatic ecosystems. Specific biogeochemical and biogeophysical transformative processes which make erosion-transported carbon a source of C emission are determined by the type of erosion (rill vs. inter-rill in hydric and saltation erosion vs. air-borne dust in aeolian erosion), soil temperature and moisture regimes, initial SOC content, texture, raindrop-stable aggregates and water repellency, crusting, slope gradient, physiography and the slope-based flow patterns, landscape position, and the attendant aerobic vs. anaerobic conditions within the landscape where the sediment-laden C is being carried by alluvial and aeolian processes. As much as 20–40% of eroded SOC may be oxidized after erosion, and erosion-induced redistribution may be a large source of C. In addition, human activities (e.g., land use and management) have altered—and are altering—the redistribution pattern of sediments and C being transported. In addition to O_2 availability, other factors affecting emissions from aquatic ecosystems include sub-surface currents and high winds, which may also affect CH_4 efflux. The transport by aeolian processes is affected by wind speed, soil texture and structure, vegetation cover, etc. Lighter fractions (SOC, clay, and fine silt) are also selectively removed in the wind-blown dust. The SOC-ER of dust originating from sand-rich soil may range from 2 to 41. A majority of the C (and nutrients) lost by aeolian erosion may be removed by saltation. Even over a short period of three seasons, wind erosion can remove up to 25% of total organic C (TOC) and total N (TN) from the top 5 cm of soil. A large proportion of C being transported by hydric and aeolian erosional processes is emitted into the atmosphere as CO_2 and CH_4, along with N_2O. While some of the C buried at the depositional site or transported deep into the aquatic ecosystems may be encapsulated within reformed soil aggregates or protected against microbial processes, even the buried SOC may be vulnerable to future loss by land use, management, alkalinity or pH, the time lag between burial and subsequent loss, mineralogical properties, and global warming.

Keywords: hydric and aeolian erosion; enrichment ratio; selective removal; erosion as a source; positive feedback to climate change; deep burial; re-aggregation; saltation; dust blown carbon; emission of greenhouse gases; stages of erosion

Citation: Lal, R. Biophysical Controls That Make Erosion-Transported Soil Carbon a Source of Greenhouse Gases. *Appl. Sci.* **2022**, *12*, 8372. https://doi.org/10.3390/app12168372

Academic Editor: Itzhak Katra

Received: 13 July 2022
Accepted: 15 August 2022
Published: 22 August 2022

Publisher's Note: MDPI stays neutral with regard to jurisdictional claims in published maps and institutional affiliations.

1. Introduction

Accelerated soil erosion, in which the magnitude of soil loss exceeds the rate of new soil formation, is a major anthropogenic-induced process. Borrelli et al. [1] estimated the global potential rate of soil erosion by water at 43 ± 9.2 Pg/y and suggested that site-specific conservation agricultural practices should be adopted to reduce the risks of accelerated soil erosion. With projections for a more vigorous hydrological cycle due to climate change, the water erosion hazard may increase +30 to +66% by 2070 [1], which will increasingly jeopardize agronomic productivity and drastically affect the global C budget [2]. In comparison with fossil fuel emissions of ~36.7 Pg CO_2 eq. per year [3], global net annual soil emissions are large [4] and are not being considered in the estimation of the global C budget. At the catchment scale, Bouchoms et al. [5] estimated that the landcover change over the last 1000 years decreased the soil organic carbon (SOC) stock of the top 1 m of soil by 31.1 Mg C/ha (32% loss) compared to the initial SOC stock. However, the global C budget [3] does not specifically consider pedological processes (e.g., accelerated soil erosion) as sources and sinks of greenhouse gases (GHGs). In this regard, the fate of soil C being transported laterally over the landscape has a strong impact which is not presently accounted for, even in the analysis by Oertel et al. [4]. Pedological processes involved as sources and sinks of GHGs differ among stages of the soil erosion process. These stages (Figure 1) include: (I) disruption of soil aggregates, detachment of soil particles, and exposure of C and soil previously encapsulated and protected against microbial processes, (II) transport of detached particles by alluvial and aeolian processes, (III) redistribution of sediments and the associated C in the landscape, (IV) deposition of sediments and C on foot slopes and other locations within the landscape, (V) possible re-aggregation of clay and C to form organo-mineral complexes which undergo burial in the subsoil, and (VI) translocation into aquatic ecosystems. Dominant pedomorphic processes depend on soil temperature and moisture regimes during specific stages, as well as the physiographic characteristics. The latter are comprised of slope shape (convex, concave, complex), aspect (north or south facing), gradient (steepness or incline), and the landscape position (summit, side slopes, toe, or foot slope).

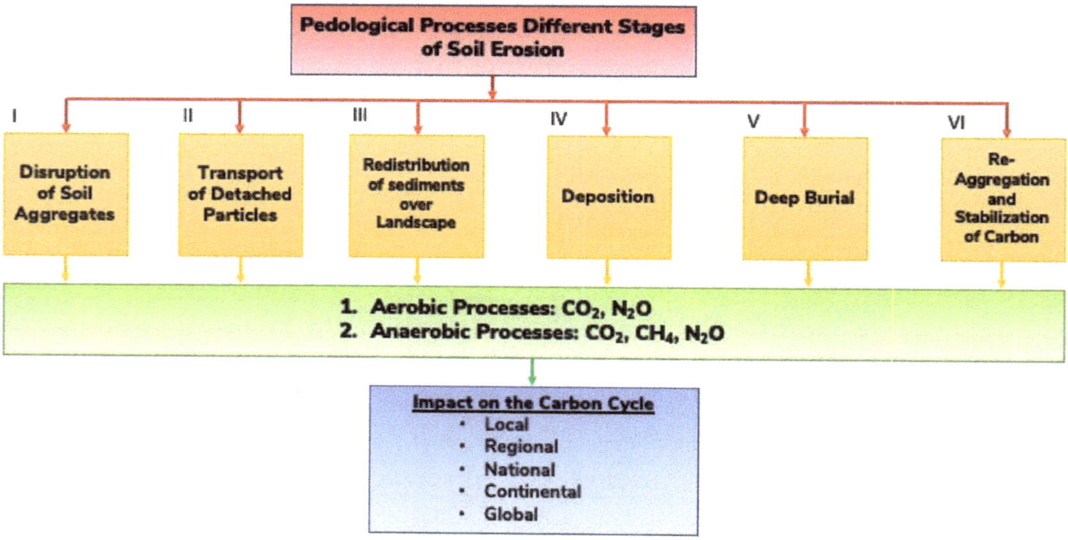

Figure 1. Fate of carbon transported by erosional processes at different stages.

The idea that soil erosion leads to sequestration of atmospheric C is contradictory to the definition of soil C sequestration. Olson et al. [6] defined soil C sequestration as, "the process of transferring CO_2 from the atmosphere into the soil of a land unit through unit plant residues and other organic solids which are derived or retained in the unit as part of the soil organic matter (SOM or humus). Retention of sequestered C in the soil (terrestrial pool) can range from short-term (not immediately released back to atmosphere) to long-term (millennia) storage. The sequestered SOC process should increase the net SOC storage during and at the end of a study to above the previous pre-treatment baseline." Thus, the deposition and burial of the lateral transport of SOC by erosion involves redistribution, and not a new sequestration, of atmospheric CO_2. Labeling the redistribution of the erosion-induced lateral transport of SOC confounds the issue.

The literature on soil erosion and the impact of land use and management on soil erosion hazard is voluminous. Therefore, this article has a limited scope of addressing specific examples of the impact of erosional processes on the fate of SOC being transported at different stages of soil erosion. The specific objective of this article is to collate and synthesize pertinent research information regarding the impact of hydric and aeolian processes at different stages on the fate of SOC being transported over the landscape and the attendant emission of GHGs into the atmosphere. The goal is to identify specific biophysical mechanisms that impact GHG emissions at different stages of the erosional process, making soil erosion an emission source, and thus create positive feedback to climate change.

The specific hypotheses tested in this review are: (a) the erosional processes at stages 1 to 3 are sources of GHG emissions, (b) the pedological processes at stages 4 to 6 six may stabilize C by deep burial and the reformation of aggregates, (c) the net effect of erosional processes at all six stages is a source of GHGs, and (d) the identification of site-specific conservation-effective measures and their implementation is a prudent strategy to reduce risks of erosion-induced emission of GHGs and minimize the positive feedback to climate change.

In view of the specific objectives and hypotheses outlined above, references cited in the article which addressed the fate of C being transported at different stages of accelerated erosion were specifically chosen. About 60 references were purposely chosen because of their focus on pedological processes at different stages and the impact on SOC dynamics and the emission of GHGs. Some articles were also chosen because these provided information on the enrichment ratio of SOC and indicated selectivity of the erosional process and preferential removal of the light colloidal fractions.

2. Criterion for Erosion-Induced Lateral Transport as a Source of Gaseous Emissions

The strategy is to identify specific biophysical processes at different stages of the erosion process which make the liberated material vulnerable to microbial processes, thus rendering the SOC released as a source of GHGs. It is important, however, that SOC mineralization during transport, which has significant implications regarding the impact of erosion on creating climate change feedback [7], is not overlooked. Similarly, reduction in mineralization at the deposition stage, deep burial, and re-aggregation must be appropriately accounted for. Indeed, the C sink/source at different stages depends on the mineralization rate of SOC along the transport pathways (stages I to III),at depositional sties (stages IV and V), and within the aquatic ecosystems (stage VI). It is, therefore, important to identify the control factors that affect the mineralization or stabilization of the SOC being transported.

The control factors at the eroding sites are those which affect agronomic productivity, the attendant dynamic replacement, and the net effect on the soil C budget. The dynamic replacement can also accentuate the priming effect and accelerate the mineralization of the exposed SOC. Vulnerability to decomposition at the eroding site is also affected by any changes in the microbial community. The mineralization of SOC during the transport is an important factor that can never be overlooked, and it is determined by the soil temperature and moisture regimes. Similarly, deep burial and reaggregation are important determinants of the fate of SOC at the depositional sites [7]. Enrichment of the sediments is an important indicator and a determinant of the fate of SOC being transported by erosional processes.

3. Enrichment of Sediments by Soil Organic Carbon

Accelerated soil erosion, due to water and wind, is a predominant process with a strong impact on the global C cycle and budget. Indeed, accelerated soil erosion, both by water and wind, affect properties of the land, atmosphere, and rivers (Figure 2), and erosional risks must be minimized through the adoption of judicious land use and of site-specific best management practices (BMPs). Climate change and the ever-increasing demands of the growing and progressively affluent human population will aggravate the soil erosion hazard. Globally, between 0.57 and 1.33 Pg C may already be transported by inter-rill soil erosion [8] because of its selectivity. Indeed, soil erosion leads to a preferential removal of the SOC fraction because it has a low bulk density (0.6–0.8 Mg/m^3), and it is concentrated in the vicinity of the soil surface [9,10].

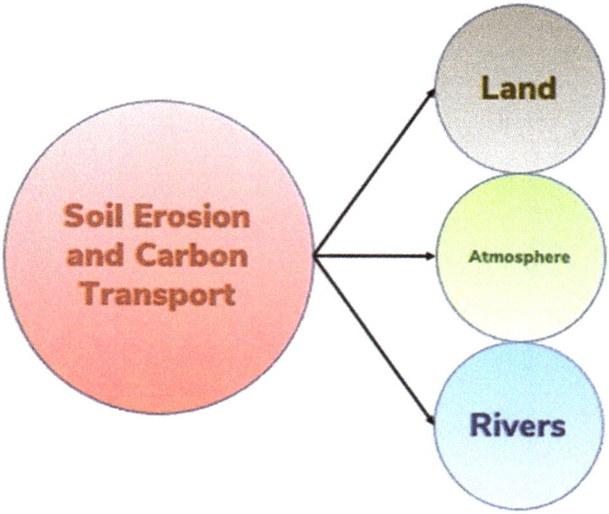

Figure 2. The soil erosion process impacting the exchange of Carbon (C) between the land, atmosphere, and rivers after the global C budget, which must be duly accounted for.

It is widely recognized that sediments have a high SOC enrichment ratio (SOC-ER) and soil organic nitrogen enrichment ratio (SON-ER), and thus are enriched in SOC/SON contents and other light fractions, which are selectively transported by hydric and aeolian processes. Based on a study of a grazed watershed of South Africa, Juarez et al. [11] observed that eroded sediments were enriched in SOC by a factor of 1.85 and in lignin by a factor of 3.7. The high lignin concentration indicated that it was less decomposed in eroded sediments than in the original soils. Further, a negative relationship was observed between the C:N ratio of eroded sediments and soil CO_2 production, indicating that the microbial biomass could be N-limited. Juarez et al. [11] concluded that the SOC in eroded sediment consists primarily of undecomposed plant material, and it is mineralized to a greater extent, regardless of the landscape position. This is an important observation indicating the mineralization of eroded SOC and the emission of GHGs.

For dryland calcareous soils, Cunliffe et al. [12] observed that eroded sediment was significantly enriched in SOC. For a study located in the Tertiary Hills, about 40 km north of Munich, Germany, Wilken et al. [13] observed that sediments are highly enriched in SOC compared to the parent soil. In northeast China, Wei et al. [14] also observed that SOC was enriched in sediments, indicating a high selectivity of the erosion processes. Holz and Augustin [15] also observed that eroded sediments were enriched in C and N by 51.3% and 50.6%, respectively, and that both elements are stored in soil fractions that are selectively eroded. Both SOC-ER and SON-ER are affected by slope gradient and soil texture. However, the C:N ratio increased by 9.9%, indicating the preferential movement of C-rich particulate organic matter (POM) compared to N-rich mineral-associated organic matter (MAOM).

4. Enrichment Ratio for Rill versus Inter-Rill Erosional Processes

Despite the overall selectivity of hydric erosion, rill and inter-rill processes may have contrasting impacts on the SOC-ER of sediment. In general, inter-rill erosion (or soil splash) may indicate preferential removal over the rill process. Based on a laboratory rainfall and flume experiment, Schiettecattle et al. [16] observed that SOC-ER was not affected by rill erosion but varied between 0.9 and 2.6 for inter-rill (splash) erosion, and it was inversely related to the unit sediment discharge. In addition, physicochemical differences between the liberated and mobilized material and the residual raindrop-stable soil aggregates can be an important determinant of the composition of sediments. Moreover, based on a rainfall simulation study, Hou et al. [17] observed that liberated particles in all tilled (plowed) soil contained higher a concentration of oxidized lignin phenols, a lower proportion of cinnamyl to vanillyl lignin, and a trend toward a higher proportion of microbially processed N, which indicates a more decomposed microbial processed SOC. These characteristics indicate the preferential removal of some constituents by inter-rill erosion.

The water repellency of aggregates, caused by SOC content, may also affect aggregate stability [18] and thus, the properties of the liberated and mobilized materials during different erosional stages. For an oxisol from Hawaii, Sutherland et al. [19] observed the preferential detachment and transport of splashed aggregates within the size range of 250 to 2000 μm. They also observed that the kinetic energy required to detach 1 g SOC varied from a median of 1870 J for the 2000–4000 μm fraction to 120 J for the 425–850 μm fraction. Crust formation by the raindrop impact is another factor that affects SOC-ER [8], and the loss of biocrust can increase the loss of SOC by erosion [20]. The SOC loss by inter-rill erosion is also affected by the initial SOC content, rainfall pattern, and the primary particle size distribution [21].

5. Soil Carbon Dynamics at Different Stages of Soil Erosion and the Attendant Emission of Greenhouse Gases

The multi-stage process of soil erosion has a variable effect on the properties and the fate of the liberated and mobilized constituents.

5.1. Stage I: Detachment and Disruption of Soil Aggregates

Raindrop-induced disruption of soil aggregates, the first stage of the erosional process (Figure 1), is the dominant cause of splash, or inter-rill soil erosion. The SOC entrapped within the aggregates is exposed to microbial processes by the disruption of aggregates and the liberation of some materials. Change in pore-size distribution and soil–water matric potential may also affect microbial activity and thus, the vulnerability of the liberated material to decomposition and emission of GHGs. Mordhorst et al. [22] indicated that the mechanical disturbance of the soil structure can alter pore systems and thus, the CO_2 effluxes. Physical influences may be overlapped by biological effects due to the higher energy supply to the microbes. Thus, the susceptibility of physical protection mechanism to SOC by the disruption of aggregates (structure) may also aggravate CO_2 effluxes [22]. Wilken et al. [23] used the event-based multi-class sediment transport model regarding a 100-year time series of rainfall observation on a 3-ha agricultural catchment in Belgium. Based on the modeling analysis, Wilken et al. [23] concluded that: (a) inter-rill erosion is a selective process which erodes primary particles, (b) rill erosion is non-selective and erodes secondary particles (aggregates), (c) particulate organic carbon (POC) is mostly encapsulated within aggregates, and (d) the export of SOC-ER is the highest in events comprised primarily of inter-rill erosion. Holz and Augustin [15] observed that the breakdown of aggregates by rainfall energy may release POM, which is then preferentially eroded. Thus, SOC and plant nutrients are selectively moved during erosion.

5.2. Stage II: Redistribution over the Landscape

The second stage, involving redistribution of sediments over the landscape, depends on physiography and the slope-based flow patterns. In the Loess hilly-gully catchment of China, Liu et al. [24] observed that although check dams served as C and N storage and sequestration structures in the Loess hilly-gully region, erosion-induced C and N redistribution may be a major source of atmospheric CO_2 and N_2O. Based on an 11.3 m check dam sediment profile in the Loess Plateau region of China, Wang et al. [25] observed that check dams intercepted 98.5 Gg of eroded soil and 172.6 Mg of SOC. Between 1960 and 1990, a total of 89.7 Mg of SOC were lost during the soil erosion process at a rate of 0.17 Mg/ha. y, and accounted for approximately 30% of the total eroded SOC exported from the eroding areas. Wang et al. [25] concluded that the soil erosion process is an important net source of GHGs. In a Mediterranean catchment, Nadeu et al. [26] observed that over a 28-year period, 26% of the eroded C remained within the catchment's hill slopes and was not homogenously distributed over the landscape. A modeling study conducted in the Nelson Farm watershed in Mississippi by Liu et al. [27] documented that those soils were consistently indicated as C sources to the atmosphere at all landscape positions. Nadeu et al. [26] observed that between 1870 and 1950, the lowest source strength was located at the eroding sites (13 to 24 gC/m^2. y), the intermediate was at ridge-top (34 gC/m^2. y), and the highest was at the depositional sites (42 to 49 gC/m^2. y). Between 1950 and 1997, the soils at all landscape positions became C sinks because of the changes in management practices (e.g., genetic improvement, fertilization). The sink strengths were the highest at eroding sites (42 to 44 gC/m^2. y), intermediate at the ridge top (35 gC/m^2. y), and the lowest at the depositional site (26 to 29 gC/m^2. y).

5.3. Stages III to V: Deposition, Deep Burial, and Re-Aggregation of Dispersed Soil

The discussions on the fate of C in stages III to V are combined because these involve inter-dependent processes of deposition, deep burial, and reaggregation of dispersed soil through the formation of organo-mineral complexes. The SOC-laden sediments are deposited on depressional sites, concave slopes, and foot slopes. The fate of SOC deposited on these sites depends on a range of factors such as soil temperature and moisture regimes, burial in the subsoil, and the degree of re-aggregation and the attendant formation of organo-mineral complexes or soil aggregates. The burial of transported C at the depositional sites is of critical importance. In the Yellow River Basin of China between 1950 and 2010, 134 ± 24.7 Pg of soil and 1.07 ± 0.15 Pg of SOC have been eroded. Approximately 63% of eroded soils were deposited in the river system and 37% were discharged into the ocean [28]. For SOC, approximately 0.53 ± 0.21 Pg (49.5%) was buried in the river systems, 0.25 ± 0.14 Pg (23.5%) was delivered into the ocean, and the remaining 0.289 ± 0.294 Pg (27%) was decomposed during the erosion/transport processes. Thus, Ran et al. [28] concluded that 20–40% of eroded SOC is oxidized after erosion, that the erosion-induced redistribution of soil is a significant source of SOC, and that human activities have and are altering its redistribution pattern.

Understanding the fate of the buried SOC is critical to compiling a credible and accurate global C budget. Wang et al. [29] used a numerical model to simulate long-term C dynamics in colluvial soils and observed that the C-burial efficiency (the ratio of current C content of the buried sediments to the original C content at the time of sedimentation) of deposited sediments may decrease over time, reaching a level of equilibrium. Half of the buried C at depositional sites may be lost over ~300 y, and the rate of depletion of SOC at the depositional site may be accelerated by agricultural management [29]. He et al. [30] studied the fate of SOC across two topo-sequences (convex and concave) over a long gentle slope in the northeast China. He et al. [30] observed that the net loss of surface soil along the eroding convex segment resulted in a greater depletion of SOC than that along the depositional concave segment. Soil re-aggregation at the deposition site can transform highly mobile-C rich fine primary particles into rather immobile soil aggregates [23]. Wang et al. [29] observed that buried SOC in colluvial profiles was more stable with a turnover time at the centennial scale. However, Ma et al. [31] observed that SOC at the depositional sites may be vulnerable to severe losses due to change in environmental factors. For example, a study in the karst graben basin of China by Lan et al. [32] showed that there was a higher concentration of SOC at the depositional sites relative to the upland soils. However, there was also a higher decomposition rate of SOC at the depositional sites than at the eroding sites, particularly under the farmland conditions. Similarly, Mao et al. [33] also observed that deposition increased C emission under both drying and wetting conditions for Ultisol and Entisol. In a Mollisol, however, C emissions increased under drying, but decreased under wetting conditions. Therefore, the effects of deposition on C emissions are regulated by soil texture and moisture regimes, and the development of macropores due to shrinkage in depositional soils increased both soil water evaporation and the attendant C emissions [33].

5.4. Stage VI: Fate of Soil Carbon in Aquatic Ecosystems

There are several factors that affect CO_2 emission from reservoirs or aquatic ecosystems. Important among these are temperature, oxygen level, properties of SOM and sediments, primary production, etc. For a study in a sub-tropical hypoxic reservoir in southeastern China, Yan et al. [34] observed that an increased supply of organic matter from storm runoff and to a lesser extent, primary production in summer, enhanced subsequent oxygen consumption. Yan et al. [34] also observed that the surface reservoir was usually a source of CO_2, except in summer, due to high primary productivity. The anaerobic respiration of OM can lead to CH_4 emission. Townsend-Small et al. [35] estimated that Lake Erie emits $1.3 \pm 0.6 \times 10^5$ kg CH_4 -C/d in late summer, with 30% of CH_4 derived from natural gas infrastructure. In addition to O_2 availability, other factors included sub-surface currents and high winds, which may also affect CH_4 efflux. The potential increase in precipitation, algal biomass, and the attendant incidence of hypoxia may aggravate the efflux of CH_4, leading to positive climate change feedback [35].

River ecosystems have also been reported to generate N_2O efflux. A study in the Yellow River of China by Ma et al. [31] indicated N_2O concentrations in the lower reaches of 8.78 to 24.26 nmol/L. Total N_2O flux from the Yellow River to the Bohai Sea was about 2.27×10^5 mol/y. Ma et al. [31] observed an 8-fold increase in N_2O flux at the beginning of the water sediment regulation, and the increase may be due to stimulation of nitrification in the water column.

Crawford et al. [36] measured CO_2 and CH_4 concentrations and emissions in a headwater stream network of interior Alaska and identified consistent areas of elevated CO_2 and CH_4. Likewise in Alaska, Ko et al. [37] also observed that watershed slope and precipitation interact to control gaseous C fluxes from the stream, reporting the highest efflux from low-gradient watersheds following rainstorms.

Riparian ecosystems and flood plains also impact the global C cycle. A study in a Thur River flood plain (northeastern Switzerland) by Shrestha et al. [38] showed that soil–water content and C availability were the key determinants controlling N pools and processes. These two factors strongly depended on the sedimentation and inundation dynamics. Shrestha and colleagues also observed locally high N_2O efflux. A study in the Prairie Pothole Region of Manitoba, Canada, by Gao et al. [39] showed that spring thaw increased the concentrations and contents of N_2O at 15 to 35 cm in the lower and riparian area before reaching the soil surface. Furthermore, the riparian area consistently exhibited the highest CH_4 emissions. Gao et al. [39] observed that these spikes in CH_4 occurred after the spring thaw N_2O emissions, with the accumulation of CH_4 and a decline in O_2 concentration. Whereas lakes are considered as sinks for sediment-laden C, Yang et al. [40] reported that lakes can be super-saturated with CO_2 and CH_4, thus emitting these GHGs. The ratio of C emission into the atmosphere to C burial into the sediment was 0.08 for Lake Donghou, China. Yang et al. [40] estimated that gaseous C emission accounted for 15% of the total C output.

6. General Discussion on the Fate of Soil Organic Carbon Transported by Hydric Erosion

The review presented indicates that the impact of erosion on SOC dynamics differs among sites of erosion and deposition, as well as during transport, along with the confounding effects of soil and crop management practices (Table 1). The important determinants of the fate of SOC being transported by erosional processes on a landscape are: (a) the detachment of sediments and SOC, (b) the net primary production at the eroding site, and (c) the changes in mineralization upon transport and deposition because of aggregate breakdown and deep burial, respectively. Thus, Kirkels et al. [41] proposed an eco-geomorphic approach. The latter includes a physical and biological approach to understand the fate of SOC transported by erosion.

Table 1. Research studies which indicate the efflux of CO_2 and other greenhouse gases at different stages of hydric erosion.

Stage	Pedological Process	References
I. Detachment	Increased access of SOC to microbial processes.	• Mordhorst et al. [22] • Wilken et al. [23] • Holz and Augustin [15]
II. Redistribution	GHG emission is affected by physiography and slope-based flow patterns.	• Liu et al. [24,27] • Nadeu et al. [26] • Wang et al. [25]
III–V. Deposition	Efflux of gases is controlled by soil temperature and moisture regimes, burial of SOC in subsoil, and re-aggregation.	• Ran et al. [28] • Wang et al. [25,29] • He et al. [30] • Ma et al. [31] • Mao et al. [33]
VI. Aquatic Ecosystems	O_2 levels, CO_2 and CH_4 concentration, temperatures, properties of sediments and SOM, watershed slope, supersaturation of lakes.	• Townsend-Small et al. [35] • Crawford et al. [36] • Ko et al. [37] • Gao et al. [39] • Yang et al. [39]

The sink/source phenomena depend on which mechanisms are dominant during different stages (Figure 1). Thus, understanding SOC dynamics during specific stages of soil erosion is essential to accurately assess the net effect of erosion on the soil C budget. There is a strong knowledge gap regarding the fate of SOC between the disruption of soil aggregates from the eroding site (stage I) through the transport and redistribution stage (stage II) and depositional/burial site (stage III). The transport of SOC subjects it to changing moisture and temperature regimes (environmental conditions) which alter the protective mechanisms (physical, chemical, biological, ecological). Thus, a thorough understanding of the fate of SOC during the transport and redistribution stages is critical to assessing the source/sink phenomena [42].

Re-aggregation at the depositional sites may also enhance the stabilization of SOC. In the Belgian Loess Belt, for example, Wang et al. [43] observed that the deposition of eroded soil material significantly increased SOC content, as well as the microbial biomass C, throughout the 2 m depth of the soil profile. Wang et al. [43] also reported reduced C mineralization that contributed to C accumulation, as was observed in depositional sites. They attributed the reduced C mineralization to the composition of the microbial community after burial of the SOC-rich material. Reduced mineralization may partly be due to encapsulation of SOC within stable aggregates, thereby reducing physical access to microbial processes [44,45].

Vandenbygaart et al. [46] reported significant effects of erosion on both the lateral and vertical distribution of soil in agricultural landscapes in different agroecosystems across

Canada over a 50-year period. The deepest aggraded profile was observed in potato fields, which contained more than 70 cm of deposited soil, accounting for 152 Mg/ha. y. The rate of deposition in some fields was as much as 4090 Mg/ha. y. Vandenbygaart et al. [46] also reported the largest stock of SOC at a 463 Mg/ha. y. to 60 cm depth, and the attendant SOC deposition rate was 2 to 4 Mg C/ha. y. across all sites. Furthermore, a large increase was observed in SOC concentration near the bottom of the A horizon, and there was also some evidence of the dynamic replacement of SOC in eroded soils. Analytical data show that the mineralization of buried SOC has been constrained since burial.

The buried SOC in alluvial depositional zones, accounting for 75% of the total SOC pool, may serve as a centennial to a millennial time scale C sink [47]. Soil alkalinity may be another factor affecting the resilience of buried SOC against microbial activities [47]. However, the quality rather than quantity of SOC is an important determinant of soil biological properties and the fate of C after its burial in depositional sites [48]. Furthermore, the time lag between burial and the subsequent loss of SOC can be an important factor affecting the mean residence time (MRT) of the deposited SOC.

Van Oost et al. [49] reported that the C buried in eroded terrestrial deposits during the agricultural expansion of the last 150 y. cannot be assumed to be inert to further destabilization, and thus, it may become a significant C source. Indeed, the buried SOC may be vulnerable to future loss by land use, management, alkalinity or pH, the time lag between burial and subsequent loss, mineralogical properties, and global warming. Mineralogical attributes of sediments are also important because of their effects on continuous soil weathering, along with mineral alterations and the breakdown of aggregates, which temporarily limit the protection of SOC by minerals and within aggregates. Continuous soil weathering can weaken the protection mechanisms and render the deposited SOC a source of emission [45]. Thus, the best option is to minimize the risks of the accelerated soil erosion of agroecosystems through the adoption of conservation-effective practices.

7. Fate of Soil Carbon Transported by Aeolian Processes

Aeolian processes are dominant in arid regions, but both hydric and aeolian processes are causes of accelerated erosion in semi-arid regions. Globally, wind erosion may affect 0.6 B ha of land [50]. Wind erosion also involves different stages: detachment, saltation, wind-blown dust, and deposition. Similar to hydric erosion, wind erosion is also a selective process which leads to a significant lateral transport of C and fine clay fraction. Wind-blown dust leads to a long-range transport on an inter-continental scale. Examples include North African dust plumes over the Atlantic Ocean, summer dust plumes from the Arabian Peninsula over the Arabian Sea and the Indian Ocean, and spring dust plumes from East Asia over the Pacific Ocean [51]. These dust plumes strongly affect the radiation budget, the hydrologic cycle, and the global C budget. Depending on the magnitude of C being transported, distributed, and deposited/buried, a significant amount may be emitted as CO_2 (and other GHGs) into the atmosphere.

SOC stocks are strongly reduced in soils prone to wind erosion. In Western Australia, Chappell and Baldock [52] estimated that the mean net SOC erosion was up to 0.2 Mg C/ha. y for the period of 1990 to 2013. Moreover, in southwestern Australia, wind erosion resulted in the loss of up to 3% of the total stock of C to 1-m depth or 3.6 Mg C/ha for the eroded soils. In the Texas High Plains, McDonald et al. [53] observed that a greater amount of CO_2-C was being assimilated by the wheat cover crop from the atmosphere than was lost from the soil, which reduced the net C losses from the soil. However, the erosion-related loss of SOC from wind erosion could be more severe during the fallow phase of the rotation cycle. Thus, farmers must be especially judicious in protecting the soil against SOC loss by wind erosion during the fallow phase in the Pacific Northwest region of the U.S. [54]. Similar to the high risks of wind erosion from croplands on plowed fields during the fallow phase, overgrazing can also reduce SOC content in the long term.

The loss of nutrients (N, P, K, and C) by wind erosion is mostly attributed to losses by suspension, because suspension preferentially removes the finest—and thus the lightest—

fractions. However, the greatest mass of the sediment is removed by saltation. Thus, the majority of the nutrients and C may also be moved by saltation. Visser and Sterk [55] observed that nutrient losses from one field during one event in the Sahel realm of Africa can be as much as 73% of the N and 100% of the P requirements of crop production. A study conducted in a desert grassland of Southern New Mexico, USA, by Li et al. [56] showed the dramatic increase in wind erosion between 75% grass cover reduction and 100% grass cover reduction, suggesting that sparsely distributed mesquites are rather ineffective at reducing wind erosion and nutrient loss compared to grasses.

8. Selectivity of the Aeolian Process

Similar to hydric erosion, gaseous emissions are also affected at different stages of wind erosion (Table 2). Lighter fractions (SOC, clay, and fine silt) are also selectivity removed in wind-blown dust. Towards an attempt to understand the magnitude and mechanism of SOC-ER in dust, using Big Spring Number Eight (BSNE) wind-vane samplers across five land types in rangelands of Western Queensland, Australia, Webb et al. [57] observed that sandy soils and finer particulate quartz-rich soils are more efficient at SOC emission and have larger SOC-ER in dust than that in clay-rich aggregated soils.

Table 2. Gaseous emissions at different stages of aeolian erosion processes.

Stage	Processes	References
I. Detachment	Sandy and particulate quartz-rich soil, local topography, wind velocity.	• Webb et al. [57] • Munch et al. [58] • Stovern et al. [59] • Touré et al. [60] Li et al. [56]
II. Saltation	Mobile sand dunes, vegetation cover, soil temperature.	• Wang et al. [61]
III–V. Deposition, Deep Burial, and Re-Aggregation	Buried SOC; climate change, anthropogenic perturbations, fallow phase of the rotation cycle.	• Chaopricha et al. [62] • Sharratt and Schillinger [54]

The SOC-ER of dust originating from sand-rich soil ranged from 2–41. Webb et al. [57] observed that two dust samples from regional transport events contained 15–20% SOC.

In addition to texture, wind speed also affects SOC-ER. Using a wind tunnel on a manured sandy soil, Munch et al. [58] observed that wind erosion caused a preferred release of manure particles at wind speeds close to the threshold of 7 m/s, with the greatest sorting effects in size, shape, and density the of particles. Munch and colleagues concluded that wind erosion immediately leads to the loss of added organic material. At higher wind speeds, Munch et al. observed that the sediment composition corresponds to the entire soil or soil-manure mixture. Depending on the wind speed and the magnitude of erosion, the loss of manure ranged from 1.7–14% of the fertilization rate.

Wind velocity interacts closely with the local topography, as a factor affecting sediment transport by wind erosion. Based on studies involving simulation or dust transport from a mine tailing using a computational fluid dynamics model, Stovern et al. [59] observed that local topography and wind velocity profiles are the major factors that influence deposition.

In a study involving sandy dunes in Southeastern Niger, Touré et al. [60] observed that saltation flux on the bare mobile dune was 2–40 times higher than that observed from cultivated fields and rangelands. Touré and colleagues measured the annual fluxes on the dune to be from 2000 to 6000 kg/m^2, due to the variability of the wind from year to year.

Soil temperature is also among the major factors controlling desert soil CO_2 flux [61]. In the Horqin Sandy Land, northeastern China, Luo et al. [63] observed that soil temperature at a depth of 20 cm was significantly correlated with soil respiration. The autotrophic respiration (R_a) was more sensitive to soil temperature than the heterotrophic respiration

(R_h), and the temperature-sensitivity coefficients (Q_{10}) were 3.39, 2.52, and 2.79 for R_a, R_h, and R_t, respectively.

Vegetation cover plays an important role in wind erosion and loss of SOC by dust-blown sediments. In a desert grassland of southern New Mexico, USA, Li et al. [56] observed that wind erosion rates and loss by dust emissions (nutrients including C) are strongly affected by plant cover. The observed dramatic increase in wind erosion between 75% grass cover reduction and 100% grass cover reduction may be due to the ineffectiveness of sparsely distributed mesquites in reducing wind erosion and nutrient loss. Over three windy seasons (March 2004 to July 2006), wind erosion removed up to 25% of total organic C (TOC) and total N (TN) from the top 5 cm of soil, and about 60% of TOC and TN during the first windy season (March–July 2004). They estimated that the lifetime of surface soil TOC and TN was about 10 y on these plots because of the 100% grass cover reduction. This drastic impact indicates the disastrous effects of accelerated wind erosion on soil resources which can occur even over short times periods [56]. Belnap et al. [64] estimated that aeolian deposition can provide up to 75% of essential plant nutrients (N, P, K, Mg, Na, Mn, Cu, Fe) through the mineralization of the deposited SOC. Therefore, understanding fate of deposited/buried SOC is critical.

9. Fate of Buried Soil Organic Carbon

Understanding the fate of C buried at depositional sites during stages III to V of the erosional process is important for determining the fate of C transported by erosional processes driven by water and wind. Knowing whether the buried soil C is a source or sink and determining the dominant processes which govern its turnover time is critical to understanding the fate of C transported by erosional processes. Despite being a globally important process, its fate under site-specific conditions is not widely understood. Chaopricha et al. [62] observed that the buried SOC concentration on a global scale can rival those of surface soils. Indeed, buried soils can contain significant regional SOC reservoirs which are not accounted for in current inventories and biogeophysical or biogeochemical models. Chaopricha et al. [62] hypothesized that buried SOC stocks may cycle biologically on annual to decade timescales because of the alteration of processes that contribute to their protection. Buried SOC can be vulnerable to emissions because of climate change and anthropogenic perturbations. The fate of buried SOC must be duly considered in terms of its role in creating positive feedback to climate change.

10. Conclusions

Accelerated erosion is among the most dominant soil degradation processes globally. It is a selective process involving preferential removal of SOC along with fine particles. Thus, erosion leads to the lateral transport of SOC over the landscape, which may be vulnerable to decomposition and the attendant gaseous emissions (CO_2, CH_4, N_2O), with positive feedback to climate change. The synthesis of the literature supports the following conclusions.

The synthesis of the literature presented herein proves the specific hypotheses that the SOC transported during stages I to III is a major source of GHGs, and the magnitude of the emission depends on site-specific temperature and moisture regimes and other physiographic and pedological conditions. The deep burial and reformation of aggregates (stages IV and V)can reduce the rate of mineralization. However, the prevalence of anaerobic conditions at depositional sites may aggravate methanogenesis, leading to the emission of CH_4 and the nitrification and denitrification processes leading to the emission of N_2O. The net effect of the erosional process, from stage I through the transport in aquatic ecosystems (stage VI), may lead to the emission of GHGs. Despite the local and global significance, the magnitude of soil C being transported by the erosional processes is not currently being accounted for in local, regional, national, and global C budgets, but it must be duly accounted for. Furthermore, the re-distribution of C over the landscape and its deep burial does not indicate a sequestration of new C within the landscape. Any redistribution of SOC or SIC over the landscape is not to be confused with sequestration.

Key conclusions from the review are as follows:

a. Sediment transported by the hydric and aeolian process are enriched with SOC and have a high SOC-ER and SON-ER;
b. Whereas erosion-induced transport of SOC is prone to mineralization and exposure to microbial processes, emissions are relatively high at stages I to III of the hydric processes;
c. Gaseous emissions are also observed at the depositional sites and in the aquatic ecosystems (e.g., river, riparian zone, lakes, and oceans, stage VI);
d. Loess deposits are also prone to mineralization and GHG emissions;
e. Re-carbonatization and deep burial of SOC may limit mineralization of SOC;
f. Lateral transport of SOC by erosional processes to deposition and burial in depressional sites (stages III to V)and aquatic ecosystems(stage VI) is redistribution rather than a new sequestration of atmospheric CO_2;
g. Buried SOC may be vulnerable to decomposition because of the climate change and other anthropogenic activities, leading to the emission of CO_2, CH_4, and N_2O;
h. Thus, a prudent strategy is to minimize the risks of soil erosion through the adoption of judicious land use and site-specific BMPs.

Funding: This research received no external funding and was funded by the CFAES Rattan Lal Center for Carbon Management and Sequestration.

Institutional Review Board Statement: Not applicable.

Informed Consent Statement: Not applicable.

Data Availability Statement: Data and Research cited in the text and tables in support of the contents have been appropriately sourced in the bibliography.

Conflicts of Interest: The author declares no conflict of interest supporting data, and appropriate references are listed in Tables 1 and 2.

References

1. Borrelli, P.; Robinson, D.A.; Panagos, P.; Lugato, E.; Yang, J.E.; Alewell, C.; Wuepper, D.; Montanarella, L.; Ballabio, C. Land use and climate change impacts on global soil erosion by water (2015–2070). *Proc. Natl. Acad. Sci. USA* **2020**, *117*, 21994–22001. [CrossRef] [PubMed]
2. Lal, R. Soil Erosion and the Global C Budget. *Environ. Int.* **2003**, *29*, 437–450. [CrossRef]
3. Friedlingstein, P.; Jones, M.W.; O'Sullivan, M.; Andrew, R.M.; Bakker, D.C.E.; Hauck, J.; Le Quéré, C.; Peters, G.P.; Peters, W.; Pongratz, J.; et al. Global Carbon Budget 2021. *Earth Syst. Sci. Data* **2022**, *14*, 1917–2005. [CrossRef]
4. Oertel, C.; Matschullat, J.; Zurba, K.; Zimmermann, F.; Erasmi, S. Greenhouse gas emissions from soils—A review. *Geochemistry* **2016**, *76*, 327–352. [CrossRef]
5. Bouchoms, S.; Wang, Z.; Vanacker, V.; Doetterl, S.; Van Oost, K. Modelling long-term soil organic carbon dynamics under the impact of land cover change and soil redistribution. *CATENA* **2017**, *151*, 63–73. [CrossRef]
6. Olson, K.R.; Al-Kaisi, M.; Lal, R.; Lowery, B. Experimental considerations, treatment and methods in determining soil organic carbon sequestration rates. *Soil Sci. Soc. Am. J.* **2014**, *78*, 348–360. [CrossRef]
7. Xiao, H.; Li, Z.; Chang, X.; Huang, B.; Nie, X.; Liu, C.; Liu, L.; Wang, D.; Jiang, J. The mineralization and sequestration of organic carbon in relation to agricultural soil erosion. *Geoderma* **2018**, *329*, 73–81. [CrossRef]
8. Kuhn, N.J.; Armstrong, E.K.; Ling, A.C.; Connolly, K.L.; Heckrath, G. Interrill erosion of carbon and phosphorus from conventionally and organically farmed Devon silt soils. *CATENA* **2010**, *91*, 94–103. [CrossRef]
9. Lal, R.; Monger, C.; Nave, L.; Smith, P. The role of soil in regulation of climate. *Philos. Trans. R. Soc. B-Biol. Sci.* **2021**, *376*, 20210084. [CrossRef]
10. Lal, R. Biophysical Controls That Make Erosion-Transported Soil Carbon a Source of Greenhouse Gases. *Appl. Sci.* 2022, 12. *in press*.
11. Juarez, S.; Rumpel, C.; Mchunu, C.; Chaplot, V. Carbon mineralization and lignin content of eroded sediments from a grazed watershed of South-Africa. *Geoderma* **2011**, *167–168*, 247–253. [CrossRef]
12. Cunliffe, A.M.; Puttock, A.K.; Turnbull, L.; Wainwright, J.; Brazier, R.E. Dryland, calcareous soils store (and lose) significant quantities of near-surface organic carbon. *J. Geophys. Res. Earth Surf.* **2016**, *121*, 684–702. [CrossRef]

13. Wilken, F.; Sommer, M.; Van Oost, K.; Bens, O.; Fiener, P. Process-oriented modelling to identify main drivers of erosion-induced carbon fluxes. *Soil* **2017**, *3*, 83–94. [CrossRef]
14. Wei, S.; Zhang, X.; McLaughlin, N.B.; Chen, X.; Jia, S.; Liang, A. Impact of soil water erosion processes on catchment export of soil aggregates and associated SOC. *Geoderma* **2017**, *294*, 63–69. [CrossRef]
15. Holz, M.; Augustin, J. Erosion effects on soil carbon and nitrogen dynamics on cultivated slopes: A meta-analysis. *Geoderma* **2021**, *397*, 115045. [CrossRef]
16. Schiettecatte, W.; Gabriels, D.; Cornelis, W.M.; Hofman, G. Enrichment of Organic Carbon in Sediment Transport by Interrill and Rill Erosion Processes. *Soil Sci. Soc. Am. J.* **2008**, *72*, 50–55. [CrossRef]
17. Hou, T.; Berry, T.; Singh, S.; Hughes, M.N.; Tong, Y.; Papanicolaou, A.T.; Wacha, K.M.; Wilson, C.G.; Chaubey, I.; Filley, T.R. Control of tillage disturbance on the chemistry and proportion of raindrop-liberated particles from soil aggregates. *Geoderma* **2018**, *330*, 19–29. [CrossRef]
18. Korenkova, L.N.; Matúš, P. Role of water repellency in aggregate stability of cultivated soils under simulated raindrop impact. *Eurasian Soil Sci.* **2015**, *48*, 754–758. [CrossRef]
19. Watung, R.; Sutherland, R.; El-Swaify, S. Influence of rainfall energy flux density and antecedent soil moisture content on splash transport and aggregate enrichment ratios for a Hawaiian Oxisol. *Soil Technol.* **1996**, *9*, 251–272. [CrossRef]
20. Chamizo, S.; Rodríguez-Caballero, E.; Román, J.R.; Cantón, Y. Effects of biocrust on soil erosion and organic carbon losses under natural rainfall. *CATENA* **2017**, *148*, 117–125. [CrossRef]
21. Parwada, C.; Van Tol, J. Influence of litter source on soil splash rates and organic carbon loss in different soil horizons. *Water SA* **2019**, *45*, 12–19. [CrossRef]
22. Mordhorst, A.; Peth, S.; Horn, R. Influence of mechanical loading on static and dynamic CO2 efflux on differently textured and managed Luvisols. *Geoderma* **2014**, *219–220*, 1–13. [CrossRef]
23. Wilken, F.; Fiener, P.; Van Oost, K. Modelling a century of soil redistribution processes and carbon delivery from small watersheds using a multi-class sediment transport model. *Earth Surf. Dyn.* **2017**, *5*, 113–124. [CrossRef]
24. Liu, C.; Li, Z.; Chang, X.; He, J.; Nie, X.; Liu, L.; Xiao, H.; Wang, D.; Peng, H.; Zeng, G. Soil carbon and nitrogen sources and redistribution as affected by erosion and deposition processes: A case study in a loess hilly-gully catchment, China. *Agric. Ecosyst. Environ.* **2018**, *253*, 11–22. [CrossRef]
25. Wang, S.; Feng, Q.; Zhao, W.; Fu, B.; Ding, J. Ecosystem service trade-offs and their influencing factors: A case study in the Loess Plateau of China. *Sci. Total. Environ.* **2017**, *607–608*, 1250–1263.
26. Nadeu, E.; Van Oost, K.; Boix-Fayos, C.; De Vente, J. Importance of land use patterns for erosion-induced carbon fluxes in a Mediterranean catchment. *Agric. Ecosyst. Environ.* **2014**, *189*, 181–189. [CrossRef]
27. Liu, S.; Bliss, N.; Sundquist, E.; Huntington, T. Modeling carbon dynamics in vegetation and soil under the impact of soil erosion and deposition. *Glob. Biogeochem. Cycles* **2003**, *17*. [CrossRef]
28. Ran, L.; Lu, X.X.; Xin, Z. Erosion-induced massive organic carbon burial and carbon emission in the Yellow River basin, China. *Biogeosciences* **2014**, *11*, 945–959. [CrossRef]
29. Wang, Z.; Van Oost, K.; Govers, G. Predicting the long-term fate of buried organic carbon in colluvial soils. *Glob. Biogeochem. Cycles* **2014**, *29*, 65–79. [CrossRef]
30. He, Y.; Hu, Y.; Gao, X.; Wang, R.; Guo, S.; Li, X. Minor topography governing erosional distribution of SOC and temperature sensitivity of CO2 emissions: Comparisons between concave and convex toposequence. *J. Soils Sediments* **2020**, *20*, 1906–1919. [CrossRef]
31. Ma, X.; Zhang, G.-L.; Liu, S.-M.; Wang, L.; Li, P.-P.; Gu, P.-P.; Sun, M.-S. Distributions and fluxes of nitrous oxide in lower reaches of Yellow River and its estuary: Impact of water-sediment regulation. *Estuarine Coast. Shelf Sci.* **2016**, *168*, 22–28. [CrossRef]
32. Lan, G.; Liu, C.; Wang, H.; Tang, W.; Wu, X.; Yang, H.; Tu, L.; Hu, B.X.; Cao, J.; Li, Q. The effect of land use change and soil redistribution on soil organic carbon dynamics in karst graben basin of China. *J. Soils Sediments* **2021**, *21*, 2511–2524. [CrossRef]
33. Mao, N.; Li, W.; Wei, X.; Shao, M. Soil moisture- and texture-dependent effects of soil deposition on evaporation and carbon emission. *Soil Tillage Res.* **2020**, *204*, 104703. [CrossRef]
34. Yan, J.; Chen, N.; Wang, F.; Liu, Q.; Wu, Z.; Middelburg, J.J.; Xie, Y.; Guo, W.; Zhang, X. Interaction Between Oxygen Consumption and Carbon Dioxide Emission in a Subtropical Hypoxic Reservoir, Southeastern China. *J. Geophys. Res. Biogeosci.* **2021**, *126*. [CrossRef]
35. Townsend-Small, A.; Disbennett, D.; Fernandez, J.M.; Ransohoff, R.W.; Mackay, R.; Bourbonniere, R.A. Quantifying emissions of methane derived from anaerobic organic matter respiration and natural gas extraction in Lake Erie. *Limnol. Oceanogr.* **2016**, *61*, S356–S366. [CrossRef]
36. Crawford, J.T.; Striegl, R.G.; Wickland, K.P.; Dornblaser, M.M.; Stanley, E.H. Emissions of carbon dioxide and methane from a headwater stream network of interior Alaska. *J. Geophys. Res. Biogeosci.* **2013**, *118*, 482–494. [CrossRef]
37. Ko, J.; Lee, J.; Kang, H. Effects of micro-topography on N2O emission from sediments in temperate streams. *Ecol. Eng.* **2020**, *155*, 105906. [CrossRef]
38. Shrestha, J.; Niklaus, P.; Frossard, E.; Samaritani, E.; Huber, B.; Barnard, R.L.; Schleppi, P.; Tockner, K.; Luster, J. Soil Nitrogen Dynamics in a River Floodplain Mosaic. *J. Environ. Qual.* **2012**, *41*, 2033–2045. [CrossRef]
39. Gao, X.; Tenuta, M.; Nelson, A.; Sparling, B.; Tomasiewicz, D.; Mohr, R.M.; Bizimungu, B. Effect of nitrogen fertilizer rate on nitrous oxide emission from irrigated potato on a clay loam soil in Manitoba, Canada. *Can. J. Soil Sci.* **2013**, *93*, 1–11. [CrossRef]

40. Yang, H.; Xing, Y.; Xie, P.; Ni, L.; Rong, K. Carbon source/sink function of a subtropical, eutrophic lake determined from an overall mass balance and a gas exchange and carbon burial balance. *Environ. Pollut.* **2008**, *151*, 559–568. [CrossRef]
41. Kirkels, F.; Cammeraat, L.; Kuhn, N. The fate of soil organic carbon upon erosion, transport and deposition in agricultural landscapes—A review of different concepts. *Geomorphology* **2014**, *226*, 94–105. [CrossRef]
42. de Nijs, E.; Cammeraat, E.L. The stability and fate of Soil Organic Carbon during the transport phase of soil erosion. *Earth-Sci. Rev.* **2020**, *201*. [CrossRef]
43. Wang, X.; Cammeraat, L.H.; Wang, Z.; Zhou, J.; Govers, G.; Kalbitz, K. Stability of organic matter in soils of the Belgian Loess Belt upon erosion and deposition. *Eur. J. Soil Sci.* **2013**, *64*, 219–228. [CrossRef]
44. Dungait, J.A.; Ghee, C.; Rowan, J.S.; McKenzie, B.M.; Hawes, C.; Dixon, E.R.; Paterson, E.; Hopkins, D.W. Microbial responses to the erosional redistribution of soil organic carbon in arable fields. *Soil Biol. Biochem.* **2013**, *60*, 195–201. [CrossRef]
45. Doetterl, S.; Cornelis, J.-T.; Six, J.; Bodé, S.; Opfergelt, S.; Boeckx, P.; Van Oost, K. Soil redistribution and weathering controlling the fate of geochemical and physical carbon stabilization mechanisms in soils of an eroding landscape. *Biogeosciences* **2015**, *12*, 1357–1371. [CrossRef]
46. VandenBygaart, A.J.; Kroetsch, D.; Gregorich, E.G.; Lobb, D. Soil C erosion and burial in cropland. *Glob. Chang. Biol.* **2011**, *18*, 1441–1452. [CrossRef]
47. Ferguson, B.; Lukens, W.E.; El Masri, B.; Stinchcomb, G.E. Alluvial landform and the occurrence of paleosols in a humid-subtropical climate have an effect on long-term soil organic carbon storage. *Geoderma* **2020**, *371*, 114388. [CrossRef]
48. Liu, X.; Sheng, H.; Jiang, S.; Yuan, Z.; Zhang, C.; Elser, J.J. Intensification of phosphorus cycling in China since the 1600s. *Proc. Natl. Acad. Sci. USA* **2016**, *113*, 2609–2614. [CrossRef]
49. Van Oost, K.; Verstraeten, G.; Doetterl, S.; Notebaert, B.; Wiaux, F.; Broothaerts, N.; Six, J. Legacy of human-induced C erosion and burial on soil–atmosphere C exchange. *Proc. Natl. Acad. Sci. USA* **2012**, *109*, 19492–19497. [CrossRef]
50. Oldeman, L. *The Global Extent of Soil Degradation*; WOS:A1994BA80F00007; Greenland, D., Szabolcs, I., Eds.; ISRIC: Wageningen, The Netherlands, 1994; pp. 99–144.
51. Choobari, O.A.; Zawar-Reza, P.; Sturman, A. The global distribution of mineral dust and its impacts on the climate system: A review. *Atmos. Res.* **2014**, *138*, 152–165. [CrossRef]
52. Chappell, A.; Baldock, J.A. Wind erosion reduces soil organic carbon sequestration falsely indicating ineffective management practices. *Aeolian Res.* **2016**, *22*, 107–116. [CrossRef]
53. McDonald, M.D.; Lewis, K.L.; Ritchie, G.L.; DeLaune, P.B.; Casey, K.D.; Slaughter, L.C. Carbon dioxide mitigation potential of conservation agriculture in a semi-arid agricultural region. *AIMS Agric. Food* **2019**, *4*, 206–222. [CrossRef]
54. Sharratt, B.; Schillinger, W.F. Windblown Dust Potential from Oilseed Cropping Systems in the Pacific Northwest United States. *Agron. J.* **2014**, *106*, 1147–1152. [CrossRef]
55. Visser, S.M.; Sterk, G. Nutrient dynamics—Wind and water erosion at the village scale in the Sahel. *Land Degrad. Dev.* **2007**, *18*, 578–588. [CrossRef]
56. Li, J.; Okin, G.; Alvarez, L.; Epstein, H. Quantitative effects of vegetation cover on wind erosion and soil nutrient loss in a desert grassland of southern New Mexico, USA. *Biogeochemistry* **2007**, *85*, 317–332. [CrossRef]
57. Webb, N.P.; Strong, C.L.; Chappell, A.; Marx, S.K.; McTainsh, G.H. Soil organic carbon enrichment of dust emissions: Magnitude, mechanisms and its implications for the carbon cycle. *Earth Surf. Process. Landforms* **2013**, *38*, 1662–1671. [CrossRef]
58. Münch, S.; Papke, N.; Leue, M.; Faust, M.; Schepanski, K.; Siller, P.; Roesler, U.; Nübel, U.; Kabelitz, T.; Amon, T.; et al. Differences in the sediment composition of wind eroded sandy soils before and after fertilization with poultry manure. *Soil Tillage Res.* **2021**, *215*, 105205. [CrossRef]
59. Stovern, M.; Felix, O.; Csavina, J.; Rine, K.P.; Russell, M.R.; Jones, R.M.; King, M.; Betterton, E.A.; Sáez, A.E. Simulation of windblown dust transport from a mine tailings impoundment using a computational fluid dynamics model. *Aeolian Res.* **2014**, *14*, 75–83. [CrossRef] [PubMed]
60. Touré, A.A.; Tidjani, A.; Rajot, J.; Marticorena, B.; Bergametti, G.; Bouet, C.; Ambouta, K.; Garba, Z. Dynamics of wind erosion and impact of vegetation cover and land use in the Sahel: A case study on sandy dunes in southeastern Niger. *CATENA* **2019**, *177*, 272–285. [CrossRef]
61. Wang, B.; Liu, J.; Zhang, X.; Wang, C. Changes in soil carbon sequestration and emission in different succession stages of biological soil crusts in a sand-binding area. *Carbon Balance Manag.* **2021**, *16*, 27. [CrossRef]
62. Chaopricha, N.T.; Marin-Spiotta, E. Soil burial contributes to deep soil organic carbon storage. *Soil Biol. Biochem.* **2014**, *69*, 251–264. [CrossRef]
63. Luo, Y.; Zhao, X.; Ding, J.; Li, Y.; Liu, X.; Wang, T.; Luo, Y. Soil respiration dynamics in a semi-fixed sand dune under the pioneer shrub Artemisia halodendron Turcz. ex Bess. in the Horqin sandy land, northeastern China. *Arid Land Res. Manag.* **2019**, *34*, 36–51. [CrossRef]
64. Belnap, J.; Phillips, S.; Duniway, M.; Reynolds, R. *Soil Fertility in Deserts: A Review on the Influence of Biological Soil Crusts and the Effect of Soil Surface Disturbance on Nutrient Inputs and Losses*; USGS, WOS: 000184355600025; Alsharhan, A., Wood, W., Goudie, A., Fowler, A., Abdellatif, E., Eds.; Taylor & Francis: Abingdo, UK, 2003; pp. 245–252.

applied sciences

MDPI

Article

The Importance of Annual Plants and Multi-Scalar Analysis for Understanding Coastal Dune Stabilization Process in the Mediterranean

Pua Bar Kutiel and Michael Dorman *

Department of Geography and Environmental Development, Ben-Gurion University of the Negev, Be'er-Sheva 84105, Israel; kutiel@bgu.ac.il
* Correspondence: dorman@bgu.ac.il

Abstract: Since ecological phenomena and patterns vary with scale, scalar analysis is a developing practice in ecology. Scalar analysis is most valuable in heterogeneous environments, since habitat heterogeneity is a key factor in determining biodiversity. One such case can be seen in the changes in annual vegetation in coastal sand dune systems. Most studies in these environments are carried out at the dune scale, comparing dunes at different stabilization states. However, a broader understanding of dune stabilization processes requires analyses at the finer scales of dune slope aspects (directions of exposure to wind) and patches (under and between woody perennial species). Here, we present the results of a study that combines the three scales (dune, slope, and patch) in the Mediterranean coastal dune systems in Israel. Through this multi-scalar analysis, we are able to describe processes at the finer patch and aspect scale and explain how they shape patterns at the dune scale. The results indicate that the dune scale exposes the differences in annual plant characteristics between mobile and fixed dunes, their slopes and patches and the reorganization and spatial distribution of annual plants within mobile and fixed dunes during the stabilization process.

Keywords: annual plants; coastal dunes; fixed dunes; mobile dunes; dune stabilization; Mediterranean; open and shrub patches; slope aspects

Citation: Bar Kutiel, P.; Dorman, M. The Importance of Annual Plants and Multi-Scalar Analysis for Understanding Coastal Dune Stabilization Process in the Mediterranean. *Appl. Sci.* **2021**, *11*, 2821. https://doi.org/10.3390/app11062821

Academic Editor: Pietro Picuno

Received: 23 February 2021
Accepted: 19 March 2021
Published: 22 March 2021

1. Introduction

As a science that discusses spatial and temporal variations and dynamics, ecology is sensitive to differences in spatial and temporal scales [1]. Since some phenomena and mechanisms are observable only at certain scales, while others may exhibit contrasting patterns at various scales, scale is a fundamental theoretical and methodological issue (e.g., [2–6]). Wiens [2] has emphasized the importance of understanding processes at finer scales for understanding patterns at broader scales. McGill [7] has demonstrated how the abundance and distribution of species are determined by variables of different spatial scales, and thus how multi-scalar analysis is sometimes essential for understanding ecological phenomena. The issue of scale becomes more important when biodiversity and landscape ecology are discussed. Biodiversity is partly maintained by habitat heterogeneity [8], which is manifested in differences between habitats at different scales, from larger scales such as climatic gradients to smaller scales such as the immediate environment of a single organism.

Wheatley and Johnson [5] have distinguished between spatial and scalar analyses. Spatial analysis deals with the locations of entities and distances between them and is independent of sizes of observational units (grains). On the other hand, scalar analysis deals with changes in the observations at different grain sizes. We suggest that scalar analysis should be further subdivided into size and entity-defined scalar analyses. In a size-defined scalar analysis, grains are defined according to their sizes (e.g., 2 or 0.5 m scales), with less regard to the entities that construct landscape heterogeneity (i.e., what is inside the grain measurement unit). In an entity-defined scalar analysis, grains are defined first based on their identity and the entities inside them (e.g., the patch or community scales; see also [2,6,9]), and less by their sizes, so measurement units at different scales may be of the same size (i.e., nested structure of plots of the same size). Intermediate designs, in which equal or almost equal emphasis on both size and entity is given, may also exist.

Sand dunes are an environment in which multiple scales of habitat heterogeneity are very obvious (e.g., [10]). Wind drift potential and vegetation cover determine sand mobility [11], and thus the percentage of vegetation cover and the stabilization state of dunes are most commonly discussed as the key independent variables [11–20]. Dunes at different stabilization states also differ in terms of their vegetation composition. Mobile dunes usually support psammophilous species, while fixed dunes usually support non-psammophile species (such as generalist species) [12,13,21]. Since variation in dune stabilization states is currently a pressing issue [16,19–31], most studies have focused on the dune scale. Studies at the dune scale commonly describe differences among dune stabilization states and discuss the processes dunes undergo as they stabilize. Ecologically, these include an increase in species richness following the establishment of non-psammohile species, and usually the loss of psammophilous species [15,18,20].

Barchans and parabolic dunes are common types of dunes in regions where the common wind blows in one direction [32]. They are comprised of three slopes that differ in their abiotic conditions: the windward, that is exposed to wind erosion; the crest, with zero net sand erosion deposition but with greater exposure to wind; and the leeward slip face subjected to sand deposition [32]. The differences in wind activity and erosion-deposition balance among the three dune aspects affect plants' ability to establish and survive. Thus, there are also apparent differences in geomorphic and ecological processes at the slope scale [14,23,28,33,34]. Consequently, plant species are not homogeneously distributed on dunes or among slopes; on fixed dunes, vegetation cover is distributed more homogeneously among slopes compared to mobile dunes [14,23,33–38].

Vegetation patches constitute a finer scale. Wind activity in mobile dunes constrains plant establishment, usually resulting in the formation of small vegetation patches (hereafter referred to as shrub patches) in an otherwise barren open matrix (open patches). The shrub patches themselves reduce wind drift potential and can facilitate the establishment of more vegetation, mainly on mobile and semi-fixed dunes [11,39,40]. The patchy distribution of vegetation is also maintained on the fixed dunes [35,41]. Studies at the patch scale usually focus on shrub patches, the effects of woody perennial plant species on their immediate surroundings and on vegetation (often annual plants) under shrub canopies [42–48]. The matrix in the open patches between perennial vegetation patches was once thought to be hardly affected by the woody perennial vegetation, but it is now evident that the effects of perennial plants can exceed the limits of their canopies and that the effects of shrubs on the vegetation differ between shrub and open patches [13,23,33,35,47,49].

Therefore, entity-defined multi-scalar analysis has the potential to reveal the complexity behind the general term "dune stabilization", providing a broad perspective of the processes that dunes undergo during stabilization at different scales, elucidating the effects of the establishment of perennial species on their immediate environment (shrub patches) and surroundings (open patches) in different dune slope aspects and stabilization states. In this paper, we demonstrate the importance of studying vegetation at multiple entity-defined scales through analyses of the annual vegetation on coastal Mediterranean sand dunes. Our baseline for analysis is the commonly discussed dune scale. We then turn to the finer scales of the slope and patch types. We demonstrate that different information is acquired from analyses at each scale and show how the information contributes to the overall description of the annual vegetation at the study site, its variability, dynamics and possible governing processes.

2. Materials and Methods

2.1. Study Site

The study was conducted at the Nizzanim Long-Term Ecological Research (LTER) site in Nizzanim Dunes Nature Reserve, located at the southern part of the Mediterranean coastal plain, Israel (31°42′–31°44′ N, 34°35′–34°36′ E), covering an area of 20 km^2 (Figure 1). The site consists of mobile, semi-fixed and fixed dunes, separated by densely vegetated inter-dune depressions [16,18]. In the present study, we focused on the mobile (hereafter M) and fixed dunes (hereafter F)—the two extreme states. The climate is Mediterranean with an annual average temperature of 20 °C and annual rainfall of 400–500 mm falling mainly during winter (November–April). The common wind direction is south-west with a very low drift potential index (147) [38].

Figure 1. A map of Israel, indicating the location of Nizzanim Long-Term Ecological Research (LTER) Nature Reserve.

The mobile parabolic dunes have 5–15% perennial vegetation cover, which is mainly distributed on the dune crest and the slipface (windward vegetation cover is 3%, crest—20%, and slipface—32%, when normalized by the slope length resulting in 5–15% total cover). Maritime grass (*Ammophila arenaria*) is characteristic only to the mobile dunes and is the dominant perennial species along with wormwood (*Artemesia monosperma*). Fixed dunes have 31–50% perennial plant cover, which is distributed almost evenly across all slopes (38%, 44% and 46%, respectively). They are dominated by desert broom (*Retama raetam*) and wormwood and have higher perennial plant species richness. They are the only dune state to have Mediterranean perennial species such as *Prasium majus* and *Asparagus horridus* [49].

2.2. Study Design and Fieldwork

The study was of a nested design: the patch scale was nested in the slope scale, which was nested in the dune scale. Six dunes were studied: three mobile dunes and three fixed dunes. We focused on annual plants as they are more susceptible to a wide spatial range of changes in soil and micro-climate properties compared to the woody perennial plants [50].

Annual plant observations were conducted during spring (March–April) in the years 2006, 2007, 2009, 2014, 2015 and 2016. For each dune, 72 quadrats of 40 × 40 cm were sampled per year, with 24 quadrats per slope aspect, 12 under the shrub and 12 in the open, adjacent to the observed shrubs. Quadrats were placed randomly but alternately into perennial bush or open patches within a 100 m^2 zone at the middle of each slope. The quadrats in the fixed dunes were placed only under *Artemisia monosperma*, a key species of the coastal sand dunes in the Levant [51], and in mobile dunes only under *Artemisia monosperma* and *Ammophila arenaria*, since these were the most common perennial species in the respective dunes. Annual plants were identified to species level, when possible, and their relative percentage cover was recorded at each quadrat.

2.3. Statistical Analysis

The effects of the environmental conditions (dune type, slope type and patch type) on total vegetation cover, species richness (i.e., number of observed species), beta diversity (i.e., distance to group centroid in ordination) and species assemblages were evaluated as follows:

1. Environmental effects on vegetation cover (first row in Figure 2) were evaluated by fitting mixed effects linear models with random effects of the dune identity (e.g., M1 = mobile dune #1, etc.) and sampling year (e.g., 2007) to take into account unmeasured sources of variation affecting plant cover across spatial locations and sampling time periods. Significance levels of differences between estimated marginal means of the groups were corrected using the Bonferroni method.
2. Environmental effects on species richness (i.e., count of unique species) (second row in Figure 2) were evaluated with the same approach as in case (1), only using mixed effects generalized linear models, with a Poisson distribution, instead of mixed effects linear models, since the dependent variable was the species count. Significance levels of differences between estimated marginal means of the groups were corrected using the Bonferroni method.
3. Environmental effects on beta diversity (third row in Figure 2) were evaluated using Anderson's PERMDISP2 procedure for the analysis of multivariate homogeneity of group dispersions [52], where the distances of group members to the group centroid are subject to ANOVA. Samples were placed in multivariate space using the β_t [53,54] measure of dissimilarity. Significance levels of differences between group means were corrected using Tukey's "Honest Significant Difference" method.
4. Environmental effects on species composition were evaluated using redundancy Analysis (RDA) ordination (Figure 3), separately for each pair of groups being compared, followed by a permutation test to assess the significance of the constraints, corrected using the Bonferroni method.

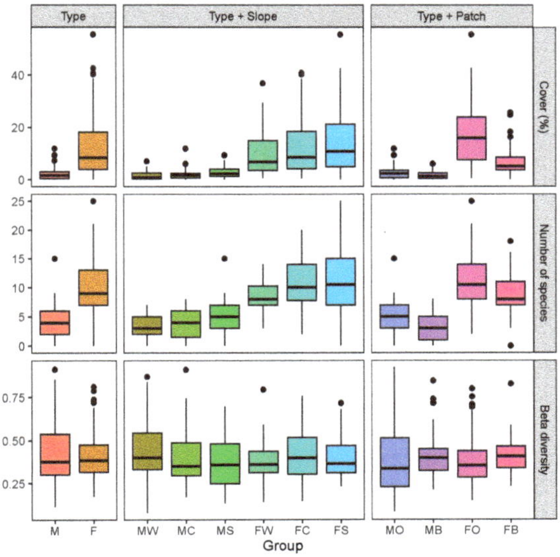

Figure 2. Distribution of vegetation cover, species richness and beta diversity among samples, separately for each dune type ("Type"), dune type and slope combination ("Type + Slope"), and dune type and patch combination ("Type + Patch") (M = mobile dune; F = fixed dune; W = windward; C = crest; S = slipface; O = open; B = bush).

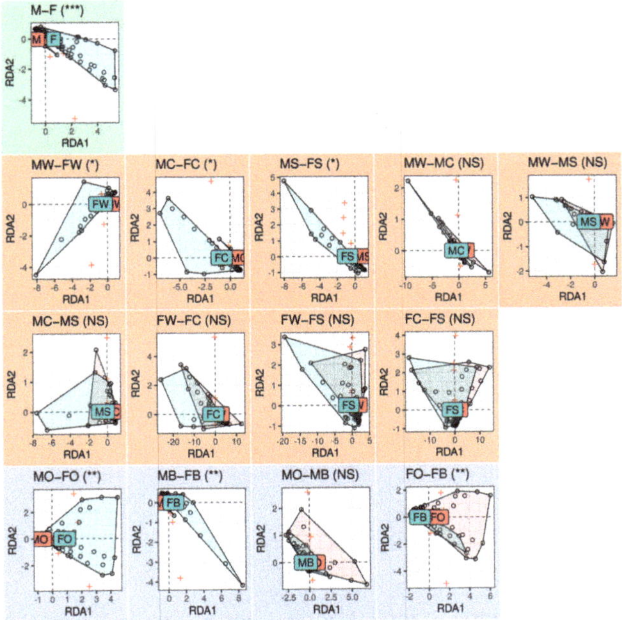

Figure 3. Redundancy analysis (RDA) of species composition, constrained by dune type (top), dune type and slope type combinations (middle), or dune type and patch type combinations (bottom). Species scores are shown in red, site scores are shown in black. Sites belonging to each of the two groups being compared are marked with a convex hull polygon (M = mobile dune; F = fixed dune; W = windward; C = crest; S = slipface; O = open; B = bush, *** $p < 0.001$, ** $p < 0.01$, * $p < 0.05$, NS = not significant).

The latter four types of models, where the dependent variables were either (1) cover, (2) richness, (3) beta diversity, or (4) plant assemblage, were fitted separately for three "levels" of observation:

1. Comparing dune types,
2. Comparing dune type/slope combinations, and
3. Comparing dune type/patch combinations.

The results regarding the environmental effects on cover, richness, beta diversity and plant assemblages are summarized in Table 1 and visualized in Figure 2 (cover, richness and beta diversity) and Figure 3 (plant assemblages).

Table 1. Estimates and significance levels when comparing vegetation cover, species richness, species assemblages (i.e., species composition) and beta diversity, among dune types ("Type"), dune type and slope combinations ("Type + Slope"), and dune type and patch combinations ("Type + Patch"). The "Effect" column specifies the pairwise comparisons; for example, an M-F cover estimate of −10.43 means that vegetation cover was higher by 10.43, on average, for fixed dunes compared to mobile dunes (M = mobile dune; F = fixed dune; W = windward; C = crest; S = slipface; O = open; B = bush). (*** $p < 0.001$, ** $p < 0.01$, * $p < 0.05$, NS = not significant).

Level	Effect	Cover	Richness	Assemblages	Beta
Type	M-F	−10.43 **	−0.90 ***	***	0.01 NS
Type + Slope	MW-FW	−8.85 **	−0.95 ***	*	0.04 NS
Type + Slope	MC-FC	−10.77 ***	−1.01 ***	*	−0.01 NS
Type + Slope	MS-FS	−11.63 ***	−0.75 ***	*	−0.02 NS
Type + Slope	MW-MC	−0.52 NS	−0.14 NS	NS	0.02 NS
Type + Slope	MW-MS	−1.32 NS	−0.44 **	NS	0.04 NS
Type + Slope	MC-MS	−0.81 NS	−0.30 NS	NS	0.03 NS
Type + Slope	FW-FC	−2.44 NS	−0.20 NS	NS	−0.04 NS
Type + Slope	FW-FS	−4.11 NS	−0.23 *	NS	−0.02 NS
Type + Slope	FC-FS	−1.67 NS	−0.04 NS	NS	0.02 NS
Type + Patch	MO-FO	−15.26 ***	−0.81 ***	**	0.00 NS
Type + Patch	MB-FB	−5.66 NS	−1.03 ***	**	0.01 NS
Type + Patch	MO-MB	1.08 NS	0.41 ***	NS	−0.02 NS
Type + Patch	FO-FB	10.68 ***	0.18 *	**	−0.02 NS

Species turnover, when moving from mobile to fixed dunes, was calculated at three levels: considering the species assemblage in all samples combined, separately for samples from each slope type and separately for samples from each patch type (Figure 4). Species turnover was defined using the β_t index, contrasting the complete list of species in mobile vs. fixed dunes [53].

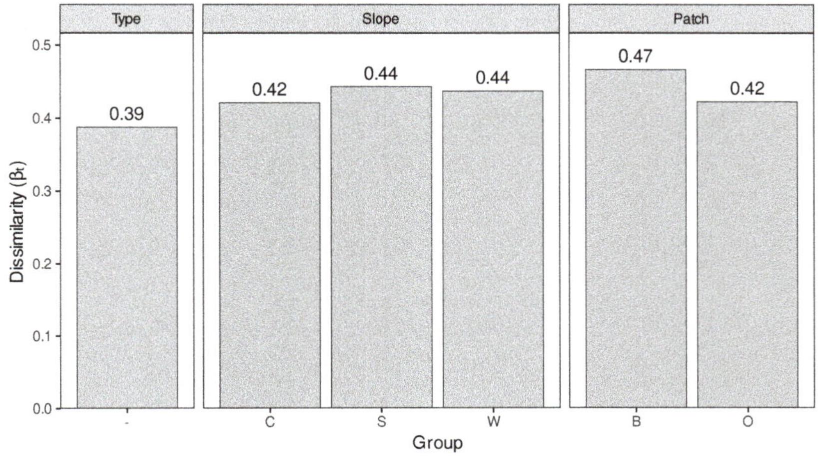

Figure 4. Species turnover (β_t) of the transition from mobile to stabilized dunes, when considering the species composition in all samples combined ("Type"), separately in each slope type ("Slope") or separately in each patch type ("Patch") (M = mobile dune; F = fixed dune; W = windward; C = crest; S = slipface; O = open; B = bush).

Indicator species were identified using the indicator value method [55], using three levels of grouping of samples: by dune types, by dune type and slope type combinations and by dune type and patch type combinations (Table 2).

Table 2. Significant ($p < 0.05$) indicator species of dune types ("Type"), dune type and slope type combinations ("Type + Slope"), and of dune type and patch type combinations ("Type + Patch") (M = mobile dune; F = fixed dune; W = windward; C = crest; S = slipface; O = open; B = bush).

Level	Ind.	Species
Type	M	*Trisetaria koelerioides*
Type	F	*Anagallis arvensis, Arenaria leptoclados, Asphodelus tenuifolius, Avena barbata, Brassica tournefortii, Bromus rigidus, Bromus sterilis, Campanula sulphurea, Daucus glaber, Erodium laciniatum, Galium aparine, Geranium robertianum, Hormuzakia* sp., *Lagurus ovatus, Lotus halophilus, Lupinus angustifolius, Lupinus palaestinus, Maresia pulchella, Ononis serrata, Phleum exaratum, Plantago sarcophylla, Rumex bucephalophorus, Rumex pictus, Torilis arvensis, Urospermum picroides*
Type + Slope	MW	*Crepis aculeata, Cutandia memphitica, Trisetaria koelerioides*
Type + Slope	MC	*Cutandia memphitica, Trisetaria koelerioides*
Type + Slope	MS	*Corynephorus articulatus, Crepis aculeata, Launaea fragilis, Lotus halophilus, Trisetaria koelerioides*
Type + Slope	FW	*Asphodelus tenuifolius, Avena barbata, Brassica tournefortii, Bromus rigidus, Corynephorus articulatus, Crepis aculeata, Daucus glaber, Galium aparine, Hormuzakia* sp., *Lotus halophilus, Maresia pulchella, Rumex bucephalophorus, Rumex pictus*

Table 2. *Cont.*

Level	Ind.	Species
Type + Slope	FC	*Anagallis arvensis, Asphodelus tenuifolius, Avena barbata, Brassica tournefortii, Bromus rigidus, Campanula sulphurea, Corynephorus articulatus, Crepis aculeata, Cutandia memphitica, Daucus glaber, Erodium laciniatum, Galium aparine, Geranium robertianum, Hormuzakia* sp., *Lotus halophilus, Lupinus palaestinus, Maresia pulchella, Plantago sarcophylla, Rumex bucephalophorus, Rumex pictus, Trisetaria koelerioides, Urospermum picroides*
Type + Slope	FS	*Anagallis arvensis, Arenaria leptoclados, Asphodelus tenuifolius, Avena barbata, Brassica tournefortii, Bromus rigidus, Campanula sulphurea, Corynephorus articulatus, Crepis aculeata, Daucus glaber, Erodium laciniatum, Galium aparine, Geranium robertianum, Hormuzakia* sp., *Lotus halophilus, Lupinus palaestinus, Ononis serrata, Rumex bucephalophorus, Rumex pictus, Trisetaria koelerioides, Urospermum picroides, Vicia sativa*
Type + Patch	MO	*Crepis aculeata, Ifloga* sp., *Lotus halophilus, Polycarpon succulentum, Trisetaria koelerioides*
Type + Patch	MB	*Launaea fragilis, Trisetaria koelerioides*
Type + Patch	FO	*Anagallis arvensis, Arenaria leptoclados, Asphodelus tenuifolius, Brassica tournefortii, Bromus rigidus, Bromus sterilis, Campanula sulphurea, Crepis aculeata, Daucus glaber, Erodium laciniatum, Galium aparine, Hormuzakia* sp., *Ifloga* sp., *Lotus halophilus, Lupinus angustifolius, Lupinus palaestinus, Maresia pulchella, Neurada procumbens, Ononis serrata, Phleum exaratum, Plantago sarcophylla, Polycarpon succulentum, Rumex bucephalophorus, Rumex pictus, Torilis arvensis, Trifolium tomentosum, Trisetaria koelerioides, Urospermum picroides*
Type + Patch	FB	*Anagallis arvensis, Asphodelus tenuifolius, Avena barbata, Brassica tournefortii, Bromus rigidus, Campanula sulphurea, Crepis aculeata, Daucus glaber, Erodium laciniatum, Galium aparine, Geranium robertianum, Hormuzakia* sp., *Lagurus ovatus, Lotus halophilus, Maresia pulchella, Plantago sarcophylla, Polycarpon succulentum, Rumex bucephalophorus, Rumex pictus, Sonchus tenerrimus, Urospermum picroides*

2.4. Software

All analyses were done using R [56]. Mixed effects linear models and mixed effects generalized linear models were fitted using the package `lme4` [57]. Marginal means were estimated and pairwise-compared using the `emmeans` package [58] (Table 1). Beta diversity (Figure 2) was estimated through the β_t index [53], using the function `betadiver` in the R package `vegan` [59]. Comparisons of beta diversity between dune types, between type/slope combinations and between type/patch combinations were done using the function `betadisper` in the R package `vegan` [59] (Table 1, Figure 2). Species turnover, also using the β_t index via the `betadiver` function, was estimated when combining all samples from each dune type into a collective species list (Figure 3). Permutation tests on redundancy analysis (RDA) ordination were done using the package `vegan` [59]. Pairwise comparisons of groups in RDA were done using the function `multiconstrained` in the R package `BiodiversityR` [60] (Table 1, Figure 3). Indicator species identification (Table 2) was done using the package `indicspecies` [61].

3. Results

3.1. Overview

Overall, 77 annual plant species were identified across all samples. Out of the 77 species, three species were observed only in mobile dunes, 40 species were observed only in fixed dunes and 34 species were observed in both mobile and fixed dunes. We refer to each scale separately, comparing the differences among the studied landscape units within each scale. Thus, the contribution of each scale can be addressed separately to understand the process of dune stabilization from mobile dunes to stabilized dunes.

3.2. Dune Scale

Annual plant cover and species richness were significantly higher for fixed dunes compared to mobile dunes (Table 1, Figure 2). The plant assemblages also differed significantly between the two dune types (Table 1, Figure 3). Beta diversity did not significantly differ between dune types (Table 1, Figure 2).

The species turnover (β_t) value for the transition from mobile to fixe dunes was 0.39, given possible values between 0–1, where 0 indicates full similarity between two habitats and 1 indicates full dissimilarity (Figure 4).

Trisetaria koelerioides was classified as a significant indicator of mobile dunes, while 25 different species were classified as significant indicators of fixed dunes (Table 2).

3.3. Slope Scale

There were no significant differences between the slopes of the same dune type considering vegetation cover, species richness, assemblages and beta diversity, excluding a significant difference in species richness between the S (slipface) slope aspect and the W (windward) slope aspect of the two dune types. However, there were significant differences in cover, richness and assemblage composition between mobile dunes and the fixed dunes in each of the three slope aspects (Table 1, Figures 2 and 3). Beta diversity among samples did not significantly differ between any of the dune types and slope aspects combinations (Table 1).

The species turnover (β_t) for the transition from mobile dunes to fixed dunes, considering samples from each of the slope types, ranged between 0.42 to 0.44 (Figure 4).

When considering dune type and slope type combinations, 13 indicator species were identified for W slopes, 22 species for C (crest) slopes and 22 for S slopes on fixed dunes, while only three, two and five species were identified as indicators of W, C and S slopes on mobile dunes, respectively (Table 2).

3.4. Patch Scale

Open patches were significantly different from patches beneath the shrubs in terms of vegetation cover (only in stabilized dunes), species richness (in both dune types) and assemblage composition (only in fixed dunes) (Table 1, Figures 2 and 3). Fixed dunes differed from mobile dunes in terms of vegetation cover, species richness and assemblage composition (Table 1, Figures 2 and 3). Beta diversity among samples did not significantly differ between any of the dune types and patch type combinations (Table 1, Figure 2).

The species turnover (β_t), for the transition from mobile to stabilized dunes, was 0.47 when considering samples from bush patches and 0.42 when considering open patches (Figure 4).

When considering dune type and patch type combinations, 28 indicator species were identified for open patches and 21 species for bush patches on fixed dunes, while only five and two species were identified as indicators of open and bush patches on mobile dunes, respectively (Table 2).

4. Discussion

4.1. Overview

Most studies on the dune stabilization process have focused on the dune scale, using the changes in perennial vegetation cover and its spatial pattern. These two measures have a significant direct correlation with dune stabilization [18,35,41]. The increase of perennial vegetation cover is followed by changes in soil properties, namely a decrease in sand mobility [33,47] and increase in soil silt and clay content [14,62], soil organic matter [13,23], soil nutrients [36] and soil moisture [13]. Therefore, the perennial plants are considered as ecosystem engineers that are responsible for the changes in the physical and geomorphic dune state from a mobile to a fixed state, thus increasing spatial heterogeneity [19]. Annual plants are facilitated by the perennial plants, which determine their composition and spatial distribution [13,49,50]. In contrast to perennial plants, annuals are more sensitive to micro-spatial and temporal changes. Annual plant compositions and their morphological and phonological properties change over small distances, sometimes even of a few centimeters, due to micro environmental changes. Perennial plants have strong impacts on the adjacent open patches, as well as the area under their own canopy—especially on sand dunes where the substrate mobility is very crucial for the plants' micro-scale [13,50,63]. Thus, annual plants represent the fine grain of the plant community scale.

4.2. Dune Scale

The transition from mobile dunes to fixed dunes is followed, as expected, by a significant increase in annual vegetation cover and species richness and a significant change in plant assemblages. The β_t species turnover, which indicates the percentage of similarity between the assemblages based on species presence/absence, was only 0.39. Only one species, *Trisetaria koelerioides*, was found to be an indicator species for mobile dunes. The Sandiness Index [41] of this species is 0.91, with index values ranging between 0 and 1, where 1 indicates the highest affinity of an annual (organism) species to mobile sand dunes. In turn, 25 indicator species were identified for fixed dunes, whereas the Sandiness Indexes for 19 species were very low due to the fact that these species can also be found in other Mediterranean habitats and not only on sandy soils. In order to assess the assemblage affinity towards mobile dunes or the similarity to mobile dune assemblages, we developed the Dune Assemblage Index (DAI) [41]. This index, calculated per dune, ranges between 0–1, where a DAI of 0 represents an assemblage in which the species are restricted to fully fixed dunes and a DAI of 1 represents an assemblage in which the species are restricted to fully mobile dunes. The average of the Dune Assemblage Index for mobile dunes at the present research site, based on annual plants, was found to be 0.66, and a score of 0.36 was found for the fixed dunes. To summarize, observations on the dune scale suggest that the transition from mobile to fixed dunes is followed by significant changes in annual plant cover, richness and composition.

4.3. Slope Scale

There were no significant differences in plant cover and assemblages between the different slope types within both dune types. The only detected difference was in the number of species, which was significantly higher in the slipface slopes compared to the windward slopes. The environmental conditions at the mobile dunes are harsh [14,17,38]: the windward slope is subject to sand erosion while the slipface is affected by sand deposition. The perennial plant cover is very low and distributed mainly at the crest and slipface slopes in a clumped pattern [35,39]. The perennial plant cover of the fixed dunes is higher than that of the mobile dunes, but almost uniform across all the slope aspects due to the improvement in the physical environmental conditions [12,50]. Still, the cover on the windward side is slightly lower compared to the other two slope aspects, probably because there is still some sand movement at the windward slope.

Comparing dune types separately for samples taken from each of the three slope types indicated that the annual plant cover, the number of species and the plant assemblages significantly differ between dune types. The β_t species turnover ranges between 0.42–0.44; i.e., there was a similarity level of about 40% among dune types, considering the various slope types.

4.4. Patch Scale

One of the most limiting factors for annual plants in the Mediterranean is light [64]. Therefore, we expect to find these plants mostly at the open patches. Moreover, the hydraulic conductivity at the dune open patches is much higher (in our case five times higher) than beneath the shrub due to the coarse texture, and absence of organic matter and allelopathy effect [35]. Therefore, higher soil moisture is available for annual plants at the first 30 cm of soil depth during the rainy season [50].

In both dune types, the number of species in the open patches was significantly higher than beneath shrubs. However, only on the fixed dunes was the difference also expressed by higher cover and different species assemblages.

4.5. Synthesis

Combining the dune, slope aspect and patch scales provides a more complete picture of the processes underlying changes in annual vegetation during dune stabilization, since different processes take place at different scales and stages. The transition from mobile to fixed dunes is characterized by the expansion of annual vegetation mainly in the slip-faces, which are relatively protected from the wind, and by the re-equalization of annual vegetation cover and species number at the patch scale. The annual vegetation cover and species number increase in both shrub and open patches, but in the latter more than in the former. The perennial plants have a stronger net facilitative effect within some distance from their canopies due to wind sheltering, while under their canopy, they have a stronger net negative effect due to competition for water and allelopathy, among other factors [40,65–71].

4.6. Implementation for Biodiversity Conservation

The dune stabilization process under a Mediterranean climate in the Eastern Mediterranean Basin, with a low drift potential, is characterized by significant geomorphological and ecological changes taking place over a period of 40–60 years [16,17]. These changes comprise the transition from parabolic dunes to fixed dunes and from perennial and annual plant assemblages composed of psammophile, desert and endemic plant species to assemblages characterized by species with high Eastern Mediterranean affinity [13,50]. In order to conserve the spatial heterogeneity simultaneously composed of mobile and stabilized dunes, it is necessary to manage the dune ecosystem by the remobilization of a certain part of the stabilized dunes. Principally, the idea sounds simple and coincides with conclusions obtained in numerous studies regarding plant succession in sand dunes (for example, [12]). However, practically, the situation is much more complicated. A study conducted on the dunes studied here for 12 years, tracking plant re-establishment on fixed dunes, from which the perennial plants were removed, showed that treatment had very little effect on the plant community. Stronger response was found in semi-fixed dune treatments, in particular for mobile dune indicator species, which showed evidence of recolonization within a few years following treatment. However, it was concluded that only long and continuous disturbance pressures, such as grazing or even controlled 4×4 vehicle activity, could lead to the desired impact [20].

5. Conclusions

We have demonstrated how entity-defined multi-scalar analysis can promote ecological understanding by integrating the unique data and knowledge that can be acquired at each scale. In this study, we followed the changes in the annual vegetation during dune stabilization on three distinct scales. Starting with the dune scale, which is the most commonly studied, we identified the general trend of increasing annual vegetation cover and species number and the change in plant assemblages during dune stabilization. When making observations on the dune scale, little could be said about processes, except for postulating that annual plants establish more successfully when perennial vegetation cover increases and wind drift potential decreases. Analysis at the slope scale revealed that stabilization does not occur evenly among the morphological units of the dune. Since annual vegetation patterns are known to be associated with exposure to wind, these patterns served as concrete evidence for the facilitative effects of perennial plants in reducing exposure to wind. Analysis at the patch scale revealed that the perennial plants also have negative effects on the annual vegetation, but that these negative effects operate at different scales, magnitudes and spatial distribution than the facilitative effects. Integrating these three scales provided a more detailed, complex and realistic picture of the processes that annual vegetation undergoes during dune stabilization.

Author Contributions: P.B.K.: Conceptualization, methodology, validation, formal analysis, investigation, resources, data curation, writing—original draft preparation, writing—review and editing, supervision, project administration, funding acquisition. M.D.: Software, validation, formal analysis, investigation, data curation, writing—original draft preparation, writing—review and editing, visualization. All authors have read and agreed to the published version of the manuscript.

Funding: This research was funded by the Israeli Nature and Park Authority.

Institutional Review Board Statement: Not applicable.

Informed Consent Statement: Not applicable.

Data Availability Statement: The data presented in this study are available on request from the corresponding author. The data are not publicly available due to being part of a large database used by the project for various research purposes.

Acknowledgments: We wish to thank the Israeli Nature and Park Authority for financing this project, the undergraduate and graduate students who did the field work and Tania Bird for organizing the data.

Conflicts of Interest: The authors declare no conflict of interest.

Abbreviations

The following abbreviations are used in this manuscript:

M	Mobile (dune)
F	Fixed (dune)
W	Windward (slope)
C	Crest (slope)
S	Slipface (slope)
O	Open (patch)
B	Bush (patch)

References

1. Levin, S.A. The problem of pattern and scale in ecology: The Robert H. MacArthur award lecture. *Ecology* **1992**, *73*, 1943–1967. [CrossRef]
2. Wiens, J.A. Spatial scaling in ecology. *Funct. Ecol.* **1989**, *3*, 385–397. [CrossRef]
3. Sale, P.F. Appropriate spatial scales for studies of reef-fish ecology. *Aust. J. Ecol.* **1998**, *23*, 202–208. [CrossRef]
4. Auestad, I.; Rydgren, K.; Økland, R.H. Scale-dependence of vegetation-environment relationships in semi-natural grasslands. *J. Veg. Sci.* **2008**, *19*, 139–148. [CrossRef]
5. Wheatley, M.; Johnson, C. Factors limiting our understanding of ecological scale. *Ecol. Complex.* **2009**, *6*, 150–159. [CrossRef]
6. Bisigato, A.J.; Villagra, P.E.; Ares, J.O.; Rossi, B.E. Vegetation heterogeneity in Monte Desert ecosystems: A multi-scale approach linking patterns and processes. *J. Arid Environ.* **2009**, *73*, 182–191. [CrossRef]
7. McGill, B.J. Matters of scale. *Science* **2010**, *328*, 575–576. [CrossRef] [PubMed]
8. Ferrier, S. Mapping spatial pattern in biodiversity for regional conservation planning: Where to from here? *Syst. Biol.* **2002**, *51*, 331–363. [CrossRef]
9. Carboni, M.; Carranza, M.L.; Acosta, A. Assessing conservation status on coastal dunes: A multiscale approach. *Landsc. Urban Plan.* **2009**, *91*, 17–25. [CrossRef]
10. Zuo, X.; Zhao, X.; Zhao, H.; Zhang, T.; Yulin, L.; Wang, S.; Li, W.; Powers, R. Scale dependent effects of environmental factors on vegetation pattern and composition in Horqin Sandy Land, Northern China. *Geoderma* **2012**, *173*, 1–9. [CrossRef]
11. Yizhaq, H.; Ashkenazy, Y.; Tsoar, H. Why do active and stabilized dunes coexist under the same climatic conditions? *Phys. Rev. Lett.* **2007**, *98*, 188001. [CrossRef] [PubMed]
12. Kutiel, P.; Danin, A.; Orshan, G. Vegetation of the sandy soils near Caesarea, Israel. I. Plant communities, environment and succession. *Isr. J. Plant Sci.* **1979**, *28*, 20–35.
13. Kutiel, P.; Danin, A. Annual-species diversity and aboveground phytomass in relation to some soil properties in the sand dunes of the northern Sharon Plains, Israel. *Vegetatio* **1987**, *70*, 45–49.

14. Danin, A.; Nukrian, R. Dynamics of dune vegetation in the southern coastal area of Israel since 1945. *Doc. Phytosociol.* **1991**, *13*, 281–296.
15. Kutiel, P. Conservation and management of the Mediterranean coastal sand dunes in Israel. *J. Coast. Conserv.* **2001**, *7*, 183–192. [CrossRef]
16. Kutiel, P.; Cohen, O.; Shoshany, M.; Shub, M. Vegetation establishment on the southern Israeli coastal sand dunes between the years 1965 and 1999. *Landsc. Urban Plan.* **2004**, *67*, 141–156. [CrossRef]
17. Levin, N.; Ben-Dor, E. Monitoring sand dune stabilization along the coastal dunes of Ashdod-Nizanim, Israel, 1945–1999. *J. Arid Environ.* **2004**, *58*, 335–355. [CrossRef]
18. Levin, N.; Kidron, G.J.; Ben-Dor, E. A field quantification of coastal dune perennial plants as indicators of surface stability, erosion or deposition. *Sedimentology* **2008**, *55*, 751–772. [CrossRef]
19. Bird, T.L.; Dorman, M.; Ramot, A.; Bouskila, A.; Bar, P.; Groner, E. Shrub encroachment effects on habitat heterogeneity and beetle diversity in a Mediterranean coastal dune system. *Land Degrad. Dev.* **2017**, *28*, 2553–2562. [CrossRef]
20. Bird, T.L.F.; Bouskila, A.; Groner, E.; Bar Kutiel, P. Can Vegetation Removal Successfully Restore Coastal Dune Biodiversity? *Appl. Sci.* **2020**, *10*, 2310. [CrossRef]
21. Martínez, M.L.; Psuty, N.P. *Coastal Dunes*; Springer: Berlin/Heidelberg, Germany, 2004.
22. Viles, H.; Spencer, T. *Coastal Problems: Geomorphology, Ecology and Society at the Coast*; Edward Arnold: London, UK, 1995; 350p, ISBN 978 0340531976.
23. Lesica, P.; Cooper, S.V. Succession and disturbance in sandhills vegetation: Constructing models for managing biological diversity. *Conserv. Biol.* **1999**, *13*, 293–302. [CrossRef]
24. Bray, H.E.; Stokes, S. Chronologies for Late Quaternary barchan dune reactivation in the southeastern Arabian Peninsula. *Quat. Sci. Rev.* **2003**, *22*, 1027–1033. [CrossRef]
25. Stokes, S.; Bailey, R.; Fedoroff, N.; O'Marah, K. Optical dating of aeolian dynamism on the West African Sahelian margin. *Geomorphology* **2004**, *59*, 281–291. [CrossRef]
26. Hugenholtz, C.H.; Wolfe, S.A. Recent stabilization of active sand dunes on the Canadian prairies and relation to recent climate variations. *Geomorphology* **2005**, *68*, 131–147. [CrossRef]
27. Thomas, D.S.; Knight, M.; Wiggs, G.F. Remobilization of southern African desert dune systems by twenty-first century global warming. *Nature* **2005**, *435*, 1218–1221. [CrossRef] [PubMed]
28. Tsoar, H. Sand dunes mobility and stability in relation to climate. *Phys. A Stat. Mech. Its Appl.* **2005**, *357*, 50–56. [CrossRef]
29. Sun, J.; Li, S.H.; Han, P.; Chen, Y. Holocene environmental changes in the central Inner Mongolia, based on single-aliquot-quartz optical dating and multi-proxy study of dune sands. *Palaeogeogr. Palaeoclimatol. Palaeoecol.* **2006**, *233*, 51–62. [CrossRef]
30. Liu, S.; Wang, T. Aeolian desertification from the mid-1970s to 2005 in Otindag Sandy Land, Northern China. *Environ. Geol.* **2007**, *51*, 1057–1064. [CrossRef]
31. Le Houérou, H.N. Man-made deserts: Desertization processes and threats. *Arid Land Res. Manag.* **2002**, *16*, 1–36. [CrossRef]
32. Goudie, A. Parabolic dunes: Distribution, form, morphology and change. *Ann. Arid Zone* **2011**, *50*, 1–7.
33. Lichter, J. Colonization constraints during primary succession on coastal Lake Michigan sand dunes. *J. Ecol.* **2000**, *88*, 825–839. [CrossRef]
34. Dech, J.P.; Maun, M.A. Zonation of vegetation along a burial gradient on the leeward slopes of Lake Huron sand dunes. *Can. J. Bot.* **2005**, *83*, 227–236. [CrossRef]
35. Kutiel, P.B.; Katz, O.; Ziso-Cohen, V.; Divinsky, I.; Katra, I. Water availability in sand dunes and its implications for the distribution of *Artemisia monosperma*. *Catena* **2016**, *137*, 144–151. [CrossRef]
36. Olff, H.; Huisman, J.; Van Tooren, B. Species dynamics and nutrient accumulation during early primary succession in coastal sand dunes. *J. Ecol.* **1993**, *81*, 693–706. [CrossRef]
37. Nichols, W.F.; Killingbeck, K.T.; August, P.V. The influence of geomorphological heterogeneity on biodiversity II. A landscape perspective. *Conserv. Biol.* **1998**, *12*, 371–379. [CrossRef]
38. Tsoar, H.; Blumberg, D.G. Formation of parabolic dunes from barchan and transverse dunes along Israel's Mediterranean coast. *Earth Surf. Process. Landf.* **2002**, *27*, 1147–1161. [CrossRef]
39. Malkinson, D.; Kadmon, R.; Cohen, D. Pattern analysis in successional communities–an approach for studying shifts in ecological interactions. *J. Veg. Sci.* **2003**, *14*, 213–222. [CrossRef]
40. Yu, S.; Bell, D.; Kutiel, P.B. Impact of microhabitats on the heterogeneity of seedling emergence in a Mediterranean coastal sand dunes community. *Ecoscience* **2009**, *16*, 369–378. [CrossRef]
41. Rubinstein, Y.; Groner, E.; Yizhaq, H.; Svoray, T.; Bar, P. An eco-spatial index for evaluating stabilization state of sand dunes. *Aeolian Res.* **2013**, *9*, 75–87. [CrossRef]
42. Holzapfel, C.; Mahall, B.E. Bidirectional facilitation and interference between shrubs and annuals in the Mojave Desert. *Ecology* **1999**, *80*, 1747–1761. [CrossRef]
43. Shumway, S.W. Facilitative effects of a sand dune shrub on species growing beneath the shrub canopy. *Oecologia* **2000**, *124*, 138–148. [CrossRef]
44. Tielbörger, K.; Kadmon, R. Temporal environmental variation tips the balance between facilitation and interference in desert plants. *Ecology* **2000**, *81*, 1544–1553. [CrossRef]

45. El-Bana, M.I.; Nijs, I.; Kockelbergh, F. Microenvironmental and vegetational heterogeneity induced by phytogenic nebkhas in an arid coastal ecosystem. *Plant Soil* **2002**, *247*, 283–293. [CrossRef]
46. Maestre, F.T.; Bautista, S.; Cortina, J. Positive, negative, and net effects in grass–shrub interactions in Mediterranean semiarid grasslands. *Ecology* **2003**, *84*, 3186–3197. [CrossRef]
47. Franks, S.J. Facilitation in multiple life-history stages: Evidence for nucleated succession in coastal dunes. *Plant Ecol.* **2003**, *168*, 1–11. [CrossRef]
48. Sternberg, M.; Yu, S.L.; Bar, P. Soil seed banks, habitat heterogeneity, and regeneration strategies in a Mediterranean coastal sand dune. *Isr. J. Plant Sci.* **2004**, *52*, 213–221. [CrossRef]
49. Perry, M. Perennial Plants Impact on Annual Plant Diversity in Sand Dunes at Different Spatial Scales. Master's Thesis, Ben Gurion University of the Negev, Be'er Sheva, Israel, 2008.
50. Kutiel, P. Annual vegetation of the coastal sand dunes of the northern Sharon, Israel. *Isr. J. Plant Sci.* **1998**, *46*, 287–298. [CrossRef]
51. Katz, O.; Bar Kutiel, P. The ecology of *Artemisia monosperma* Delile: A desert keystone species in sandy ecosystems in the Middle East. In *Artemisia: Classification, Cultivation and Use*; Roe, J.M., Ed.; Nova Science Publishers: New York, NY, USA, 2020; pp. 1–29.
52. Anderson, M.J.; Ellingsen, K.E.; McArdle, B.H. Multivariate dispersion as a measure of beta diversity. *Ecol. Lett.* **2006**, *9*, 683–693. [CrossRef]
53. Koleff, P.; Gaston, K.J.; Lennon, J.J. Measuring beta diversity for presence–absence data. *J. Anim. Ecol.* **2003**, *72*, 367–382. [CrossRef]
54. Wilson, M.V.; Shmida, A. Measuring beta diversity with presence-absence data. *J. Ecol.* **1984**, *72*,1055–1064. [CrossRef]
55. Dufrêne, M.; Legendre, P. Species assemblages and indicator species: The need for a flexible asymmetrical approach. *Ecol. Monogr.* **1997**, *67*, 345–366. [CrossRef]
56. R Core Team. *R: A Language and Environment for Statistical Computing*; R Foundation for Statistical Computing: Vienna, Austria, 2020.
57. Bates, D.; Mächler, M.; Bolker, B.; Walker, S. Fitting Linear Mixed-Effects Models Using lme4. *J. Stat. Softw.* **2015**, *67*, 1–48. [CrossRef]
58. Lenth, R. *emmeans: Estimated Marginal Means, aka Least-Squares Means*; 2020; R Package Version 1.5.2-1. Available online: https://rdrr.io/cran/emmeans/ (accessed on 20 February 2021)
59. Oksanen, J.; Blanchet, F.G.; Friendly, M.; Kindt, R.; Legendre, P.; McGlinn, D.; Minchin, P.R.; O'Hara, R.B.; Simpson, G.L.; Solymos, P.; et al. *Vegan: Community Ecology Package*; 2020; R Package Version 2.5-7. Available online: https://cran.r-project.org/web/packages/vegan/index.html (accessed on 20 February 2021)
60. Kindt, R.; Coe, R. *Tree Diversity Analysis: A Manual and Software for Common Statistical Methods for Ecological and Biodiversity Studies*; World Agroforestry Centre (ICRAF): Nairobi, Kenya, 2005; ISBN 92-9059-179-X.
61. De Caceres, M.; Legendre, P. Associations between species and groups of sites: Indices and statistical inference. *Ecology* **2009**, *90*, 3566–3574.
62. Danin, A. Plant species diversity and plant succession in a sandy area in the Northern Negev. *Flora* **1978**, *167*, 409–422. [CrossRef]
63. Wang, X.; Jiang, J.; Lei, J.; Zhang, W.; Qian, Y. Distribution of ephemeral plants and their significance in dune stabilization in Gurbantunggut Desert. *J. Geogr. Sci.* **2003**, *13*, 323–330.
64. Cid-Benevento, C.R.; Werner, P.A. Local distributions of old-field and woodland annual plant species: Demography, physiological tolerances and allocation of biomass of five species grown in experimental light and soil-moisture gradients. *J. Ecol.* **1986**, *74*, 857–880.
65. Halligan, J.P. Toxicity of Artemisia californica to four associated herb species. *Am. Midl. Nat.* **1976**, *95*, 406–421. [CrossRef]
66. Weaver, T.; Klarich, D. Allelopathic effects of volatile substances from Artemisia tridentata Nutt. *Am. Midl. Nat.* **1977**, *97*, 508–512. [CrossRef]
67. Melkania, N.; Singh, J.; Bisht, K. Allelopathic potential of *Artemisia vulgaris* L. and Pinus roxburghii Sargent: A bioassay study. *Proc. Indian Nat. Sci. Acad. B* **1982**, *48*, 685–688.
68. Kil, B.S.; Yun, K.W. Allelopathic effects of water extracts of Artemisia princeps var. orientalis on selected plant species. *J. Chem. Ecol.* **1992**, *18*, 39–51. [CrossRef]
69. Lydon, J.; Teasdale, J.R.; Chen, P.K. Allelopathic activity of annual wormwood (*Artemisia annua*) and the role of artemisinin. *Weed Sci.* **1997**, *45*, 807–811. [CrossRef]
70. Escudero, A.; Albert, M.J.; Pita, J.M.; Pérez-García, F. Inhibitory effects of Artemisia herba-alba on the germination of the gypsophyte *Helianthemum squamatum*. *Plant Ecol.* **2000**, *148*, 71–80. [CrossRef]
71. Barney, J.N.; Hay, A.G.; Weston, L.A. Isolation and characterization of allelopathic volatiles from mugwort (*Artemisia vulgaris*). *J. Chem. Ecol.* **2005**, *31*, 247–265. [CrossRef]

applied sciences

MDPI

Article

Asynchrony Drives Plant and Animal Community Stability in Mediterranean Coastal Dunes

Tania L.F. Bird [1,*], Pua Bar (Kutiel) [1], Elli Groner [2] and Amos Bouskila [3]

1 Department of Geography and Environmental Development, Ben Gurion University of the Negev, Beer-Sheva 8410501, Israel; kutiel@bgu.ac.il
2 The Dead Sea-Arava Science Center, Mitzpe Ramon, Dead-Sea Mobile Post 86910, Israel; elli@adssc.org
3 Department of Life Sciences, Ben Gurion University of the Negev, Beer-Sheva 8410501, Israel; bouskila@bgu.ac.il
* Correspondence: birdt@post.bgu.ac.il

Abstract: Substantial evidence now suggests that a positive diversity–stability relationship exists. Yet few studies examine the facets of biodiversity that contribute to this relationship, and empirical research is predominantly conducted on grassland communities under controlled conditions. We investigate the roles of species richness, environmental condition (vegetation cover), asynchrony, and weighted population stability in driving community stability across multiple taxa. We used data from a Long-term Ecological Research project to investigate temporal stability of annual plants, beetles, reptiles, and rodents in Nizzanim Coastal Sand Dune Nature Reserve in Israel. All four taxa had a strong positive relationship between asynchrony and community stability. Only rodents showed a positive richness–stability relationship. Perennial plant cover had a significant relationship with community stability for three taxa, but the direction of the correlation varied. Asynchrony had a stronger relationship with perennial plant cover than it did with richness for both plants and beetles. We suggest that community stability is driven by asynchrony for flora as well as fauna. Stability appears to be determined by species' interactions and their responses to the environment, and not always by diversity. This has important consequences for understanding the effects of environmental degradation on ecosystem stability and productivity, which have destabilizing consequences beyond biodiversity loss.

Keywords: community stability; covariance effect; biodiversity; population stability; species richness; species synchrony; diversity–stability relationship; coastal dunes; multi-taxa; cross-taxa congruence

Citation: Bird, T.L.F.; Bar (Kutiel), P.; Groner, E.; Bouskila, A. Asynchrony Drives Plant and Animal Community Stability in Mediterranean Coastal Dunes. *Appl. Sci.* **2021**, *11*, 6214. https://doi.org/10.3390/app11136214

Academic Editor: Stefano Loppi

Received: 1 June 2021
Accepted: 28 June 2021
Published: 5 July 2021

1. Introduction

Substantial evidence has demonstrated that diversity stabilizes ecosystem functioning over time [1–7]. Meanwhile, the current degradation of ecosystems and resultant losses of biodiversity reported globally could be reducing the capacity of ecosystems to maintain stable levels of function and productivity [4,8–11]. Thus, understanding the drivers of temporal stability has become critically important for conserving biodiversity and ecosystem function.

Diversity–stability relationship (DSR) theories suggest that species richness can contribute to community stability through various mechanisms, such as statistical averaging [12], overyielding [4], species asynchrony and other species interactions [3,4,13–18]. Nevertheless, species-rich communities may be more susceptible to disruption of key interactions and could, therefore, be less stable at the population level [7,19,20].

Community stability can be modelled as a function of species population stability weighted by their dominance, and the covariance (synchrony) between species [18]. The relative importance of different facets of biodiversity underlying the diversity–stability relationship remains unclear [21]. Asynchrony has played a key part in theoretical studies to predict DSR in modelled communities [4,18,21–24]. Yet only a handful studies have investigated the relationship between asynchrony and stability empirically [21,25–27]. The dominance of a species is also likely to affect its contribution to overall stability [26,28,29]. Weighted population stability and asynchrony should therefore both affect community stability. We investigate the balance between them in the context of DSR.

Stability cannot be understood outside the context of the environmental condition, and examining DSR in real -world ecosystems has been recognized as a key research need in ecology [30–32]. Yet most empirical studies of DSR focus on controlled, short-term, and small-scale experiments, under standardized environmental conditions with constant community composition, often only considering the role of species richness in driving community stability [27,33].

Empirical support for the DSR in natural terrestrial ecosystems comes overwhelmingly from plants in grasslands and trees in mixed forests, due to the relative simplicity of the communities in these habitats [8,32,34–37]. The insights gained in considering cross-taxon congruence have been highlighted in a range of studies, from the effect of spatial heterogeneity on species distributions (e.g., [38,39], to anthropogenic disturbance and fragmentation on species composition [40–42], and functional diversity [43]. Nevertheless, few studies examine multiple taxa in natural systems in the context of DSR [27].

Studying stability in natural systems and across multiple taxa is challenging. Data from Long-Term Ecological Research (LTER) networks (see https://www.ilter.network/ accessed on 28 June 2021) help to overcome this challenge [8,36,44,45]. We take advantage of a multi-taxa coastal dune LTER site with a natural gradient of increasing perennial vegetation to examine the mechanisms through which environmental condition and richness could affect the community stability of four taxonomic groups.

Coastal dune systems provide a convenient system for studying the impact of environmental gradients on DSR because they are highly dynamic, relatively simple in terms of species diversity and complexity, and for Mediterranean coasts, most species are well documented [46]. In addition, coastal dunes are subject to a gradient of disturbance and stress, coupled with increasing Perennial Plant Cover (PPC), and disturbance has been shown to affect plant and animal community stability [47–50].

Given that shrub encroachment is a major issue for biodiversity loss in coastal systems across the world [51–53], an investigation of the impact of perennial plant cover on stability is important. Dunes with greater PPC are expected to have greater species diversity and therefore, create more stable communities [54,55]. Conversely, shrub encroachment can reduce species diversity and community stability in sand dune habitats [56,57]. Alternatively, intermediate levels of plant cover could confer the highest levels of diversity [58,59], and therefore, could be the most stable.

Only two research groups have explicitly investigated stability in relation to diversity within coastal dunes [60–62], and both found evidence for a positive DSR for plant communities (dominated by perennial species with >10% annuals). However, the taxonomic scope in these studies was limited to plants only. Moreover, these studies did not investigate the underlying mechanisms of stability, such as asynchrony among species.

In this article, we empirically investigate the roles of species richness and environmental condition (vegetation cover), and compare asynchrony and weighted population stability in driving the community stability across multiple taxonomic groups.

2. Materials and Methods

2.1. Study Site

Biodiversity trends in Nizzanim coastal sand dune Nature reserve (NDNR) in Israel have been monitored across four major taxonomic groups as part of a Long-Term Ecological Research (LTER) project since 2004. The Nizzanim LTER is an ongoing collaborative project that monitors plant, arthropod, rodent, and reptile diversity [53,57,63–67], and their responses to restoration practices [57,63]. NDNR (31°42′–31°44′ N, 34°35′–34°36′ E) is the largest remaining natural coastal dune system in Israel, covering an area of 20 km^2. The climate is Mediterranean, with an annual average temperature of 20 °C and annual rainfall of 450 mm falling almost exclusively between November–April.

The LTER site consists of coastal sand dunes classified into three categorical states: mobile, semi-fixed, and fixed dunes, which are separated by densely vegetated inter-dune depressions [68–70]. The classification of fixation states is based on perennial plant cover (PPC), sand movement, and visual indicators such as dune geomorphic structure, perennial plant distribution, dominant perennial species, and soil characteristics [68,71–75]. Fixation state is a stronger explanatory variable than perennial plant cover alone [53,66] likely due to these local scale factors [76]. However, PPC can be used as a continuous environmental gradient that may affect stability dynamics, and so was the preferred parameter for this study.

2.2. Collection Methods

The Nizzanim LTER has been monitoring 10 undisturbed dunes, including three mobile, four semi-fixed, and three fixed, since 2005 (here we present 2006–2017). Each dune was sampled yearly for all taxa (dunes for plants, beetles, reptiles, and rodents) using various methods [57]: perennial plants were sampled using 100 m transects, annuals were sampled in 40 × 40 cm quadrats, beetles were sampled with dry pitfall traps, rodents were mark-recaptured using Sherman traps and their abundance was estimated using the Lincoln–Petersen Index [77], and reptiles were sampled using a combination of methods and later combined to give a rank abundance measure between 1 and 5. Full sampling methods are described in Supplementary Materials in Section S1.

There may be heterogeneity within dunes themselves; however, we did not consider intra-dune differences here and all the data within a dune were pooled to provide a single measure of stability/richness for each dune per taxa. Herewith, a "sample" refers to the monitoring of a single dune in one year for a given taxa.

2.3. Measures of Stability

The most commonly used measure of community stability is temporal variability [2,13,32,78–80]. A frequently used measure for temporal variability is the Coefficient of Variation (CV). This is the temporal standard deviation over the temporal mean of abundance, such that a higher CV implies lower stability [81]. We used Community variability ($CVcomm$) as an inverse measure of stability for each dune separately, referring to a community as the assemblage of species in a single taxonomic group found on a single dune.

$CVcomm$ ranges from 0 to 1 and is a measure of the standard deviation (*s.d.*) over the mean (μ):

$$CVcomm = s.d_{comm} / \mu_{comm} \quad (1)$$

$CVcomm$ can also be derived using Thibaut and Connolly's [18] function:

$$CVcomm = \sqrt{\phi} * CVpop\_av.weighted \quad (2)$$

where ϕ (or Phi) is Loreau's synchrony [14] and $CVpop\_av.weighted$ is the CV for each species, weighted by its relative abundance and averaged across all species [18]. Synchrony (ϕ) measures covariance among species within a community and ranges from 0 (highly asynchronous) to 1 (complete synchrony) [14]. Note that a value of $\phi = 0.5$ suggests that any synchrony between species is random.

Dominant species contribute to $CVpop\_av.weighted$, and dominance has been shown to have an important role in regulating community stability [8,26,35,36,82]. Meanwhile, communities with high levels of asynchrony could have higher community stability, irrespective of the stability of individual populations. Unless species abundances are perfectly synchronous, $CVcomm$ must be always be smaller than $CVpop\_av.weighted$ [27].

We tested the relationships between parameters using $CVcomm$, synchrony, and $CVpop\_av.weighted$ as the inverse measures of community stability, asynchrony, and weighted population stability, respectively, in order to maintain the calculations given in Equations (1) and (2). However, when discussing the significance of relationships, we refer to stability and asynchrony, because conceptually they are easier to understand in relation to DSR theory. Community stability is considered $CVcomm^{-1}$, population stability is $CVpop\_av.weighted^{-1}$, and asynchrony equates to $1 - \phi$.

To exemplify the relationship between the various measures of stability, Figure 1 represents the simulated data for simple two-species communities, showing how community stability can be affected by the balance between population stability and asynchrony (modelled data is provided in Supplementary Materials Table S1). The temporal mean for overall (community) abundance was $\mu = 45$ in all cases. Low population stability coupled with low asynchrony (when species fluctuations are synchronized) created the most unstable community (Figure 1a). The combination of high population stability coupled with low asynchrony (Figure 1b) achieved the same degree of community stability as low population stability coupled with high asynchrony (Figure 1c). High population stability together with high asynchrony produced the most stable community in terms of overall abundance (Figure 1d).

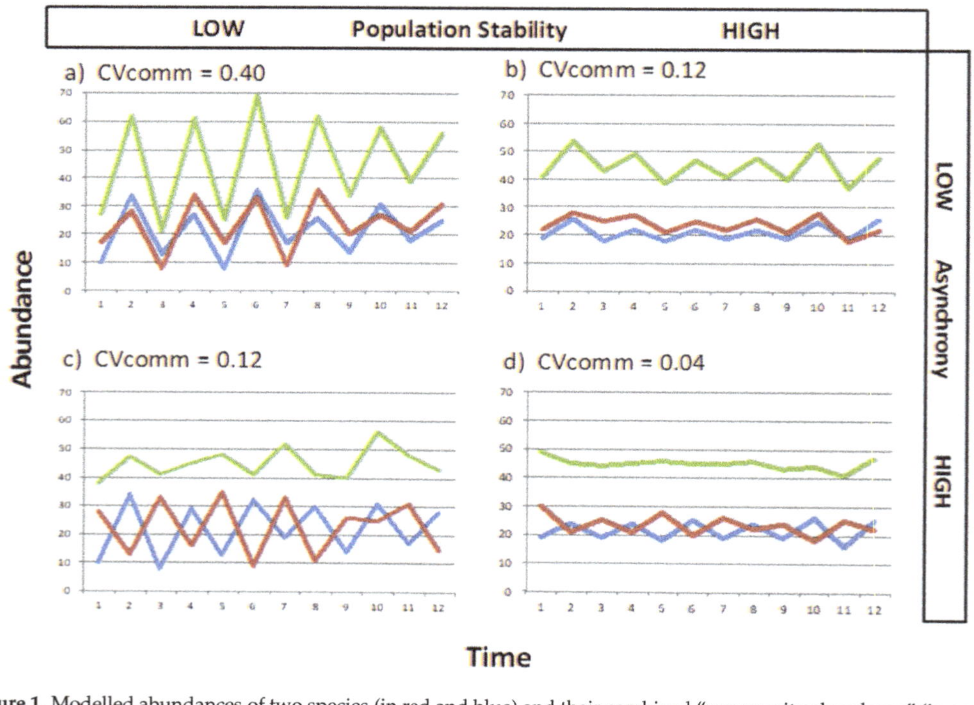

Figure 1. Modelled abundances of two species (in red and blue) and their combined "community abundance" (in green), showing how community stability can be affected by the balance between population stability and asynchrony. (**a**) Low population stability and low asychrony, (**b**) high population stability and low asynchrony, (**c**) low population and stability high asynchrony, (**d**) high population stability and high asynchrony. *CVcomm* can range from 0 to 1 and is the inverse of community stability. The Temporal mean for each community (μ, pooled abundance) was 45. The Standard deviation ranged between 3.0 and 3.5 for high population stability and between 9.50 and 10.0 for low population stability. Asynchrony is measured as $1 - \phi$, with high asynchrony ranging between 0.8 and 0.9 and low asynchrony ranging between 0.1 and 0.2. Data used for these models can be found in Supplementary Materials Table S1.

2.4. Statistical Analysis

We used R version 3.4 [83] for all analyses. We ignored rare (low-abundance) species, because they can create artefacts in the analyses despite not contributing a lot to community stability. Species that were sampled less than five years across the entire data set were removed. Second, we only selected species that were present in any given dune at least five times consecutively, or where any absences were not consecutive.

Due to the fact that stability is a single measurement across many years, our large dataset was reduced to a single data point for each dune. This precluded the use of structural equation models for our data. Instead, we used a series of linear regressions of Y~X, where Y is dependent on X, using the stats package [83]. For annual plants, we found that the relationships appeared unimodal for most parameters, so we also tested $Y \sim X^2$, where Y~X was not significant. Regression lines and standard error ranges for graphs were plotted with a linear model smoothing function using ggplot2 [84].. We compared the adjusted R^2 of each linear regression across different independent (X) parameters for each given dependent (Y) parameter for each taxon. We did not compare the relative contribution of X parameters to Y; rather, the X parameter with the highest adjusted R^2 for a given Y was regarded as the parameter that explained the most variation of Y, and was therefore considered the strongest predictor of Y.

3. Results

In total, data points were collected in 4176 quadrats for annuals, 4140 pitfall trap-nights for beetles, and 5022 Sherman trap-nights for rodents, as well as 704 activity-transects, 352 track-transects, and 1760 pitfall trap-nights for reptiles. Overall, species richness was highest for annual plant species (n = 63), of which 41 species were included in the analysis after the exclusion of rare species. Beetles were the second most diverse group across all dunes (n = 48 morphospecies), of which 32 were included in the analysis. Twenty reptile species were recorded, of which 15 were included. Finally, a total of five rodent species were found in Nizzanim LTER, of which three species were included.

Figure 2 shows the regressions for PPC and Average species richness (Richness) against *CVcomm* for each taxon. Richness was not significantly correlated to community variability (*CVcomm*) for any taxon except rodents (Figure 2c), such that increased richness was associated with a decrease in *CVcomm* (an increase in community stability). Annual plants also showed a unimodal trend for PPC, with intermediate cover associated with the highest degree of community variability (Figure 2e). *CVcomm* for both rodents and beetles significantly increased with increasing PPC (Figure 2f,g). Reptile stability showed no correlation with either parameter (Figure 2d,h).

Plots of the regressions of *CVpop_av.weighted* and synchrony against community variability are given in Figure 3. As can be seen above, only beetles had a positive relationship between *Cvpop_av.weighted* and *CVcomm* (Figure 3b), while all four taxa had a significant positive relationship between synchrony and *CVcomm* (Figure 3e–h).

Since there were many correlations between all the different variables across four taxonomic groups, we provide a summary for each taxa in Figure 4, showing all significant correlations between parameters (including those shown in Figures 2 and 3), with directionality depicted as arrows from the independent (X) to the dependent (Y) parameter Regression outputs for all taxa are reported in Supplementary Materials Table S2. As seen in Figure 4, each taxonomic group showed different strengths and directions for the relationships among parameters.

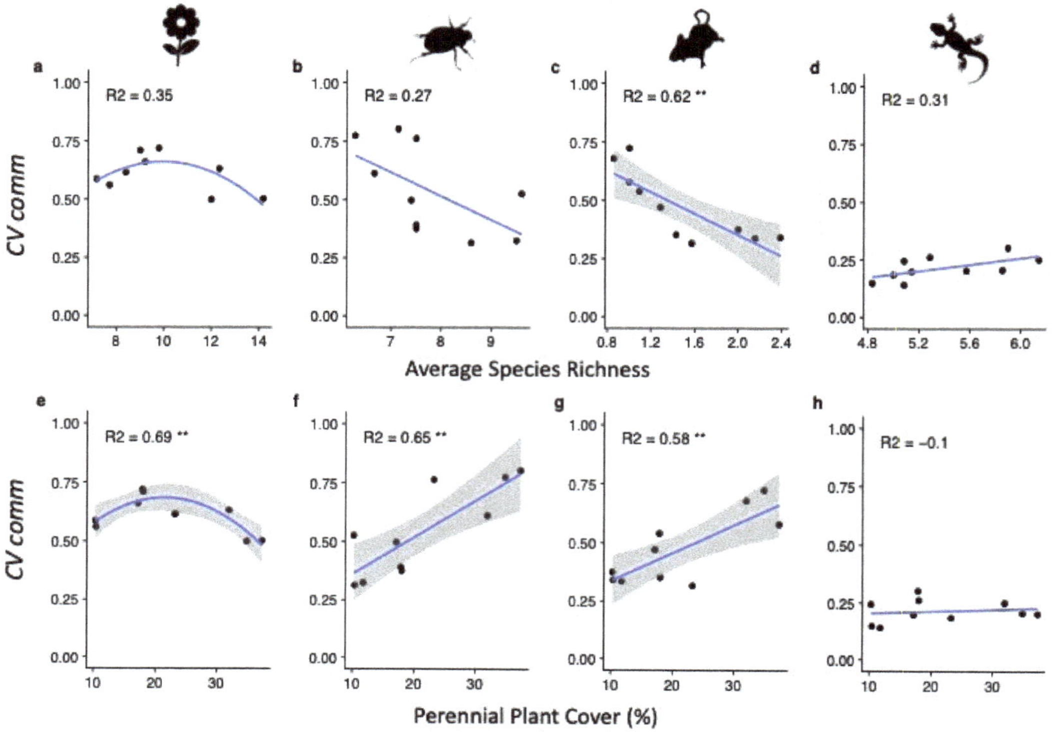

Figure 2. Regressions of (**a–d**) average species richness and (**e–h**) of perennial plant cover (PPC) against community variability (*CVcomm*) for each taxon. Standard error ranges are shown for all significant regressions. Regression lines were plotted using a linear model smoothing function. NB: an increase in *CVcomm* equates to a reduction in community stability. Explained variation (R^2) can only be compared across parameters within each taxon. ** $p < 0.01$.

Overall, no two taxa shared the same directionality in relationships across all parameters. PPC had little effect on weighted population variability (*CVpop_av.weighted*) except for annuals, where a unimodal bell-shaped curve was observed. For annuals, richness also presented a bell-shaped relationship with *CVpop_av.weighted*, and the latter had a similar bell-shaped curve with *CVcomm*.

Perennial plant cover (PPC) had a different relationship with synchrony depending on the taxon. A positive relationship was found for beetles and rodents, whereas a U-shaped response was present for annual plants, with intermediate cover associated with the lowest degree of synchrony. Reptiles did not show any significant relationship between synchrony and PPC; synchrony remained relatively high across all dunes, ranging between 0.7 and 0.8.

All four taxa were found to have a significant relationship between synchrony and richness, but the directionality differed between taxa; a negative relationship was found for beetles and reptiles, a positive correlation for reptiles, and a U-relationship for annuals.

Within each taxon, the adjusted R^2 ($adj.R^2$) values for each regression were used as an indication of which parameter explained the most variation for a given dependent variable. Thus, PPC explained more variation than richness for asynchrony of annuals ($adj.R^2$ = 0.75 and $adj.R^2$ = 0.56, respectively), and beetles ($adj.R^2$ = 0.67 and 0.55, respectively). However, PPC explained less than richness for rodent synchrony ($adj.R^2$ = 0.70 and 0.91, respectively) and reptile synchrony ($adj.R^2$ = 0.01 and 0.56, respectively).

Beetles were the only taxon to show a significant positive correlation between *CVpop_av.weighted* and *CVcomm*, but even for this taxon, there was a stronger adjusted R^2 for synchrony and *CVcomm*.

All taxa displayed different relationships and directionality among parameters, except for the relationship between synchrony and *CVcomm*. The key finding in our study was that despite the differences in strengths and directions of relationships between other parameters, the strongest positive relationship for all four taxa was consistently between synchrony and *CVcomm* (as shown by the highest $adj.R^2$ for a positive correlation in a given taxon as depicted by the star ★ in Figure 4a–d, and see Figure 3e–h). In other words, a high degree of asynchrony was associated with the highest levels of community stability.

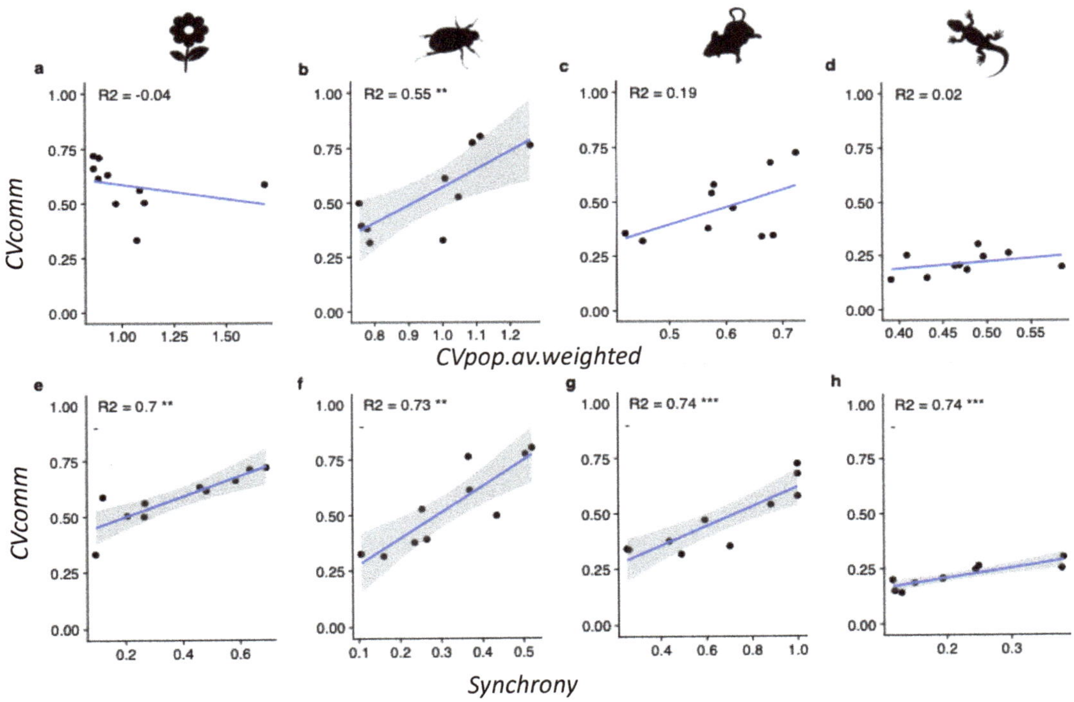

Figure 3. Regressions of (**a**–**d**) weighted average population Coefficient of Variation (*CVpop_av.weighted*), and (**e**–**h**) Synchrony (ϕ), against community variability (*CVcomm*) for each taxon. Standard error ranges are only shown for significant regressions. Regression lines were plotted using a linear model smoothing function. Note an increase in *CVcomm* equates to a reduction in community stability. Explained variation (adjusted R^2) can be compared across parameters within each taxon. ** $p < 0.01$, *** $p < 0.001$.

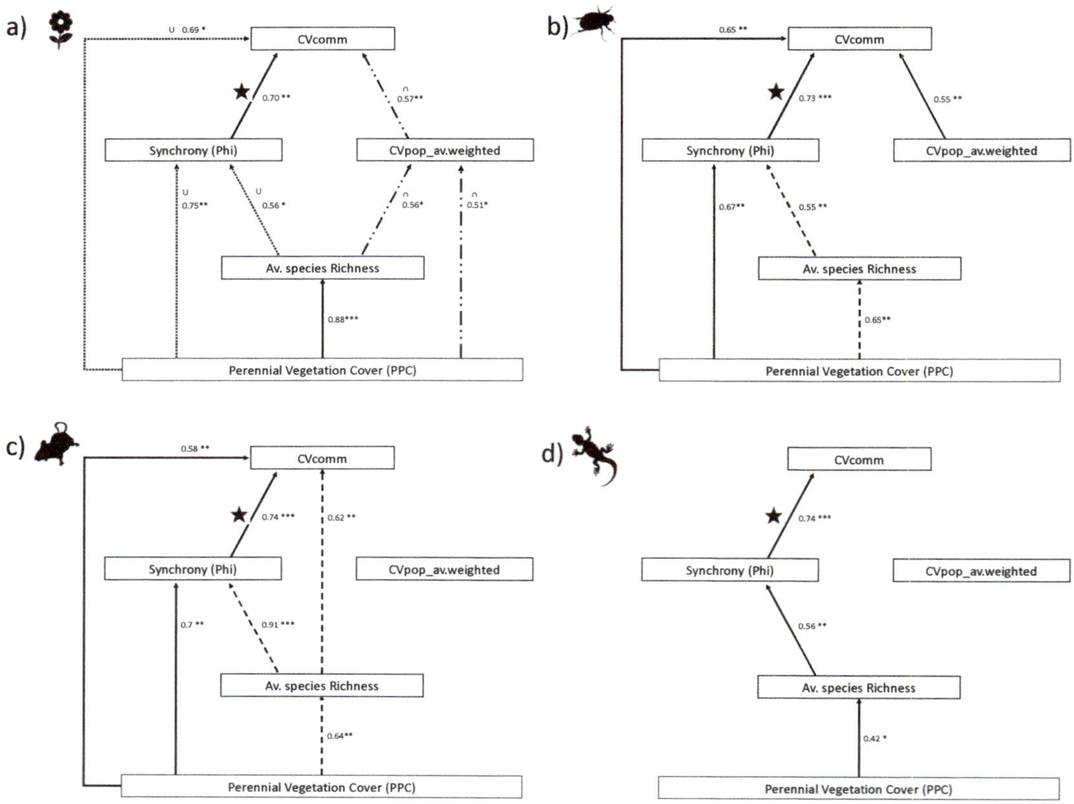

Figure 4. Linear regressions lm(Y~X^2) for (**a**) annual plants, and lm(Y~X) for (**b**) beetles (**c**) rodents and (**d**) reptiles. Arrows are drawn from independent (X) to dependent (Y) for all significant regressions (* $p < 0.05$, ** $p < 0.01$, *** $p < 0.001$); Adjusted R^2 values are given and patterns depict directions of correlation: Solid = positive-linear, dashed= negative-linear, dotted = U shaped unimodal, double-dashed = bell-shaped unimodal. Full regression outputs are given in Supplementary Materials Table S2. ★ Depicts the strongest (highest R^2) positive correlation for each taxon and is presented in full in Figure 3e–h.

4. Discussion

Diversity–stability relationship studies often focus exclusively on richness and stability, without considering the environmental factors or interspecific relationships that may drive stability [21]. We investigated the roles of species richness, environmental condition (perennial vegetation cover), synchrony, and weighted population variability in driving community variability across multiple taxonomic groups. Our empirical data consistently presented synchrony as the most strongly and positively correlated parameter for community variability (*CVcomm*) for all taxa. In addition, synchrony was a better predictor from community variability than weighted population variability, despite both parameters being mathematically linked to community variability [18]. In stability terms (as the inverse of variability), the asynchrony–community stability relationship was always significant and positive across all taxa; a high degree asynchrony appeared to confer more stable communities no matter which taxon and despite the different directionalities and strengths in the relationships among all other parameters. In contrast, a positive diversity–stability relationship (DSR) was only supported in rodents (a negative richness–*CVcomm* correlation equates to a positive richness–community stability relationship).

Asynchrony has been recognized as a key driver in theoretical and modelled DSR [18,24,85]. In empirical studies, asynchrony can play a key role in stabilizing real plant communities in grasslands, but other studies had conflicting results [8,21,25,26,29,86,87]. Despite some of the founding theory of DSR being focused on fluctuations of animal populations [88,89] few studies have looked at animal stability in relation to asynchrony. Blüthgen et al. [27] considered asynchrony in relation to community stability for plants, arthropods, birds, and bats, and found (as we did) a consistent positive relationship between asynchrony and community stability for animal and plant taxa in both grassland and forest systems. As with our findings, they too suggest that asynchrony was a stronger driver of community stability than richness directly.

In terms of asynchrony and richness, several empirical studies have also reported a positive relationship in experimental [85] and naturally assembled grasslands [90], whereas Blüthgen et al. [27] reported the opposite for grassland plants. We found support for a positive asynchrony–richness relationship for beetles and rodents (i.e., a negative synchrony–richness correlation), whereas the asynchrony–richness relationship was negative for reptiles and unimodal for annuals.

Asynchrony could be expected to increase with increasing richness due to processes such as niche partitioning at the evolutionary scale or competitive exclusion at the ecological scale [14,18,91]. The mechanisms for this relationship thus remain unclear. Asynchrony can reflect either heterogeneity in species (functional) responses to environmental conditions (response diversity), or simply their demographic stochasticity [8,12,27,29,32].

The degree to which asynchrony is linked to diversity has been shown to be influenced by environmental condition in both empirical [86,90] and modelled [29] systems. Disturbance and stochastic environments are known to have direct as well as indirect effects on the DSR via the changes they create in species richness, community composition, and species traits [92,93]. Spatial heterogeneity in natural systems increases species diversity and stability of birds [94], and riverine fish [95]. In our findings, the direction of the relationship between asynchrony and PPC was dependent on the taxon, but all taxa had a positive asynchrony–community stability relationship.

Environmental conditions can also be more important than richness in predicting community stability across a range of ecosystems [33]. Eutrophication weakened the stabilizing effect of richness on a grassland community by increasing the temporal variation of productivity and decreasing species asynchrony in more diverse communities, rather than by reducing diversity per se [34]. It is perhaps not surprising, then, that the DSR was not supported in our findings for most taxa, whereas a significant effect of PPC on community stability was found for three out of four taxa (all except reptiles).

The species richness gradient of annuals in the Nizzanim dunes was similar to other sites along the coast of Israel, with the highest richness in fixed dunes, where the PPC was also highest [57]. PPC had a significant U-shaped relationship with annuals' community stability (depicted as a bell shape with *CVcomm* in Figures 2e and 3a), whereas the richness–stability relationship was weak. Species asynchrony was also found to have a unimodalrelationship with PPC, such that the lowest levels of asynchrony and stability were found in semi-fixed dunes. Kuiters [62] found that plant stability in coastal dunes was largely explained by diversity rather than by abiotic factors, but he mostly considered perennial plants, which are not as vulnerable to environmental conditions.

Rodents were the only taxon to show a significant positive DSR in Nizzanim, and it was a stronger predictor than PPC. It is apparent that the stabilizing effect operated through the effect of richness on asynchrony; rodent communities in Nizzanim Dunes Nature Reserve have very low overall richness and are almost exclusively dominated by two species: the Greater Egyptian gerbil, *Gerbillus pyramidum*, and Allenby's gerbil, *Gerbillus andersoni allenbyi*, comprising 98.5% of rodent captures (with three other species presenting 1.5% of the captures). An increase from one to two species was positively associated with a significant increase in rodent community stability. These species are known to have low covariance due to spatio-temporal niche differentiation [96,97].

Reptiles did not appear to be as influenced by PPC as other taxa, both in terms of asynchrony and stability. Asynchrony was in general quite high across all sites for reptiles, which could infer a high degree of niche differentiation for this taxon irrespective of plant cover. Alternatively, there may be some dampening of the range of values for covariance due to the ranking methodology that was applied.

Dominance of species in their communities has been shown to be an important role in regulating population and community stability [8,26,35,36,82]. Surprisingly, we found no support for any effect of weighted population stability ($CVpop\_av.weighted^{-1}$) on community stability except for beetles. For the latter, a positive relationship between population stability and community stability was found despite the theoretical expectations of a negative relationship [7,18,98]. Nevertheless, for beetles the asynchrony–community stability relationship was stronger than population stability–community stability, consistent with other taxa.

5. Conclusions

Our findings suggest that in sand dunes, community stability is driven by asynchrony rather than population stability. Furthermore, our results provided only equivocal evidence for a relationship between diversity and stability in coastal dune systems. Each of the four taxonomic groups in this study appeared to operate under a different mechanism in terms of the community's response to richness and plant cover (a proxy for environmental condition), yet consistently demonstrated a positive asynchrony–stability relationship.

If this pattern is applicable to other systems, this would suggest that the emergent property of the community is the interactions among species (asynchrony), rather than the number of species present. This in turn infers a deterministic formation of community assemblages, rather than random stochasticity [99–101]. Hence, although individual populations may fluctuate, competition and niche differentiation allow species to fluctuate in complementary ways, maintaining certain characteristics of stability. From a generalist predator's point of view (or an anthropocentric view of ecosystem services), resource availability remains stable as the decline of one species is compensated by an increase in another. This could also explain the contradictory theories of stability (e.g., [7,88,102,103]; diversity can beget stability [88], even if population stability is negatively correlated to diversity [102] when the community is highly asynchronous.

Understanding how biotic mechanisms confer stability in variable environments is a fundamental quest in ecology, and one that is becoming increasingly urgent due to global change [36]. If stability can be determined by collective responses of species to each other and to their environment, rather than by the richness or dominance of species present, this has important consequences for understanding the effects of environmental degradation on ecosystem function and productivity. Anthropogenic changes to our planet may be having a two-fold impact on ecosystems, by diminishing ecosystem stability in addition to the direct effects on biodiversity (species) loss.

More attention is needed to bridge our understanding of ecological theory with conserving ecosystem function [104]. Long-term studies such as those within the International LTER network are particularly well suited to a metadata analysis, which could examine DSR in natural systems and across multiple taxa. We need to better understand the drivers and effects of asynchrony among species both theoretically and with empirical data in natural systems; interspecific interactions and mechanisms that determine asynchrony and stability are far more complex in natural systems than under controlled conditions. A focus on asynchrony as a stabilizing mechanism could better inform conservation management regarding the risk of environmental degradation on ecosystem function.

Supplementary Materials: The following are available online at https://www.mdpi.com/article/10.3390/app11136214/s1, Section S1: Sampling methods for data collection across 5 taxonomic groups, Section S2: Supporting data, Table S1: Modelled data for demonstrating the combine effects of population stability and asynchrony on community stability as depicted in Figure 1, Table S2: Regression results for all parameters summarized in Figures 2–4.

Author Contributions: Conceptualization, T.L.F.B., P.B., E.G. and A.B.; methodology, T.L.F.B., P.B., E.G. and A.B.; formal analysis, T.L.F.B.; investigation, T.L.F.B., P.B., E.G. and A.B.; resources, P.B., E.G. and A.B.; data curation, T.L.F.B.; writing—original draft preparation, T.L.F.B.; writing—review and editing, P.B., E.G. and A.B.; supervision, P.B., E.G. and A.B.; project administration, T.L.F.B.; funding acquisition, P.B., E.G. and A.B. All authors have read and agreed to the published version of the manuscript.

Funding: This study and the Nizzanim LTER are funded by the Israel Nature and Parks Authority, (INPA) and the International Arid Land Consortium (IALC).

Institutional Review Board Statement: Not applicable.

Informed Consent Statement: Not applicable.

Data Availability Statement: For full datasets contact the corresponding author.

Acknowledgments: Special thanks to Michael Dorman, who provided substantial time and advice on the initial analyses and use of R coding. Great thanks are due to Yael Zilka, Boaz Shacham, Zehava Siegal, Adi Ramot, Meirav Perry, Ittai Renan, Gal Vine, Oz Rittner, Arnon Tsairi, and the numerous team heads and graduate and undergraduate students from Ben Gurion University for their assistance in collecting monitoring data across the years. We thank the entomologists at the Steinhardt Museum of Natural History for their help with identifying beetles, in particular Chicatunov and Laibale Friedman. Thanks also to the Shikmim Field Study Center (Society for the Protection of Nature in Israel) for their hospitality during our fieldwork sessions over the years. We also acknowledge contributions from the Ministry of Science and Technology (MOST) and the International Arid Land Consortium (IALC). Finally, we thank Yehoshua Shkedi, Yariv Malihi, and the Israel Nature & Parks Authority (INPA) rangers for their continuous support and assistance.

Conflicts of Interest: The authors declare no conflict of interest. The funders had no role in the design of the study; in the collection, analyses, or interpretation of data; in the writing of the manuscript, or in the decision to publish the results.

References

1. Pimm, S.L. The complexity and stability of ecosystems. *Nature* **1984**, *307*, 321–326. [CrossRef]
2. Tilman, D.; Downing, J.A. Biodiversity and stability in grasslands. *Nature* **1994**, *367*, 363–365. [CrossRef]
3. McCann, K.S. The diversity–stability debate. *Nature* **2000**, *405*, 228–233. [CrossRef]
4. Loreau, M.; de Mazancourt, C. Biodiversity and ecosystem stability: A synthesis of underlying mechanisms. *Ecol. Lett.* **2013**, *16*, 106–115. [CrossRef] [PubMed]
5. Kéfi, S.; Domínguez-García, V.; Donohue, I.; Fontaine, C.; Thébault, E.; Dakos, V. Advancing our understanding of ecological stability. *Ecol. Lett.* **2019**, *22*, 1349–1356. [CrossRef] [PubMed]
6. Schindler, D.E.; Hilborn, R.; Chasco, B.; Boatright, C.P.; Quinn, T.P.; Rogers, L.; Webster, M.S. Population diversity and the portfolio effect in an exploited species. *Nature* **2010**, *465*, 609–612. [CrossRef] [PubMed]
7. May, R.M. *Stability and Complexity in Model Ecosystems*; Princeton University Press: Princeton, NJ, USA, 2001; ISBN 9780691088617.
8. Grman, E.; Lau, J.A.; Schoolmaster, D.R.; Gross, K.L. Mechanisms contributing to stability in ecosystem function depend on the environmental context. *Ecol. Lett.* **2010**, *13*, 1400–1410. [CrossRef] [PubMed]
9. Naeem, S. Ecosystem consequences of biodiversity loss: The evolution of a paradigm. *Ecology* **2002**, *83*, 1537–1552. [CrossRef]
10. Domínguez-García, V.; Dakos, V.; Kéfi, S. Unveiling dimensions of stability in complex ecological networks. *Proc. Natl. Acad. Sci. USA* **2019**, *116*, 25714–25720. [CrossRef]
11. Thebault, E. Uncertain predictions of species responses to perturbations lead to underestimate changes at ecosystem level in diverse systems. *Peer Community Ecol.* **2020**, *1*, 100063. [CrossRef]
12. Doak, D.F.; Bigger, D.; Harding, E.K.; Marvier, M.A.; O'Malley, R.E.; Thomson, D. The statistical inevitability of stability-diversity relationship in community ecology. *Am. Nat.* **1998**, *151*, 264–276. [CrossRef]
13. Lehman, C.L.; Tilman, D. Biodiversity, stability, and productivity in competitive communities. *Am. Nat.* **2000**, *156*, 534–552. [CrossRef] [PubMed]
14. Loreau, M.; de Mazancourt, C. Species Synchrony and Its Drivers: Neutral and Nonneutral Community Dynamics in Fluctuating Environments. *Am. Nat.* **2008**, *172*, E48–E66. [CrossRef] [PubMed]
15. Tilman, D.; Reich, P.B.; Knops, J.M.H. Biodiversity and ecosystem stability in a decade-long grassland experiment. *Nature* **2006**, *441*, 629–632. [CrossRef] [PubMed]
16. Hector, A.; Hautier, Y.; Saner, P.; Wacker, L.; Bagchi, R.; Joshi, J.; Scherer-Lorenzen, M.; Spehn, E.M.; Bazeley-White, E.; Weilenmann, M.; et al. General stabilizing effects of plant diversity on grassland productivity through population asynchrony and overyielding. *Ecology* **2010**, *91*, 2213–2220. [CrossRef]

17. Lhomme, J.-P.; Winkel, T. Diversity–stability relationships in community ecology: Re-examination of the Portfolio effect. *Theor. Popul. Biol.* **2002**, *279*, 271–279. [CrossRef]
18. Thibaut, L.M.; Connolly, S.R. Understanding diversity-stability relationships: Towards a unified model of portfolio effects. *Ecol. Lett.* **2013**, *16*, 140–150. [CrossRef]
19. Mrowicki, R.J.; O'Connor, N.E.; Donohue, I. Temporal variability of a single population can determine the vulnerability of communities to perturbations. *J. Ecol.* **2016**, *104*, 887–897. [CrossRef]
20. White, L.; O'Connor, N.E.; Yang, Q.; Emmerson, M.C.; Donohue, I. Individual species provide multifaceted contributions to the stability of ecosystems. *Nat. Ecol. Evol.* **2020**, *4*, 1594–1601. [CrossRef]
21. Craven, D.; Eisenhauer, N.; Pearse, W.D.; Hautier, Y.; Isbell, F.; Roscher, C.; Bahn, M.; Beierkuhnlein, C.; Bönisch, G.; Buchmann, N.; et al. Multiple facets of biodiversity drive the diversity–stability relationship. *Nat. Ecol. Evol.* **2018**, *2*, 1579–1587. [CrossRef]
22. Wang, S.; Loreau, M. Ecosystem stability in space: Alpha, beta and gamma variability. *Ecol. Lett.* **2014**, *17*, 891–901. [CrossRef]
23. Gouhier, T.C.; Guichard, F. Synchrony: Quantifying variability in space and time. *Methods Ecol. Evol.* **2014**, *5*, 524–533. [CrossRef]
24. Morin, X.; Fahse, L.; de Mazancourt, C.; Scherer-Lorenzen, M.; Bugmann, H. Temporal stability in forest productivity increases with tree diversity due to asynchrony in species dynamics. *Ecol. Lett.* **2014**, *17*, 1526–1535. [CrossRef] [PubMed]
25. Sasaki, T.; Lu, X.; Hirota, M.; Bai, Y. Species asynchrony and response diversity determine multifunctional stability of natural grasslands. *J. Ecol.* **2019**, *107*, 1862–1875. [CrossRef]
26. Sasaki, T.; Lauenroth, W.K. Dominant species, rather than diversity, regulates temporal stability of plant communities. *Oecologia* **2011**, *166*, 761–768. [CrossRef] [PubMed]
27. Blüthgen, N.; Simons, N.K.; Jung, K.; Prati, D.; Renner, S.C.; Boch, S.; Fischer, M.; Hölzel, N.; Klaus, V.H.; Kleinebecker, T.; et al. Land use imperils plant and animal community stability through changes in asynchrony rather than diversity. *Nat. Commun.* **2016**, *7*, 10697. [CrossRef] [PubMed]
28. Maestre, F.T.; Castillo-Monroy, A.P.; Bowker, M.A.; Ochoa-Hueso, R. Species richness effects on ecosystem multifunctionality depend on evenness, composition and spatial pattern. *J. Ecol.* **2012**, *100*, 317–330. [CrossRef]
29. Tredennick, A.T.; de Mazancourt, C.; Loreau, M.; Adler, P.B. Environmental responses, not species interactions, determine synchrony of dominant species in semiarid grasslands. *Ecology* **2017**, *98*, 971–981. [CrossRef]
30. Worm, B.; Duffy, J.E. Biodiversity, productivity and stability in real food webs. *Trends Ecol. Evol.* **2003**, *18*, 628–632. [CrossRef]
31. Hillebrand, H.; Langenheder, S.; Lebret, K.; Lindström, E.; Östman, Ö.; Striebel, M. Decomposing multiple dimensions of stability in global change experiments. *Ecol. Lett.* **2018**, *21*, 21–30. [CrossRef]
32. Ives, A.R.; Carpenter, S.R. Stability and diversity of ecosystems. *Science* **2007**, *317*, 58–62. [CrossRef] [PubMed]
33. Brose, U.; Hillebrand, H. Biodiversity and ecosystem functioning in dynamic landscapes. *Philos. Trans. R. Soc. B Biol. Sci.* **2016**, *371*, 20150267. [CrossRef] [PubMed]
34. Hautier, Y.; Seabloom, E.W.; Borer, E.T.; Adler, P.B.; Harpole, W.S.; Hillebrand, H.; Lind, E.M.; MacDougall, A.S.; Stevens, C.J.; Bakker, J.D.; et al. Eutrophication weakens stabilizing effects of diversity in natural grasslands. *Nature* **2014**, *508*, 521–525. [CrossRef] [PubMed]
35. Xu, Z.; Ren, H.; Li, M.-H.; van Ruijven, J.; Han, X.; Wan, S.; Li, H.; Yu, Q.; Jiang, Y.; Jiang, L. Environmental changes drive the temporal stability of semi-arid natural grasslands through altering species asynchrony. *J. Ecol.* **2015**, *103*, 1308–1316. [CrossRef]
36. Hallett, L.M.; Hsu, J.S.; Cleland, E.E.; Collins, S.L.; Dickson, T.L.; Farrer, E.C.; Gherardi, L.A.; Gross, K.L.; Hobbs, R.J.; Turnbull, L.; et al. Biotic mechanisms of community stability shift along a precipitation gradient. *Ecology* **2014**, *95*, 1693–1700. [CrossRef]
37. Seabloom, E.W. Compensation and the stability of restored grassland communities. *Ecol. Appl.* **2007**, *17*, 1876–1885. [CrossRef]
38. Duan, M.; Liu, Y.; Yu, Z.; Baudry, J.; Li, L.; Wang, C.; Axmacher, J.C. Disentangling effects of abiotic factors and biotic interactions on cross-taxon congruence in species turnover patterns of plants, moths and beetles. *Sci. Rep.* **2016**, *6*, 23511. [CrossRef]
39. Gossner, M.M.; Getzin, S.; Lange, M.; Pašalić, E.; Türke, M.; Wiegand, K.; Weisser, W.W. The importance of heterogeneity revisited from a multiscale and multitaxa approach. *Biol. Conserv.* **2013**, *166*, 212–220. [CrossRef]
40. Rooney, R.C.; Azeria, E.T. The strength of cross-taxon congruence in species composition varies with the size of regional species pools and the intensity of human disturbance. *J. Biogeogr.* **2015**, *42*, 439–451. [CrossRef]
41. Yong, D.L.; Barton, P.S.; Okada, S.; Crane, M.; Lindenmayer, D.B. Birds as surrogates for mammals and reptiles: Are patterns of cross-taxonomic associations stable over time in a human-modified landscape? *Ecol. Indic.* **2016**, *69*, 152–164. [CrossRef]
42. Maes, D.; Bonte, D. Using distribution patterns of five threatened invertebrates in a highly fragmented dune landscape to develop a multispecies conservation approach. *Biol. Conserv.* **2006**, *133*, 490–499. [CrossRef]
43. Flynn, D.F.B.; Gogol-Prokurat, M.; Nogeire, T.; Molinari, N.; Richers, B.T.; Lin, B.B.; Simpson, N.; Mayfield, M.M.; DeClerck, F. Loss of functional diversity under land use intensification across multiple taxa. *Ecol. Lett.* **2009**, *12*, 22–33. [CrossRef]
44. Vihervaara, P.; D'Amato, D.; Forsius, M.; Angelstam, P.; Baessler, C.; Balvanera, P.; Boldgiv, B.; Bourgeron, P.; Dick, J.; Kanka, R.; et al. Using long-term ecosystem service and biodiversity data to study the impacts and adaptation options in response to climate change: Insights from the global ILTER sites network. *Curr. Opin. Environ. Sustain.* **2013**, *5*, 53–66. [CrossRef]
45. Turner, M.G.; Collins, S.L.; Lugo, A.L.; Magnuson, J.J.; Rupp, T.S.; Swanson, F.J. Disturbance dynamics and ecological response: The contribution of Long-Term Ecological Research. *Bioscience* **2003**, *53*, 46. [CrossRef]
46. Carboni, M.; Zelený, D.; Acosta, A.T.R. Measuring ecological specialization along a natural stress gradient using a set of complementary niche breadth indices. *J. Veg. Sci.* **2016**, *27*, 892–903. [CrossRef]

47. Bonte, D.; Baert, L.; Maelfait, J.-P. Spider assemblage structure and stability in a heterogeneous coastal dune system (Belgium). *J. Arachnol.* **2002**, *30*, 331–343. [CrossRef]
48. Ruocco, M.; Bertoni, D.; Sarti, G.; Ciccarelli, D. Mediterranean coastal dune systems: Which abiotic factors have the most influence on plant communities? *Estuar. Coast. Shelf Sci.* **2014**, *149*, 213–222. [CrossRef]
49. Ciccarelli, D.; Bacaro, G.; Chiarucci, A. Coastline dune vegetation dynamics: Evidence of no stability. *Folia Geobot.* **2012**, *47*, 263–275. [CrossRef]
50. Van Der Wurff, A.W.G.; Kools, S.A.E.; Boivin, M.E.Y.; Van Den Brink, P.J.; Van Megen, H.H.M.; Riksen, J.A.G.; Doroszuk, A.; Kammenga, J.E. Type of disturbance and ecological history determine structural stability. *Ecol. Appl.* **2007**, *17*, 190–202. [CrossRef]
51. Isermann, M. Expansion of Rosa rugosa and Hippophae rhamnoides in coastal grey dunes: Effects at different spatial scales. *Flora* **2008**, *203*, 273–280. [CrossRef]
52. Pye, K.; Blott, S.J.; Howe, M.A. Coastal dune stabilization in Wales and requirements for rejuvenation. *J. Coast. Conserv.* **2014**, *18*, 27–54. [CrossRef]
53. Bird, T.L.F.; Dorman, M.; Ramot, A.; Bouskila, A.; Bar Kutiel, P.; Groner, E. Shrub encroachment effects on habitat heterogeneity and beetle diversity in a Mediterranean coastal dune system. *Land Degrad. Dev.* **2017**, *28*, 2553–2562. [CrossRef]
54. Zuo, X.; Zhao, X.; Wang, S.; Li, Y.; Lian, J.; Zhou, X. Influence of dune stabilization on relationship between plant diversity and productivity in Horqin Sand Land, Northern China. *Environ. Earth Sci.* **2012**, *67*, 1547–1556. [CrossRef]
55. Rosenzweig, M.L.; Abramsky, Z. How are diversity and productivity related? In *Species Diversity in Ecological Communities: Historical and Geographical Perspectives*; Ricklefs, R., Schluter, D., Eds.; University of Chicago Press: Chicago, IL, USA, 1993; pp. 52–65.
56. Báez, S.; Collins, S.L. Shrub Invasion Decreases Diversity and Alters Community Stability in Northern Chihuahuan Desert Plant Communities. *PLoS ONE* **2008**, *3*, e2332. [CrossRef]
57. Bird, T.L.F.; Bouskila, A.; Groner, E.; Bar Kutiel, P.; Kutiel, P.B. Can Vegetation Removal Successfully Restore Coastal Dune Biodiversity? *Appl. Sci.* **2020**, *10*, 2310. [CrossRef]
58. Huston, M.A. Disturbance, productivity, and species diversity: Empiricism vs. logic in ecological theory. *Ecology* **2014**, *95*, 2382–2396. [CrossRef]
59. Ferreira, S.M.; van Aarde, R.J. Maintaining diversity through intermediate disturbances: Evidence from rodents colonizing rehabilitating coastal dunes. *Afr. J. Ecol.* **2000**, *38*, 286–294. [CrossRef]
60. Kuiters, A.T.; Kramer, K.; Van der Hagen, H.G.J.M.; Schaminée, J.H.J. Plant diversity, species turnover and shifts in functional traits in coastal dune vegetation: Results from permanent plots over a 52-year period. *J. Veg. Sci.* **2009**, *20*, 1053–1063. [CrossRef]
61. Isermann, M. Patterns in species diversity during succession of coastal dunes. *J. Coast. Res.* **2011**, *274*, 661–671.
62. Kuiters, A.T. Diversity-stability relationships in plant communities of contrasting habitats. *J. Veg. Sci.* **2013**, *24*, 453–462. [CrossRef]
63. Bar, P. Restoration of coastal sand dunes for conservation of biodiversity: The Israeli experience. In *Restoration of Coastal Dunes*; Springer Series on Environmental Management; Springer: Berlin/Heidelberg, Germany, 2013; pp. 173–185.
64. Shacham, B. Dune Management and Reptiles—Implications for Habitat Reconstruction and Conservation Strategies. Ph.D. Thesis, Ben-Gurion University of the Negev, Be'er Sheva, Israel, 2010.
65. Shacham, B.; Bouskila, A. Vegetation removal as a management tool in Nizzanim dunes (Israel): Preliminary assessment of effects on reptile and mammal populations. In Proceedings of the International Conference on Management and Restoration of Coastal Dunes (ICCD), Santander, Spain, 3–5 October 2007; pp. 144–145.
66. Perry, M. Perennial Plants Impact on Annual Plant Diversity in Sand Dunes at Different Spatial Scales. Master's Thesis, Ben Gurion University of the Negev, Be'er Sheva, Israel, 2008.
67. Ramot, A. Effect of Plant Cover on Arthropod Community in Nizzanim Coastal Dunes. Master's Thesis, Ben Gurion University of the Negev, Be'er Sheva, Israel, 2007.
68. Levin, N.; Kidron, G.J.; Ben-Dor, E. A field quantification of coastal dune perennial plants as indicators of surface stability, erosion or deposition. *Sedimentology* **2008**, *55*, 751–772. [CrossRef]
69. Kutiel, P.; Cohen, O.; Shoshany, M.; Shub, M. Vegetation establishment on the southern Israeli coastal sand dunes between the years 1965 and 1999. *Landsc. Urban Plan.* **2004**, *67*, 141–156. [CrossRef]
70. Tsoar, H.; Blumberg, D.G. Formation of parabolic dunes from barchan and transverse dunes along Israel's Mediterranean coast. *Earth Surf. Process. Landf.* **2002**, *27*, 1147–1161. [CrossRef]
71. Kutiel, P.; Danin, A.; Orshan, G. Vegetation of the sandy soils near Caesarea, Israel. I. Plant communities, environment and succession. *Isr. J. Bot.* **1979**, *28*, 20–35.
72. Kutiel, P. Annual vegetation of the coastal sand dunes of the northern Sharon, Israel. *Isr. J. Plant Sci.* **1998**, *46*, 287–298. [CrossRef]
73. Kutiel, P. Conservation and management of the Mediterranean coastal sand dunes in Israel. *J. Coast. Conserv.* **2001**, *7*, 183–192. [CrossRef]
74. Tsoar, H. Sand dunes mobility and stability in relation to climate. *Phys. A Stat. Mech. Its Appl.* **2005**, *357*, 50–56. [CrossRef]
75. Rubinstein, Y.; Groner, E.; Yizhaq, H.; Svoray, T.; Bar (Kutiel), P. An eco-spatial index for evaluating stabilization state of sand dunes. *Aeolian Res.* **2013**, *9*, 75–87. [CrossRef]
76. Fenu, G.; Carboni, M.; Acosta, A.T.R.; Bacchetta, G. Environmental factors influencing coastal vegetation pattern: New insights from the Mediterranean Basin. *Folia Geobot.* **2013**, *48*, 493–508. [CrossRef]

77. Brittain, S.; Böhning, D. Estimators in capture–recapture studies with two sources. *AStA Adv. Stat. Anal.* **2009**, *93*, 23–47. [CrossRef]
78. Carpenter, S.; Walker, B.; Anderies, J.M.; Abel, N. From metaphor to measurement: Resilience of what to what? *Ecosystems* **2001**, *4*, 765–781. [CrossRef]
79. Cottingham, K.L.; Brown, B.L.; Lennon, J.T. Biodiversity may regulate the temporal variability of ecological systems. *Ecol. Lett.* **2001**, *4*, 72–85. [CrossRef]
80. Elmqvist, T.; Folke, C.; Nyström, M.; Peterson, G.; Bengtsson, J.; Walker, B.; Norberg, J. Response diversity, ecosystem change, and resilience. *Front. Ecol. Environ.* **2003**, *1*, 488–494. [CrossRef]
81. Tilman, D. Biodiversity: Population versus ecosystem stability. *Ecology* **1995**, *77*, 350–363. [CrossRef]
82. Lepš, J. Variability in population and community biomass in a grassland community affected by environmental productivity and diversity. *Oikos* **2004**, *107*, 64–71. [CrossRef]
83. R Core Team. *R: A Language and Environment for Statistical Computing*; R Foundation for Statistical Computing: Vienna, Austria; Available online: https://www.r-project.org/ (accessed on 10 October 2018).
84. Wickham, H. *ggplot2: Elegant Graphics for Data Analysis*; Springer: New York, NY, USA, 2009.
85. de Mazancourt, C.; Isbell, F.; Larocque, A.; Berendse, F.; De Luca, E.; Grace, J.B.; Haegeman, B.; Wayne Polley, H.; Roscher, C.; Schmid, B.; et al. Predicting ecosystem stability from community composition and biodiversity. *Ecol. Lett.* **2013**, *16*, 617–625. [CrossRef] [PubMed]
86. Wilcox, K.R.; Tredennick, A.T.; Koerner, S.E.; Grman, E.; Hallett, L.M.; Avolio, M.L.; La Pierre, K.J.; Houseman, G.R.; Isbell, F.; Johnson, D.S.; et al. Asynchrony among local communities stabilises ecosystem function of metacommunities. *Ecol. Lett.* **2017**, *20*, 1534–1545. [CrossRef] [PubMed]
87. Yang, H.; Jiang, L.; Li, L.; Li, A.; Wu, M.; Wan, S. Diversity-dependent stability under mowing and nutrient addition: Evidence from a 7-year grassland experiment. *Ecol. Lett.* **2012**, *15*, 619–626. [CrossRef] [PubMed]
88. MacArthur, R. Fluctuations of animal populations and a measure of community stability. *Ecology* **1955**, *36*, 533–536. [CrossRef]
89. Elton, C.S. *The Ecology of Invasions by Plants and Animals*; Methuen: London, UK, 1958.
90. Hautier, Y.; Tilman, D.; Isbell, F.; Seabloom, E.W.; Borer, E.T.; Reich, P.B. Anthropogenic environmental changes affect ecosystem stability via biodiversity. *Science* **2015**, *348*, 336–340. [CrossRef]
91. Isbell, F.I.; Polley, H.W.; Wilsey, B.J. Biodiversity, productivity and the temporal stability of productivity: Patterns and processes. *Ecol. Lett.* **2009**, *12*, 443–451. [CrossRef]
92. Collins, S.L. Disturbance frequency and community stability in native tallgrass prairie. *Am. Nat.* **2000**, *155*, 311–325. [CrossRef]
93. Donohue, I.; Hillebrand, H.; Montoya, J.M.; Petchey, O.L.; Pimm, S.L.; Fowler, M.S.; Healy, K.; Jackson, A.L.; Lurgi, M.; McClean, D.; et al. Navigating the complexity of ecological stability. *Ecol. Lett.* **2016**, *19*, 1172–1185. [CrossRef] [PubMed]
94. Hovick, T.J.; Elmore, R.D.; Fuhlendorf, S.D.; Engle, D.M.; Hamilton, R.G. Spatial heterogeneity increases diversity and stability in grassland bird communities. *Ecol. Appl.* **2015**, *25*, 662–672. [CrossRef]
95. Oberdorff, T.; Hugueny, B.; Vigneron, T. Is assemblage variability related to environmental variability? An answer for riverine fish. *Oikos* **2001**, *93*, 419–428. [CrossRef]
96. Ziv, Y.; Smallwood, J.A. Gerbils and Heteromyids—Interspecific Competition and the Spatio-Temporal Niche. In *Activity Patterns in Small Mammals*; Springer: Berlin/Heidelberg, Germany, 2000; pp. 159–176.
97. Kotler, B.P.; Brown, J.S.; Subach, A. Mechanisms of Species Coexistence of Optimal Foragers: Temporal Partitioning by Two Species of Sand Dune Gerbils. *Oikos* **1993**, *67*, 548. [CrossRef]
98. Campbell, V.; Murphy, G.; Romanuk, T.N. Experimental design and the outcome and interpretation of diversity-stability relations. *Oikos* **2011**, *120*, 399–408. [CrossRef]
99. Anderson, K.J. Temporal patterns in rates of community change during succession. *Am. Nat.* **2007**, *169*, 780–793. [CrossRef]
100. Connor, E.F.; Simberloff, D. The Assembly of Species Communities: Chance or Competition? *Ecology* **1979**, *60*, 1132. [CrossRef]
101. Cornell, H.V.; Lawton, J.H. Species Interactions, Local and Regional Processes, and Limits to the Richness of Ecological Communities: A Theoretical Perspective. *J. Anim. Ecol.* **1992**, *61*, 1. [CrossRef]
102. May, R.M. Will a large complex system be stable? *Nature* **1972**, *238*, 413–414. [CrossRef] [PubMed]
103. Tilman, D. The ecological consequences of changes in biodiversity: A search for general principles. *Ecology* **1999**, *80*, 1455–1474. [CrossRef]
104. Harvey, E.; Gounand, I.; Ward, C.L.; Altermatt, F. Bridging ecology and conservation: From ecological networks to ecosystem function. *J. Appl. Ecol.* **2017**, *54*, 371–379. [CrossRef]

applied sciences

MDPI

Article

Wind-Tunnel Studies on Sand Sedimentation around Wind-Break Walls of Lanxin High-Speed Railway II and Its Prevention

Hongchao Dun [1,2,†,‡], Guowei Xin [1,2], Ning Huang [1,3,4], Guangtian Shi [3,4] and Jie Zhang [1,2,*,‡]

1 Key Laboratory of Mechanics on Disaster and Environment in Western China, Lanzhou University, Lanzhou 730000, China; dunhc@lzu.edu.cn (H.D.); xingw17@lzu.edu.cn (G.X.); huangn@lzu.edu.cn (N.H.)
2 College of Civil Engineering and Mechanics, Lanzhou University, Lanzhou 730000, China
3 Key Laboratory of Service Environment and Intelligent Operation and Maintenance of Rail Transit, Lanzhou Jiaotong University, Lanzhou 730070, China; zxaazxy@163.com
4 School of Mechanical Engineering, Lanzhou Jiaotong University, Lanzhou 730070, China
* Correspondence: zhang-j@lzu.edu.cn
† Current address: No. 222 South Tianshui Road, Lanzhou 730000, China.
‡ These authors contributed equally to this work.

Abstract: Wind-break walls along Lanxin High-Speed Railway II were studied and approved as effective measures to reduce strong wind damage to the high-speed trains. The results show that sand sedimentation on the leeward sides of wind-break walls along the railway within Gobi Desert could significantly threaten the operation safety of running trains. Different from the current sand sedimentation prevention measures without adequate consideration of the deposition process of airborne sand particles, this study revealed the mechanism of sand sedimentation on the leeward sides of three wind-break walls within different terrains. A series of wind-tunnel experiments were carried out to measure the horizontal velocity, number density, transport flux, and deposition rate of sand particles, and it was found that the horizontal speed of sand particles was first increased and then decreased on the railway track, and the peak speed over the concave subgrade was much smaller than those over convex and flat subgrades. The number density and horizontal sand flux were largest over the concave subgrade, and were the smallest over the convex subgrade. The sand particle deposition rate and distribution were also the largest within the concave subgrade, and some measures were also proposed to prevent sand sedimentation on the leeward sides of wind-break walls.

Keywords: sand particle; wind-break wall; terrain; railway; wind-tunnel experiment

Citation: Dun, H.C.; Xin, G.W.; Huang, N.; Shi, G.T.; Zhang, J. Wind-Tunnel Studies on Sand Sedimentation around Wind-Break Walls of Lanxin High-Speed Railway II and Its Prevention. *Appl. Sci.* **2021**, *11*, 5989. https://doi.org/10.3390/app11135989

Academic Editor: Itzhak Katra

Received: 19 April 2021
Accepted: 17 June 2021
Published: 27 June 2021

1. Introduction

Within China's extensive high-speed railway system, the Lanxin High-Speed Railway II from Lanzhou to Xinjiang in western China is the first known high-speed railway crossing a long wind area (more than 462.4 km) with extremely high wind speeds [1–3]. Because there are 208 days every year with wind speeds of higher than 20 m/s and the highest wind speed is over 60 m/s [4,5], strong wind will cause train overturning, vehicle runaway, window glass shattering, power supplying equipments damage, etc., which further leads to the loss of safety in train operation, maintenance, and transportation [6–9].

In order to prevent damage to trains from strong winds, many wind-proof constructions, such as subgrade wind-break walls, bridge windscreens, and wind-proof tunnels, were built along Lanxin High-Speed Railway II [2]. Because wind-break walls are the most common wind-proof structure with low cost and simple construction [10], many wind-break walls with a height of 2.0 to 4.3 m have been built in wind zones to significantly reduce both the wind speed and the overturning force to vehicles.

However, as shown in Figure 1c,d, wind-break walls can easily cause sand sedimentation around railway tracks, which threatens the safe operation of high-speed trains. For example, in the Gobi Desert with dry loose soil and abundant surface particles [11–13], strong wind is usually accompanied by sand and dust flow, which will be deposited on the railway track [14]. Therefore, the high-speed train can only cross Lanxin High-Speed Railway II during the daytime in order to clear the deposited sand during the night [15]. The mechanism of sand sedimentation on the leeward sides of wind-break walls has recently been given serious attention. Cheng analyzed the characteristics of sand accumulation of three different types of wind-break walls by three-dimensional numerical simulations and wind-tunnel experiments and found that the wind-break wall with bottom-openings achieved improved sand dredging at high wind speeds [16]. Huang explored the characteristics of wind-blown sand flow around the wind-break wall and the causes of sand accumulation by setting up a turbulent flow-sand particle-terrain coupling model using the Lagrangian particle-tracing model and found that sand particles will be blown up and rolled back to the subgrade under the effect of vertical wind velocity and reflux [15]. Similar phenomena had been found in highways and civil constructions in arid and semi-arid regions [17–19].

Figure 1. (**a**) Convex subgrade and (**b–d**) concave subgrades. Note: The wind-break wall is different from the sand fence (see Appendix A).

Up to present, the sand accumulation around the wind-break wall is mainly cleared manually with very low efficiency. The current measures to prevent sand deposition are practiced without adequate consideration of the deposition process of airborne sand particles and topographic factors. However, the varied terrain along the Lanxin High-Speed Railway II includes convex, flat, and concave topography as shown in Figure 1a (convex) and Figure 1b (concave). Therefore, studies and optimization on the original wind-break wall structures within different terrains to reduce sand sedimentation are essentially urgent.

In this study, according to the sand sedimentation caused by wind-break walls and different terrain and railway track characteristics of Lanxin High-Speed Railway II, three models of wind-break walls were established and examined through the wind-tunnel experiment with the reproduced movement process of windblown sand. In the wind-tunnel experiments, the particle image velocimetry (PIV) system was used to collect sand movement images on the leeward side, and PTV technology was employed to obtain the sand particle movement laws. The changes in wind speed, sand particle speed, sand particle number density, sand transport flux, and sand particle deposition rate on the railway track as well as the flow field around wind-break walls (numerical simulation) were analyzed systematically to understand the sand sedimentation around wind-break walls under different terrain, which will provide an excellent reference for train-running safety in strong wind areas with sand damage.

2. Experimental Setup and Methods

2.1. Wind-Tunnel Experiment

All wind-tunnel experiments were carried out in a blow-down wind tunnel at Lanzhou University. The wind tunnel was 55 m long and included a powerful fan system, a rectification section, a working section, and a diffuser. The length of its working section was about 22 m, and the cross-sectional area of its working section was 1.3 m (width) × 1.45 m (height). The wind tunnel was controlled by a computer, and the wind speed could be adjusted between 3 and 40 m/s as the normal wind speed in nature [20,21].

As shown in Figure 2, in the experimental setup, roughness elements were in the front of the work section to generate a turbulent boundary layer with a thickness of 0.3 to 0.4 m. A monitoring Pitot tube was used to measure the inlet wind speed. A measuring Pitot tube was fixed to a frame to measure wind speeds around three models at the 17 points of 2.0, 4.0, 7.0, 10.0, 15.0, 20.0, 30.0, and 45.0 cm above the surface. The obtained data were used to analyze the spatial and temporal distributions of the wind speed. Particle image velocimetry (PIV) was used to measure the speed of sand particle on the railway track.

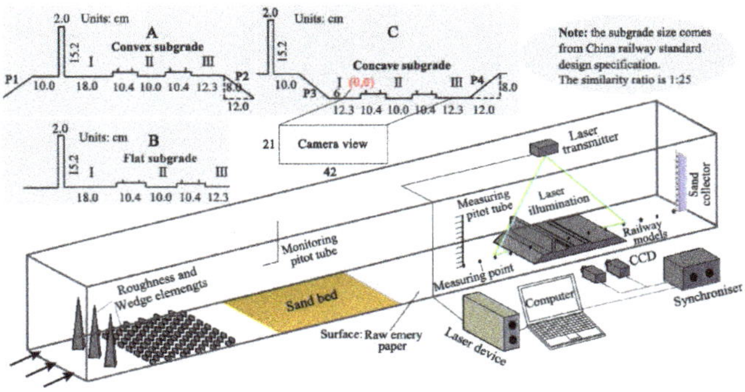

Figure 2. Configuration of the wind-tunnel experiment, in which downstream of the roughness-element is the test surface, the model (100 cm wide and 100 cm long) is 9 m away from the sand bed, the sand bed (130 cm wide, 300 cm long and 5 cm deep) is placed after the roughness elements, A is the convex wind-break wall with P1 and P2 as its two slopes, B is the flat wind-break wall, C is the concave wind-break wall with P3 and P4 as its two slopes, and the coordinate origin (0, 0) of the camera view is as shown in the figure. Turbulence intensity is 0.05.

The PIV system had a laser and transmitter, a high-resolution charge coupled device (CCD) camera, a synchronizer, and a computer with an image acquisition card. The laser emitter was on the top of the wind tunnel, and its laser beam coincided with the axis section of the model. The frequency of the laser was 5 Hz and the pulse energy was 120 mJ. The high-resolution CCD camera and the pulsed laser were synchronized with a pulse delay generator. Two CCD cameras with a pixel resolution of 2048 × 2048 and a sampling rate of 500 frames per second were at the side of the wind tunnel. The field of these two cameras was about 21 cm (height) × 42 cm (width). The images of particle tracking velocimetry (PTV) were processed to obtain the sand particle velocity and number density.

Although there are many similarity requirements in wind-tunnel experiments, such as geometric, motion, dynamic, and thermodynamic similarity [22,23], with sand flow it is difficult to perfectly simulate the shrinkage ratio of the model size, boundary layer size, and particle size in the real environment. Therefore, in this study, some compromises must be adopted [24]. Only the geometric similarity of the model, the motion similarity and dynamic similarity of the wind field, and the corresponding wind-tunnel results were mainly used in the mechanical studies as a key problem to be solved in current wind-sand physics [23,25,26].

2.2. Numerical Simulation

Due to the complicated structure of the flow field in the leeward side of wind-break walls, it is difficult to obtain flow information accurately with a Pitot tube as the only one-dimensional and unidirectional speed measuring device. Therefore, numerical simulation on computational fluid dynamics (CFD) was used to study the wind field information under the corresponding conditions of the wind-tunnel experiments (Figure 3). Meanwhile, the wind speed comparison between the wind-tunnel experiment and the numerical simulation was used to verify the reliability of the numerical simulation. Results show that the three-dimensional channel size of the numerical simulation was consistent with the working section of the wind tunnel. The wind speed profiles of wind-tunnel experiments were collected at the inlet with the velocity-inlet as the inlet boundary condition, the outflow as the outlet boundary, and the walls as other boundaries [27–29].

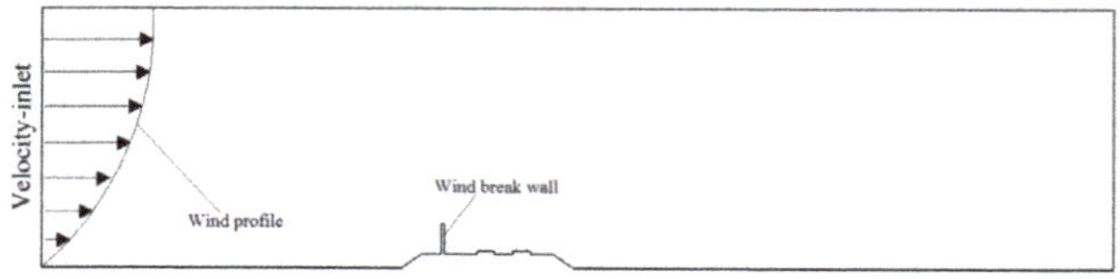

Figure 3. Configuration of numerical simulation. The dimensional channel size of the numerical simulation was consistent with the working section of the wind tunnel.

2.3. PTV Image Processing

The PTV is a non-contact, transient whole-field particle velocity measuring method to identify the movement of sparse particles [30,31]. Figure 4 shows the post-processing process of particle motion caught by the PIV system [32]. Within the square region (0.01 m × 0.01 m) of the PIV image, the number of sand particles was counted to calculate the velocity and number density of sand particles according to the following Equations (1) and (2).

$$v_i = \frac{\min(\sqrt{(x_{2i} - x_{1i})^2 + (y_{2i} - y_{1i})^2})}{\Delta t}, \tag{1}$$

$$\frac{S_{2i}}{S_{1i}} \leq 3, \tag{2}$$

where $\min(\sqrt{(x_{2i} - x_{1i})^2 + (y_{2i} - y_{1i})^2})$ represents the minimum distance between all particles in the second frame and the i th particle in the first frame, v_i represents the velocity of the i th particle in the first frame image, Δt represents the time interval between two frames, S_{1i} is the sand particle area in the first frame, and S_{2i} is the sand particle area in the second frame.

The sand particle number density N_z, horizontal particle velocity v_x, and horizontal sand flux Q_x around the three wind-break walls were calculated out from the obtained images according to the following Equations (3)–(5) [33].

$$N_z = \frac{n_z}{\Delta x \Delta y \Delta z}, \tag{3}$$

$$v_x = \frac{L_x}{\Delta t}, \tag{4}$$

$$Q_x = \frac{\pi}{6} \rho d^3 N_z v_x, \tag{5}$$

where L_x is the moving distance of a particle along x-direction; Δt is the time interval of 110 μs ($u_* = 0.9701$ m/s), 170 μs ($u_* = 0.6595$ m/s) and 250 μs ($u_* = 0.4756$ m/s) between two laser pulses; n_z is the total number of sand particles in a cuboid; Δx, Δy, and Δz are the three sides of the cuboid, in which both Δx and Δy are 0.01 m, and Δz (1 mm) is the depth of the laser; d (∼160 μm) is the average sand particle diameter in the railway track; and ρ is the particle density (2650 $\text{kg} \cdot \text{m}^{-3}$).

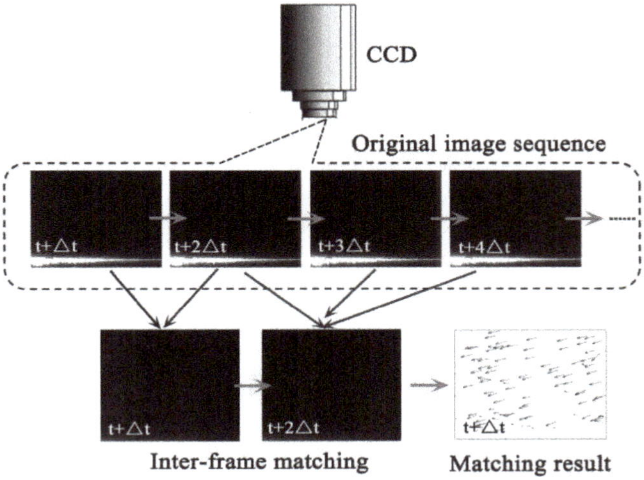

Figure 4. The post-processing procedure.

After the sedimentation of sand particles within I, II, and III on the railway track was weighed, the sand deposition rate was determined according to the following Equation (6).

$$\lambda = \frac{m_z i}{t \cdot s_i}, \tag{6}$$

where λ is the deposition rate, m is the total sedimentation of sand particles, t is the experimental time, and s_i is the area of I, II, and III.

3. Results

3.1. Wind Profiles and Size Distribution of Sand Particles

Figure 5 shows the size distribution of sand particles in the wind-tunnel experiment. The sand particle size followed a normal distribution, and the mean particle size was about 220 μm.

Figure 6 shows the horizontal wind speeds at different heights over the surface without the Pitot tube. The fitting parameters of the wind profiles, such as the friction velocity (u_*) and the roughness length (z_0), are shown in the label of Figure 6. The results show that the wind profiles perfectly obey a logarithmic distribution.

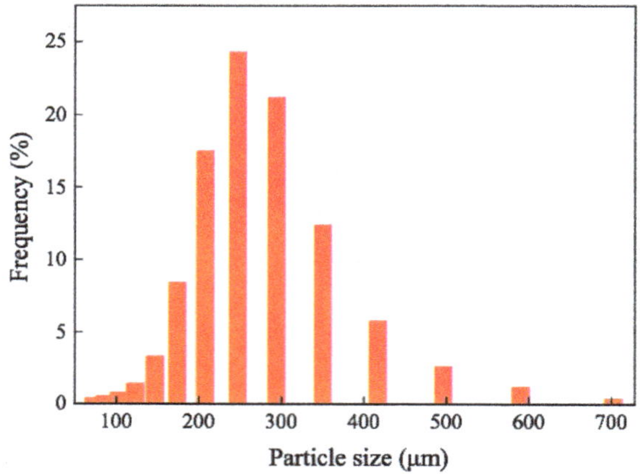

Figure 5. The size distribution of sand in the sand bed.

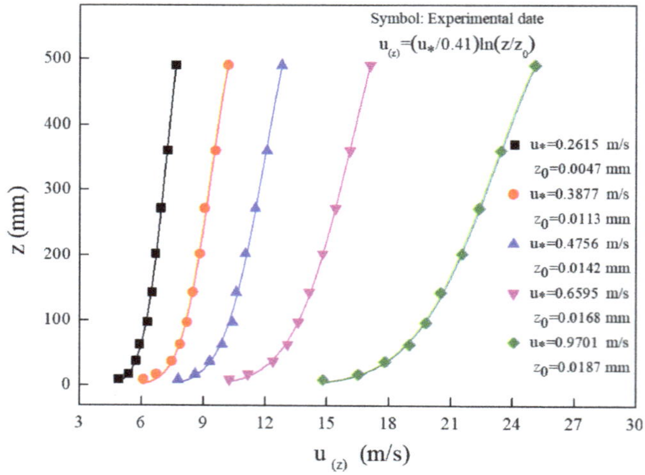

Figure 6. Wind profiles with experimental data and the profile curves fitted to the logarithmical law.

3.2. Wind Field around Three Wind-Break Walls

The horizontal wind speeds in different terrains around three wind-break walls measured with the Pitot tube are shown with black, blue, and red lines in Figure 7. The wind-break wall is located at the zero point of the x-axis, and the measuring height in Figure 7a is 15 cm, and 30 cm in Figure 7b. A wind speed of $u_* = 0.9701$ m/s was utilized as an example to analyze the wind speed variation. As shown in Figure 7, the horizontal wind speeds in three terrains were decreased then increased with x distance. The horizontal wind speed reached its peak on the leeward side and was gradually increased with x distance, and the peak of the concave subgrade was much lower than that of the convex and flat subgrade. Meanwhile, as shown in Figure 7a, the minimum speed of convex, flat, and concave subgrade on the leeward side was 2.73, 2.23, and 0 m/s, respectively. Although, according to numerical simulations, the speed on the leeward side of the concave subgrade should be negative at the height of 15 cm, the speeds of convex and flat were both positive because the Pitot tube measurements were all positive and could not be negative in wind-tunnel experiments. Therefore, the numerical simulation was applied to supplement the complete flow field information around wind-break walls.

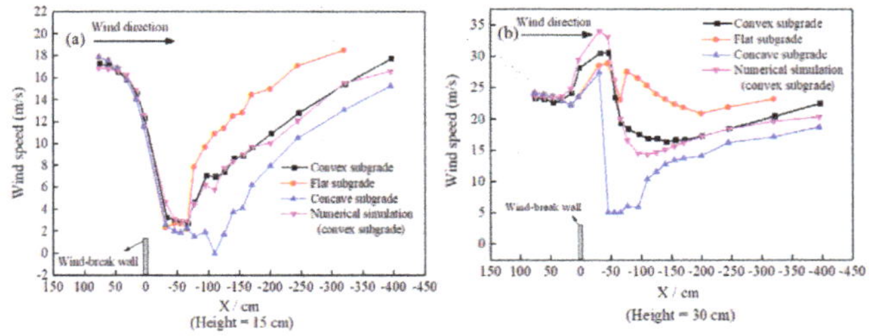

Figure 7. Wind speeds around three wind-break walls ($u_* = 0.9701$ m/s), and the measuring heights are: (**a**) 15 cm and (**b**) 30 cm.

In order to verify the reliability of numerical simulations in this study, numerical wind speeds of the convex wind-break wall at different heights were compared with those from the wind-tunnel experiments, as shown with the pink line in Figure 7. The results show that all simulation results are basically consistent with those collected in the wind tunnel.

The major stream-wise wind fields around these three wind-break walls under different terrains are shown in Figure 8a–c (symmetry plane in x-z direction). The results show that the acceleration and vortex zone appear successively, and a large vortex zone was formed on the leeward side of the wind-break walls, which is consistent with reported results [28]. As shown in Figure 8, when the wind rolled back to the track under the action of the vortex zone, the wind speed was reduced due to the obstruction of convex slope P2. However, concave slope P4 could increase the wind speed towards the railway track, and slope P3 could slow down the wind speed that carried away a lot of sand particles, indicating that a large number of sand particles would be deposited in the concave subgrade. The wind speed on the leeward side of the flat subgrade was between those of concave and convex terrains.

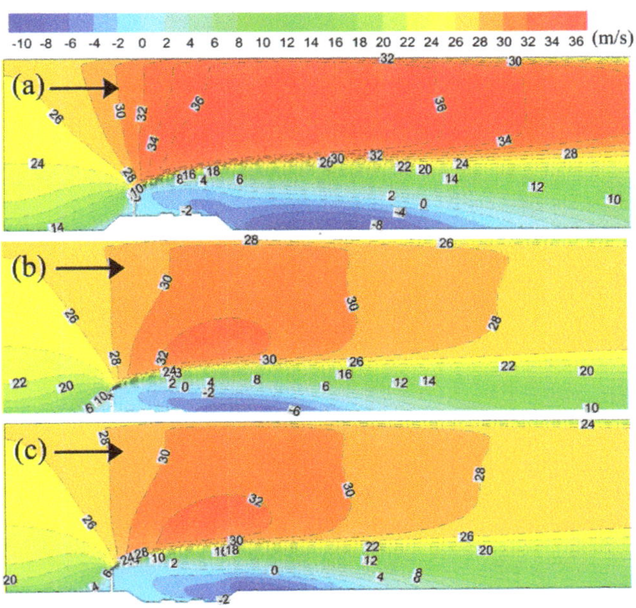

Figure 8. Wind fields around wind-break walls ($u_* = 0.9701$ m/s): (**a**) convex subgrade, (**b**) flat subgrade, and (**c**) concave subgrade.

3.3. Horizontal Speeds, Number Density, and Fluxes of Sand Particles

Figure 9 shows the changes in horizontal speeds of sand particles along the x-direction on the leeward side with different terrains when the incoming wind speed u_* is 0.9701 m/s. The results show that the horizontal speeds of sand particles were increased first and then decreased negatively with x distance on the leeward side. The speeds reached their peaks within 0.4–0.5 m and were gradually decreased after 0.5 m. The increase of horizontal speed of sand particles was the largest in the case of C, which were -11.2 m/s at 0.05 m and -10.2 m/s at 0.1 m, respectively. Compared with that in case of C, in the cases of A and B the horizontal speeds of sand particles were respectively reduced by 33% and 13% at 0.05 m, and by 22% and 8% at 0.10 m, because the horizontal speeds of sand particles within 0.4–0.5 m were affected by the vortex zone, and the negative feedback effect on sand particles was significant. Moreover, the horizontal speeds of sand particles were reduced due to the obstruction of convex slope P2 and increased due to the concave slope P4. Therefore, the horizontal speed of sand particles in the case of C was higher than that in A or B.

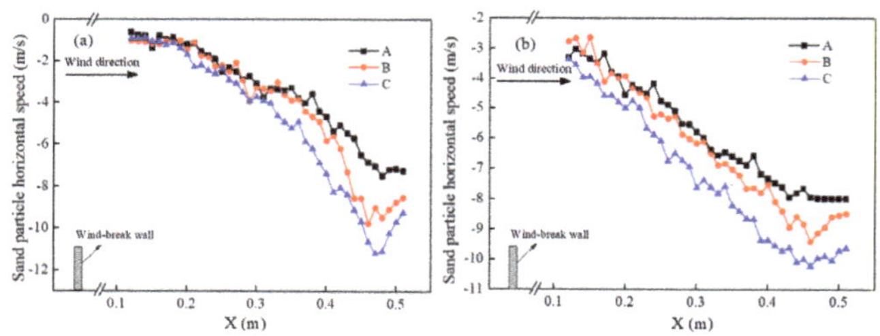

Figure 9. Sand particle's horizontal speed along the leeward side ($u_* = 0.9701$ m/s). (**a**) H = 0.05 m (above the track). (**b**) H = 0.10 m (above the track).

Figure 10 shows the change in sand particle number density along the x-direction at different heights. The results show that the sand particle number density of the three wind-break walls was first increased and then decreased. The sand particle number density of case C was the highest. Compared with that in case C, the peaks of case B were decreased by 21% and 47% at heights 0.05 and 0.10 m, respectively; and those of case A were decreased by 50% and 65% at heights 0.05 and 0.10 m, respectively. With the increase in height, the sand particle number density was gradually decreased and the amplitude of these changes was increased, indicating that the sand particle concentration was larger at a lower height and smaller at a higher location. Additionally, the sand particle number density was smaller on the side than that in the middle because the laser intensity on both sides was weaker, and the corresponding sand particle information captured by PIV was reduced.

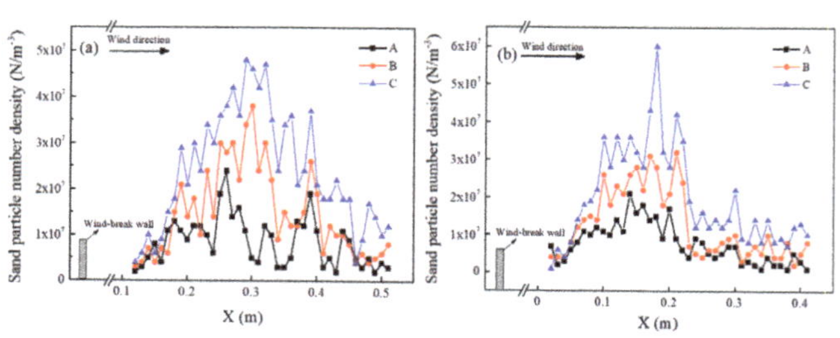

Figure 10. Number density of sand particles ($u_* = 0.9701$ m/s). (**a**) H = 0.05 m (above the track). (**b**) H = 0.10 m (above the track).

As shown in Figure 11, the horizontal sand flux in case C on the railway track was the largest, that in case A was the smallest, and that in case B was between the two.

The horizontal sand flux at different heights showed a wave-like change. At the height of 0.05 m, the maximum horizontal sand flux in cases of A, B, and C all appeared at 0.39 m along the x-direction. The horizontal sand flux in case C was the largest, and those in cases of A and B were lower by 70% and 47%, respectively, than that in case C. At the height of 0.10 m, the horizontal sand flux in cases of A and B were lower by 76% and 52%, respectively, than that in case C.

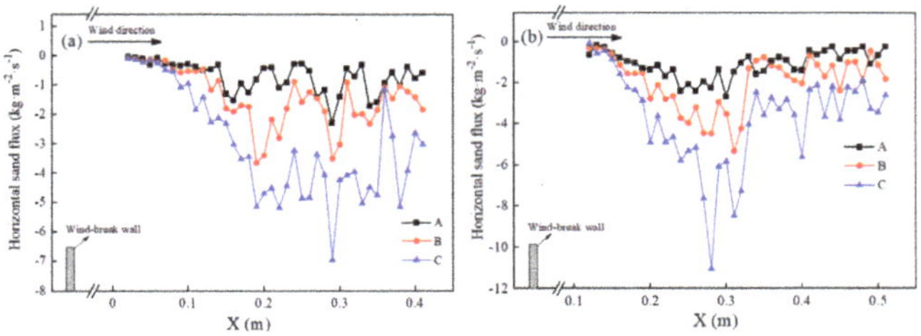

Figure 11. Horizontal sand flux ($u_* = 0.9701$ m/s). (**a**) H = 0.05 m (above the track). (**b**) H = 0.10 m (above the track).

3.4. Sand Particle Deposition Rate on Railway Track

As shown in Figure 12a, the sand particle deposition rate of I was the highest. On the whole railway track, the sand particle deposition rate in case C was higher by 89% and 87% than those in cases of A and B subgrades, indicating that the sand sedimentation in case C on the railway track was the largest. Figure 11b shows the total sand particle deposition rate on the leeward side of three wind-break walls at different wind speeds. As the incoming wind speed was increased, the total sand particle deposition rate was gradually increased, and the increase rate in case C was higher.

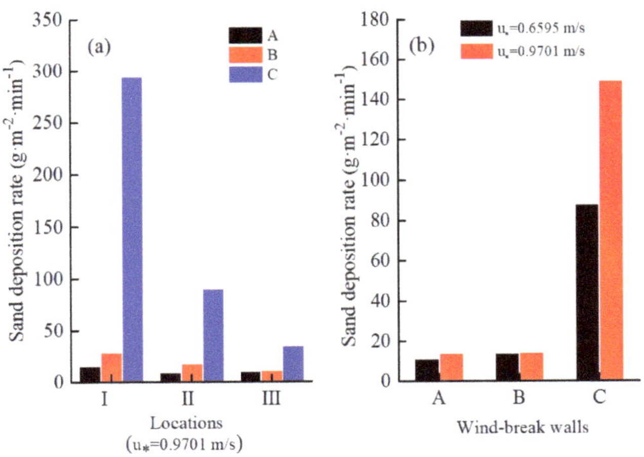

Figure 12. Sand particle deposition rates along three wind-break walls: (**a**) deposition rates at different locations and (**b**) total deposition rates at different wind speeds.

Figure 13 shows the sand particle distribution on the leeward sides of three wind-break walls, and the wind direction is shown with arrows. The results show that the final sand particle distribution was consistent with all obtained results.

Figure 13. Sand particle sedimentation on the leeward sides of three wind-break walls: (**a**) convex subgrade, (**b**) flat subgrade, and (**c**) concave subgrade.

4. Discussion

In this study, the results show that sand sedimentation on the railway track of the concave subgrade was the most serious, and was about 10.7 and 4.4 times ($u_* = 0.9701$ m/s) that of convex and flat subgrade. However, the tall length of concave wind-break wall was shorter than that of convex wind-break wall that was most of the 462.4 km wind zone railway of Lanxin Railway II. Therefore, in order to balance the influence of sand sedimentation on the whole railway track, it was necessary to take measures to reduce sand sedimentation along convex wind-break walls with the longest mileage and concave wind-break walls with serious sand sedimentation harm.

According to the flow field and sand sedimentation, some of the sand particles were carried back to the subgrade due to the existence of the backflow vortex. Therefore, in addition to setting up sand-fixing and sand-fence measures on the windward side of wind-break walls, it might be an effective measure to optimize the structure of wind-break walls or set up corresponding measures in the leeward backflow areas to reduce the sand deposition caused by backflow.

Along convex and flat wind-break walls, a second retaining wall at the top of slope P2 could be added in order to block the backflow of sand particles and make sand deposit behind the second retaining wall (Figure 14a,b). Because the distance between the wind-break walls and railway track is long (8.5 m) along concave wind-break walls, an inclined board could be set at a certain angle with the wind direction on top of the existing wind-break wall to make the original backflow area move backwards. A second retaining wall could then be added on top of slope P4 (Figure 14c). However, the inclined board could not be applied to convex and flat wind-break walls because the distance between the wind-break wall and railway track was short (4.5 m), and it would affect the overhead catenary system and other power supplying equipment of the high-speed train, such as the overhead contact line and cantilever structures [8,9].

Through this study, the distribution laws of flow field and sand sedimentation on the leeward sides of three wind-break walls under different terrains were established, and some preliminary protective measures were accordingly proposed. Further studies on protective measures of wind-break walls within different terrains, such as the angle of the inclined board and the physical parameters of the second retaining wall, are continued and will be reported in the near future.

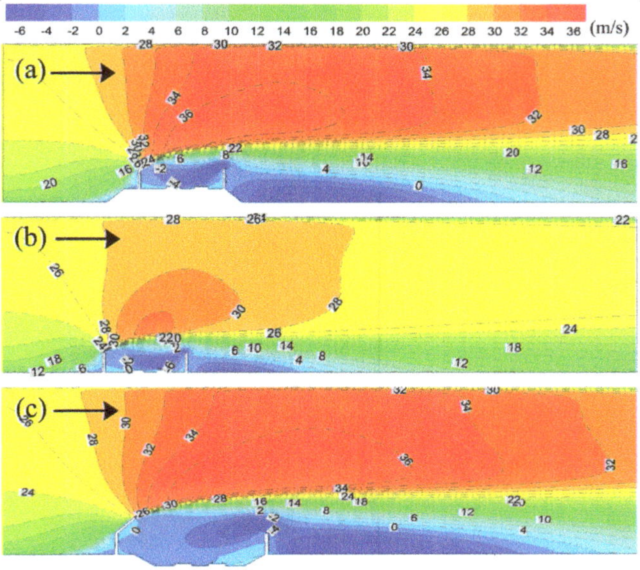

Figure 14. Sand sedimentation prevention measures along wind-break walls: (a) convex subgrade, (b) flat subgrade, and (c) concave subgrade.

5. Conclusions

In this study, three wind-break walls in different terrains, including convex, flat, and concave, were examined in a wind tunnel to determine the flow field, sand particle horizontal velocity, number density, sand flux, and deposition rate on the railway track.

The results show that the horizontal wind speeds along the three wind-break walls were first decreased and then increased with x distance at different heights, and the peak wind speed of concave subgrade was much lower than those of convex and flat subgrades. The horizontal speed of sand particles was first increased and then decreased with x distance on the leeward side at heights of 0.05 and 0.10 m. The number density and horizontal sand flux along three wind-break walls showed that they were the largest in concave subgrade and the smallest in convex subgrade, with that in the flat terrain being between these two, indicating that the sand sedimentation of concave subgrade on railway tracks was the largest, which could be additionally confirmed with the sand deposition rate.

Meanwhile, some preliminary sand sedimentation prevention measures along the three wind-break walls were suggested according to the experimental results. Although these measures did not change the structure of existing wind-break walls and facilitate construction, they could effectively reduce the sand sedimentation on railway tracks and ensure the operation safety of high-speed trains in the strong wind area.

Not limited to wind-tunnel experiments and numerical simulations, more real-size field experiments should be carried out to verify theoretical assumptions in the future. Sand sedimentation caused by strong wind is widely distributed around the globe, and our research methods and results can be extended to highway and civil construction regions to resist sand sedimentation.

Author Contributions: Conceptualization, H.D. and G.X.; methodology, G.X.; software, G.X.; validation, J.Z.; formal analysis, H.D.; investigation, G.S.; resources, N.H.; data curation, G.X.; writing—original draft preparation, G.X.; writing—review and editing, H.D. and J.Z.; supervision, G.S.; project administration, N.H.; funding acquisition, N.H. All authors have read and agreed to the published version of the manuscript.

Funding: This research was funded by the State Key Program of National Natural Science Foundation of China OF FUNDER grant number 41,931,179, the National Natural Foundation of China OF FUNDER grant number 11,772,143 and 11,702,163, and the Fundamental Research Funds for the Central Universities OF FUNDER grant number lzujbky-2021-pd09 and lzujbky-2020-cd06.

Institutional Review Board Statement: This study did not involve humans or animals.

Informed Consent Statement: This study did not involve humans.

Data Availability Statement: This study did not report any data.

Conflicts of Interest: The authors declare no conflicts of interest.

Appendix A

This study was mainly focused on the wind-break wall rather than the sand fence, because the sand fence was mainly used to reduce the sand particles from passing through the railway line and control the sand accumulation in the leeward sand-fixing barriers. As shown in Figure A1a–g, the sand fence with a porosity of about 30–60% and a height of about 1.5–2.0 m was set at about 100–150 m away from the railway line on the windward side.

However, the wind-break wall was mainly used to prevent damage to trains from strong winds. As shown in Figure A1h and Figure 1, the wind-break wall with no holes, but with a height of about 3.5–4.3 m, was generally set at 4–5 m away from the railway tracks. Initially, the wind-break wall was not believed to cause the sand particles to accumulate on the railway tracks. After the construction, it was found that the wind-break wall could reduce the speed of sand particles and cause sand particles to accumulate on the railway tracks (Figure A1c,d). The deposited sand particles would seriously affect the wheels, bogies, and electric motors of high-speed trains.

Up to the present, there have been a large number of reports on sand fences, but studies on sand accumulation caused by wind-break walls and the corresponding prevention measures are rarely reported.

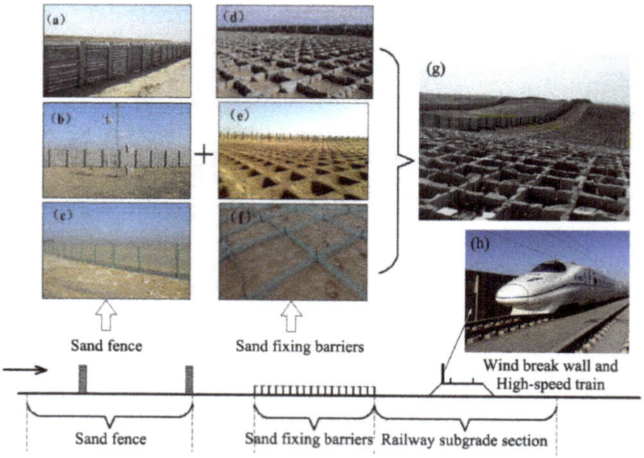

Figure A1. (**a–c**) Sand fences, (**d–g**) sand-fixing barriers, and (**h**) wind-break walls.

References

1. Zhang, J.; Wang, Y.; Jiang, F. Numerical Analysis on the Features of Sand Flow Movement around the Embankment of Lan-Xin Railway in Gobi Region. *China Railw. Sci.* **2011**, *32*, 14–18.
2. Cheng, J.J.; Lei, J.Q.; Li, S.Y.; Wang, H.F. Disturbance of the inclined inserting-type sand fence to wind–sand flow fields and its sand control characteristics. *Aeolian Res.* **2016**, *21*, 139–150. [CrossRef]

3. Wang, T.; Qu, J.; Ling, Y.; Liu, B.; Xiao, J. Shelter effect efficacy of sand fences: A comparison of systems in a wind tunnel. *Aeolian Res.* **2018**, *30*, 32–40. [CrossRef]
4. Cheng, J.J.; Lei, J.Q.; Li, S.Y.; Wang, H.F. Effect of hanging-type sand fence on characteristics of wind-sand flow fields. *Wind. Struct. Int. J.* **2016**, *22*, 555–571. [CrossRef]
5. He, Y.W. Design of the Wind-resistant Gallery in Lanzhou-Xinjiang High Speed Railway. *J. Railw. Eng. Soc.* **2017**, *34*, 55–59.
6. Cheng, J.J.; Jiang, F.Q.; Xue, C.X.; Xin, G.W.; Li, K.C.; Yang, Y.H. Characteristics of the disastrous wind-sand environment along railways in the Gobi area of Xinjiang, China. *Atmos. Environ.* **2015**, *102*, 344–354. [CrossRef]
7. Xiao, J.; Yao, Z.; Qu, J.; Jiang, F. Characteristics and formation mechanism of extreme wind in hundred-li wind zone along Lanxin Railway. *China Railw. Sci.* **2016**, *37*, 130–137.
8. Rechena, D.; Infante, V.; Sousa, L.; Baptista, R. Design and failure modes of a standard railway catenary cantilever support. *Eng. Fail. Anal.* **2019**, *107*, 104217. [CrossRef]
9. Yang, S.A.; Zw, B.; Zl, C.; Rw, D. A spatial coupling model to study dynamic performance of pantograph-catenary with vehicle-track excitation. *Mech. Syst. Signal Process.* **2021**, *151*, 107336.
10. Pan, X.; Xiuqing, M.A.; Jie, X.U. Analysis and Evaluation About Anti-wind Efficiency of Windbreak Experimental Section in Lan-Xin High Railway. *J. Arid Meteorol.* **2019**, *37*, 496–499.
11. Qu, J.J.; Huang, N.; Lei, J.Q.; Dong, Z.B.; Liu, X.W.; Xue, X.; Zu, R.P.; Zhang, K.C. Structural characteristics of Gobi sand-drift and its significance. *Adv. Earth Ences* **2005**, *20*, 19–23.
12. Li, J.; Kandakji, T.; Lee, J.A.; Tatarko, J.; Collins, J.D. Blowing dust and highway safety in the southwestern United States: Characteristics of dust emission "hotspots" and management implications. *Sci. Total Environ.* **2018**, *621*, 1023–1032. [CrossRef] [PubMed]
13. Tian, Y.; Wang, Z.; Pan, X.; Li, J.; Wang, Z. Influence of the morphological change in natural Asian dust during transport: A modeling study for a typical dust event over northern China. *Sci. Total Environ.* **2020**, *739*, 139791. [CrossRef] [PubMed]
14. Zhou, D.; Yuan, X.X.; Yang, M.Z.; Xu, L.Z.; Zhang, L. Research on the anti-wind/sand effect of windbreak wall on high-speed train. *J. Exp. Fluid Mech.* **2012**, *26*, 63–67.
15. Huang, N.; Gong, K.; Xu, B.; Zhao, J.; Xin, G. Investigations into the law of sand particle accumulation over railway subgrade with wind-break wall. *Eur. Phys. J. E* **2019**, *42*, 145. [CrossRef]
16. Cheng, J.J.; Xin, G.W.; Zhi, L.Y.; Jiang, F.Q. Unloading Characteristics of Sand-drift in Wind-shallow Areas along Railway and the Effect of Sand Removal by Force of Wind. *Sci. Rep.* **2017**, *7*, 41462. [CrossRef] [PubMed]
17. Yan, Y.X.; Yan, X. Origin Analysis on Farmers' Housing Environment Adaptability Research in the Northwest Settlement of Dry Oasis Edge. *Appl. Mech. Mater.* **2012**, *209–211*, 248–251. [CrossRef]
18. Zeng, Q.L.; Zhen-Shan, L.I.; Fu-An, L.U.; Wang, Y.L. A Computational Study of Wind-break Performance of Ventilated Wind-break Wall in Different Locations on the Highway. *J. Desert Res.* **2012**, *32*, 1542–1550.
19. Lu, B.; Shen, A.Q.; Wei, Z.X.; Highway, S.O.; University, C. Comparison of In-situ Observation Effect of Highway Windbreak Structures and Simulation of Installation Optimization. *J. Highw. Transp. Res. Dev.* **2016**, *33*, 39–45.
20. Xiaohui, L.U.; Huang, N.; Ding, T. Wind tunnel experiments on natural snow drift. *Sci. China Technol. Sci.* **2012**, *55*, 927–938.
21. Sun, W.; Huang, N.; He, W. Turbulence burst over four micro-topographies in the wind tunnel. *CATENA* **2017**, *4*, 017. [CrossRef]
22. Tetsuya, K.; Yamagishi, Y.; Kimura, S.; Sato, K. Aerodynamic Behavior of Snowflakes on an Uneven Road Surface during a Snowstorm. *Open J. Fluid Dyn.* **2017**, *7*, 696–708. [CrossRef]
23. Xie, S.; Qu, J.; Han, Q.; Pang, Y. Wind Dynamic Environment and Wind Tunnel Simulation Experiment of Bridge Sand Damage in Xierong Section of Lhasa–Linzhi Railway. *Sustainability* **2020**, *12*, 5689. [CrossRef]
24. Dong, Z.; Qinan, G.; Lu, P.; Luo, W.; Wang, H. Turbulence fields in the lee of two-dimensional transverse dunes simulated in a wind tunnel. *Earth Surf. Process. Landf.* **2010**, *34*, 204–216. [CrossRef]
25. White, B.R. Laboratory Simulation of Aeolian Sand Transport and Physical Modeling of Flow Around Dunes. *Ann. Arid Zone* **1996**, *35*, 187–213.
26. Musick, H.B.; Trujillo, S.M.; Truman, C.R. Wind-Tunnel Modelling of the Influence of Vegetation Structure On Saltation Threshold. *Earth Surf. Process. Landf.* **2015**, *21*, 589–605. [CrossRef]
27. Yan, Y.; Shao, Y. Numerical simulations of flow and pollution dispersion in urban atmospheric boundary layers. *Environ. Model. Softw.* **2008**, *23*, 906–921.
28. Shi, F.; Huang, N. Computational simulations of blown sand fluxes over the surfaces of complex microtopography. *Environ. Model. Softw.* **2010**, *25*, 362–367.
29. Liu, S.; Pan, W.; Zhao, X.; Zhang, H.; Cheng, X.; Long, Z.; Chena, Q. Influence of surrounding buildings on wind flow around a building predicted by CFD simulations. *Build. Environ.* **2018**, *140*, 1–10. [CrossRef]
30. Wang, D.; Wang, Y.; Yang, B. An Algorithm in PIV Measurement of wind-sand two-phase flow. *Chin. J. Theor. Appl. Mech.* **2006**, *3*, 302–308.
31. Sun, W.; Huang, N. A study on velocity of aeolian sand particles over leeward slope by particle image velocimetry. *Geomorphology* **2018**, *317*, 157–169. [CrossRef]

32. Zhang, Y.; Wang, Y.; Yang, B.; Jia, P. Measurement of sand creep on a flat sand bed using a high-speed digital camera: Mesoscopic features of creeping grains. *Sedimentology* **2015**, *63*, 629–644. [CrossRef]
33. Kang, L.; Zhang, J.; Zou, X.; Zhang, C.; Hong, C. Characteristics of Sand Transport during Initiation Process of Aeolian Sand Transport. *J. Desert Res.* **2017**, *37*, 1051–1058.

applied sciences

Article

Limiting Wind-Induced Resuspension of Radioactively Contaminated Particles to Enhance First Responder, Early Phase Worker and Public Safety—Part 1

Hadas Raveh-Amit [1,*,†], Avi Sharon [2,*,†], Itzhak Katra [3], Terry Stilman [4], Shannon Serre [5], John Archer [6] and Matthew Magnuson [6]

1 Department of Chemistry, Nuclear Research Centre Negev, P.O. Box 9001, Beer Sheva 8419000, Israel
2 Environmental Research Unit, Nuclear Research Centre Negev, P.O. Box 9001, Beer Sheva 8419000, Israel
3 Department of Geography and Environmental Development, Ben-Gurion University, P.O. Box 653, Beer Sheva 8410501, Israel; katra@bgu.ac.il
4 EPA Region 4, Atlanta, GA 30303, USA; stilman.terry@epa.gov
5 EPA Office of Land and Emergency Management, Washington, DC 20002, USA; serre.shannon@epa.gov
6 EPA Office of Research and Development, Cincinnati, OH 45268, USA; archer.john@epa.gov (J.A.); magnuson.matthew@epa.gov (M.M.)
* Correspondence: hadasa@nrcn.gov.il (H.R.-A.); avish@nrcn.gov.il (A.S.)
† These authors contributed equally to this work.

Citation: Raveh-Amit, H.; Sharon, A.; Katra, I.; Stilman, T.; Serre, S.; Archer, J.; Magnuson, M. Limiting Wind-Induced Resuspension of Radioactively Contaminated Particles to Enhance First Responder, Early Phase Worker and Public Safety—Part 1. *Appl. Sci.* **2022**, *12*, 2463. https://doi.org/10.3390/app12052463

Academic Editor: Angeles Sanroman Braga

Received: 26 January 2022
Accepted: 24 February 2022
Published: 26 February 2022

Abstract: An accidental radiological release or the operation of a radiological dispersal device (RDD) may lead to the contamination of a large area. Such scenarios may lead to health and safety risks associated with the resuspension of contaminated particles due to aeolian (wind-induced) soil erosion and tracking activities. Stabilization technologies limiting resuspension are therefore needed to avoid spreading contamination and to reduce exposures to first responders and decontamination workers. Resuspension testing was performed on soils from two sites of the Negev Desert following treatment with three different stabilization materials: calcium chloride, magnesium chloride, and saltwater from the Dead Sea in Israel. Two and six weeks post-treatment, resuspension was examined by inducing wind-driven resuspension and quantitatively measuring particle emission from the soils using a boundary-layer wind tunnel system. Experiments were conducted under typical wind velocities of this region. Treating the soils reduced resuspension fluxes of particulate matter < 10 μm (PM_{10}) and saltating (sand-sized) particles to around background levels. Resuspension suppression efficiencies from the treated soils were a minimum of 94% for all three stabilizers, and the Dead Sea salt solution yielded 100% efficiency over all wind velocities tested. The impact of the salt solutions (brine) was directly related to the salt treatment rather than the wetting of the soils. Stabilization was still observed six weeks post-treatment, supporting that this technique can effectively limit resuspension for a prolonged duration, allowing sufficient time for decision making and management of further actions.

Keywords: soil resuspension; stabilization material; soil contamination; Dead Sea saltwater; dust emission; soil erosion

1. Introduction

An accidental radiological release or the operation of a Radiological Dispersal Device (RDD) may lead to the contamination of a large area with radioactive materials. During the immediate emergency phase of a response, life-saving operations and securing of critical infrastructure must be conducted for the safety of the public and first responders [1,2]. During the operations, emergency responders, as well as decontamination workers assisting with the response, may be further exposed due to inhalation of resuspended particles and direct contact, owing to the tracking of contamination from the contaminated areas, i.e., roads, other construction materials and soils. Containment of the contaminated area to prevent resuspension could reduce the overall exposure for emergency responders and decontamination workers and also reduce the spread of contamination. Hence, stabilization technologies and methodologies to minimize this exposure are needed [1,2].

Aeolian (wind-induced) soil erosion, and the following process of dust emission, results in the resuspension of soil-derived particles to the atmosphere and air pollution [3–5]. Stabilization technologies are designed to prevent the spread of particles (such as by resuspension) and are routinely used in industries, such as road constructions and mining sites, for dust control [2]. The application of rapidly available and easily applied stabilization technologies has the potential for accomplishing multiple goals following the release of radioactive particles from a radiological contamination event. Primarily, the application of a stabilization material may reduce exposures to first responders and decontamination workers assisting with the response due to tracking. In addition, such technologies would limit the wind-induced spread of contamination to other non-contaminated, less-contaminated, or recently decontaminated areas, subsequently reducing the time and resources needed for additional decontamination operations [2].

The United States Environmental Protection Agency (EPA) previously conducted work on stabilization technologies [1,2,6,7]. From these studies, the list below presents some options recommended by stakeholders and experts that may be suitable for stabilization:

- Soil2O®1 dust control wetting agent (available in the US);
- Calcium chloride ($CaCl_2$);
- Phos-Chek®MVP-F3 fire retardant (available in the US);
- Locally available firefighting foam;
- Capping with locally available gravel, mulch, sand or clay;
- Misting with water or saltwater (brine), with the possible addition of additives;
- Application of a polymer coating/gel.

There is a lack of fundamental research examining the applicability of stabilization materials required in an event leading to the contamination of a large area with radioactive materials. Stabilization materials suitable for large areas of contaminated soils are expected to be cheap, easily applied and highly effective in limiting wind-induced contamination dispersal. Previous works showed the potential of specific brines to reduce dust emission from unpaved roads of different soils [3,8,9], with low environmental salinization risk [10]. The current study aimed to test the effectiveness of different brines to stabilize arid soils that may be subjected to soil contamination and are already associated with natural dust emission.

2. Materials and Methods

2.1. Soil Sampling and Physicochemical Characterization

Soils were sampled from two sites that are undisturbed and associated with dust emission in the field: Ze'elim sandy area (31.16° E/34.53° N) at the western Negev Desert [11], and the Yamin plateau (31.04° E/35.08° N) at the northeastern Negev Desert in Israel [12]. The soil samples were analyzed for elemental composition by X-ray fluorescence (XRF) using an Axios spectrometer (PANanalytical, Malvern, UK). Mineralogical phase identification was performed by X-ray powder diffraction (XRPD) using an Empyrean Philips 1050/70 diffractometer (PANanalytical, Malvern, UK). Particle size distribution (PSD) was performed by laser diffraction using Analysette 22 MicroTec Plus (Fritsch International, Idar-Oberstein, Germany). XRF, XRPD and PSD analyses were performed at the Ben-Gurion University of the Negev in Beer-Sheva. pH was measured using a Metrohm pH meter (Metrohm, Herisau, Switzerland). Water content in soils was measured gravimetrically. Total organic content (TOC) was determined by titration of the dissolved organics with ammonium iron sulfate using an 848 Titrino plus (Metrohm, Herisau, Switzerland) at the Geological Survey of Israel.

2.2. Soil Stabilization

2.2.1. Stabilization Materials

Three brine solutions were tested in this study: magnesium chloride ($MgCl_2$), calcium chloride ($CaCl_2$) and a solution sampled from the Dead Sea in Israel. $MgCl_2$ and $CaCl_2$ salts were purchased from Carlo Erba Reagents, Italy. $MgCl_2$ and $CaCl_2$ solutions were freshly prepared in deionized water at concentrations of 30% *w*/*v* and 35% *w*/*v*, respectively, with similar concentrations reported previously [2,3]. Samples of Dead Sea solution were collected in 3 L containers a few days prior to the experiments (Figure 1). Samples were analyzed for Na, K, Ca, Mg and Sr by ICP-AES (Optima 3300, Perkin Elmer, Seer Green, UK) and Br by ICP-MS (NexION 300D, Perkin Elmer, Seer Green, UK). Cl concentrations were calculated by subtracting Br concentrations, analyzed by ICP-MS, from the total Br and Cl concentrations obtained using AgNO3 titration. Titrations were performed using an 848 Titrino plus (Metrohm, Herisau, Switzerland). Chemical analysis was performed at the Geological Survey of Israel. The chemical composition of the Dead Sea salt solutions is presented in Table 1.

Figure 1. Samples of Dead Sea salt solutions collected in 3 L containers (left side) and trays of Ze'elim soil treated with different brines (surface area of 0.5 m × 1.0 m and height of 0.02 m).

Table 1. Chemical composition of the stabilization solutions.

Ion/Element	Stabilization Material		
	$CaCl_2$	$MgCl_2$	Dead Sea Salt Water
Na^+	-	-	8700
K^+	-	-	13,600
Ca^{++}	127,000	-	28,800
Mg^{++}	-	76,600	74,000
Cl^-	223,000	223,000	291,000
Br^-	-	-	8400

2.2.2. Application of Stabilizers

Soils were placed in trays customized to fit the wind tunnel dimensions (surface area of 0.5 m × 1.0 m and height of 0.02 m) (Figure 1). Brine solutions were applied to the soils by spraying the soil using a sprayer at equal volume to surface area ratios (1.5 L m^{-2}). As controls, soils were either untreated or sprayed with tap water (clean drinking water). After applying the solutions and prior to the wind-tunnel experiments, the trays were left in the laboratory in order to avoid any environmental effect on the soils (e.g., wind-induced resuspension). Table 2 summarize the stabilization experimental matrix.

Table 2. Stabilization experimental matrix.

Soil	Non-Stabilized Control (Untreated)	Tap Water	Stabilization Material		
			$CaCl_2$	$MgCl_2$	Dead Sea Salt
Yamin	A	B	C	C	C
Ze'elim	A	B	C	C	C

[A] Testing of untreated soils; [B] Testing of soils 2 weeks post-treatment; [C] Testing of soils 2 weeks and 6 weeks post-treatment.

2.3. Boundary-Layer Wind Tunnel Experiments: Resuspension Testing and Calculations

Resuspension testing was performed at the Aeolian Simulation Laboratory, Ben-Gurion University of the Negev, using a boundary-layer wind tunnel [13]. Untreated and treated soils were tested following either 2 weeks or 6 weeks beginning from the day of treatment. The different times were chosen to represent different periods of aging following an incident. Experiments were conducted under four wind velocities, 5.3, 6.8, 8.1, and 9.6 m s^{-1}, representing typical natural winds associated with dust emission in this region. PM_{10} dust concentrations were recorded by light-scattering laser photometers DustTrak DRX 8534 (TSI Inc., Shoreview, MN, USA) placed 25 cm above the tunnel bed. Before placing the soil trays in the wind tunnel, PM_{10} background levels of up to 20 μg m^{-3} were recorded. Background levels were subtracted from the PM_{10} measurements, which were taken at different wind velocities. Each sample was measured for a duration of 30 s, at 1 s intervals. This short duration is enough to determine the dust emission patterns in controlled experiments [3,5]. Mass flux values of PM_{10} resuspended from the ground (g m^{-2} s^{-1}), expressed as $F(PM_{10})$, were calculated according to the following [13]:

$$F(PM_{10}) = \frac{C(PM_{10}) \times Vt}{Ap \times t} \tag{1}$$

where $C(PM_{10})$ is the recorded PM_{10} concentration (μg m^{-3}), Vt is the air volume in the wind tunnel (3.43 m^3), Ap is the area of the experimental plot (0.25 m^2) and t is time (in seconds).

Mean mass flux values of PM_{10} ($\overline{F}(PM_{10})$) were calculated by averaging all FPM_{10} results per sample, i.e., 30 calculated flux values obtained over 30 s per wind velocity.

Saltating particles associated with the initiation of the dust emission process from soils [4,5] were collected by traps placed 2.5 to 10.5 cm above the tunnel bed and along the wind direction. Collected particles were weighted at the end of each experiment. Mean mass flux values of saltating particles (g m^{-2} s^{-1}), expressed as $\overline{F}$(saltation), were calculated according to the following:

$$\overline{F}(\text{saltation}) = \frac{m(\text{saltation})}{Ap \times t} \tag{2}$$

where m(saltation) is the measured weight of the saltating particles (g), At is the cross-sectional area of the traps (0.02 m^{-2}) and t is time (in seconds).

Suppression efficiencies (SE) of PM_{10} or saltating particles (in percentage) were calculated for each stabilizer and soil type at each wind velocity according to the following:

$$SE = \left(1 - \frac{\overline{F}}{\overline{F}(\text{control})}\right) \times 100 \tag{3}$$

where $\overline{F}$ is the mean mass flux values of PM_{10} or saltating particles (see above) and $\overline{F}$(control) is the mean flux of the control sample (untreated) for the same wind velocity and soil type.

3. Results

3.1. Physicochemical Characteristics of the Soils

Soils were collected from two sites. The first sampling site was the Yamin plateau at the northern Negev Desert in Israel, and the second site was the Ze'elim sandy area at the western Negev Desert in Israel. Both soils are mainly composed of quartz (SiO_2), silicate minerals (anorthite ($CaAl_2Si_2O_8$), sanidine ($CaAl_2Si_2O_8$)), carbonate minerals (dolomite ($CaMg(CO_3)_2$) and calcite ($CaCO_3$)) and clay-sized minerals (hematite (Fe_2O_3)), as characterized by XRF and XRD analyses (Table 3, Figure 2). Additional analysis showed the soils were alkaline and contained low water and organic matter contents (Table 4), which are typical characteristics of desert soils.

Table 3. X-ray fluorescence (XRF) measurements of soils from the Ze'elim area and the Yamin Plateau in Israel.

Compound	Ze'elim Soil (wt%)	Yamin Soil (wt%)
SiO_2	89	84
Al_2O_3	5	3
CaO	2	8
Fe_2O_3	2	2
K_2O	1	1
SiO_2	89	84

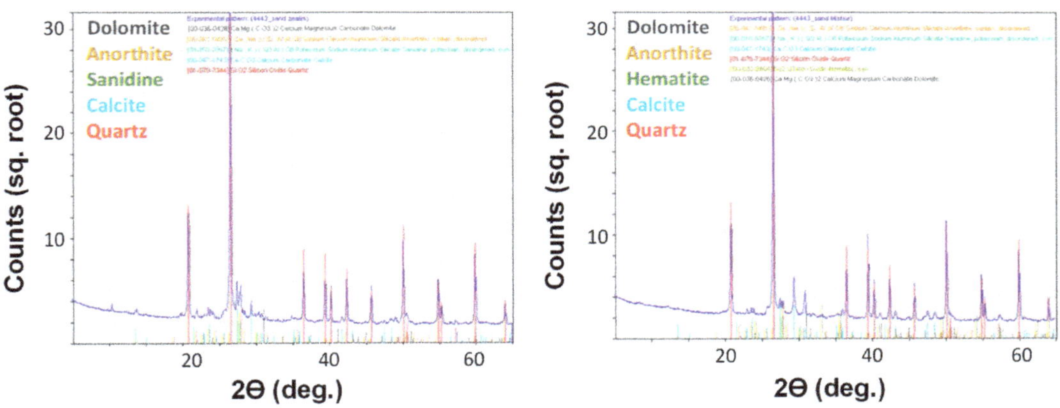

Figure 2. X-ray diffraction (XRD) patterns of the Ze'elim (**left** side) and Yamin soils (**right** side).

Table 4. Soil properties of the Ze'elim and the Yamin soils in Israel.

Properties	Ze'elim Soil (wt%)	Yamin Soil (wt%)
pH	7.9	7.8
Water content (wt%)	<1	<1
Total organic content (wt%)	<0.3	<0.3

PSD analysis showed different characteristics in grain size, whereas the Ze'elim soil was classified as sand, the Yamin soil was classified as silt loam (Figure 3). The Ze'elim soil demonstrated a higher mean grain size (170 μm vs. 50 μm) and a lower PM_{10} content (3% vs. 28%) than the Yamin soil (Table 5). It was found that the Ze'elim soil is mainly composed of fine and medium sand fractions, while silt and fine sand are the main fractions in the Yamin soil.

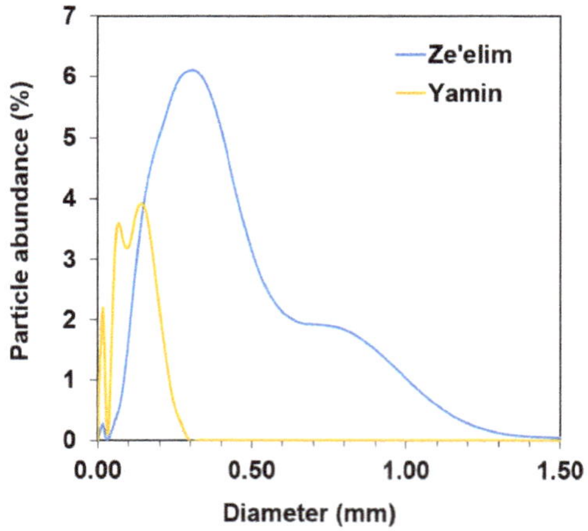

Figure 3. Particle size distribution of the Ze'elim and the Yamin soils in Israel.

Table 5. Particle size fractions of the Ze'elim and the Yamin soils in Israel.

Size Fraction		Ze'elim Soil (%)	Yamin Soil (%)
Clay (<0.002 mm)		1	7
Silt (0.002–0.063 mm)		6	51
Sand	Fine (0.063–0.25 mm)	37	41
	Medium (0.25–0.5 mm)	39	1
	Coarse (0.5–2.0 mm)	17	0
PM_{10} (<0.01 mm)		3	28

3.2. Effectiveness of Brine Stabilizers on Resuspension Suppression from the Ze'elim Soil

To test the impact of the brine stabilizers on the resuspension from the soils, soils were treated with different stabilizers, left to dry for two weeks, and then tested for wind-induced dust emission. Untreated soils served as non-stabilized controls (NSCs). Soils were treated with either of the following stabilizers: $CaCl_2$, $MgCl_2$ and saltwater from the Dead Sea in Israel. Soils were also treated with tap water in order to control for the impact of wetting (Table 2).

PM_{10} concentrations recorded during the wind tunnel experiment, representing wind-induced dust emissions from the Ze'elim soil, are presented in Figure 4. Higher wind velocities resulted in higher PM_{10} resuspension levels from the untreated soil (control).

Resuspension was slightly reduced from soils sprayed with tap water (followed by drying) at all wind velocities tested, with a significant reduction at the lowest wind velocity. Extremely low resuspension levels were detected in brine treated soils, demonstrating that the soils were effectively stabilized following the treatments. The most effective dust suppressor was the Dead Sea salt treatment, yielding average PM_{10} concentrations similar to background levels (~20 $\mu g/m^3$).

Based on the PM_{10} concentrations recorded during the wind tunnel experiment and the mass measurements of the collected salting particles, mean PM_{10} fluxes and mean saltation fluxes were calculated, respectively. Figure 5 show the mean PM_{10} fluxes and mean saltation fluxes from the Ze'elim soil under different treatment conditions, tested under four wind velocities. From these results, it was evident that the resuspension fluxes of saltating particles were significantly lower (by at least an order of magnitude) than dust particles, supporting that PM_{10} are the major resuspension contributors under natural conditions.

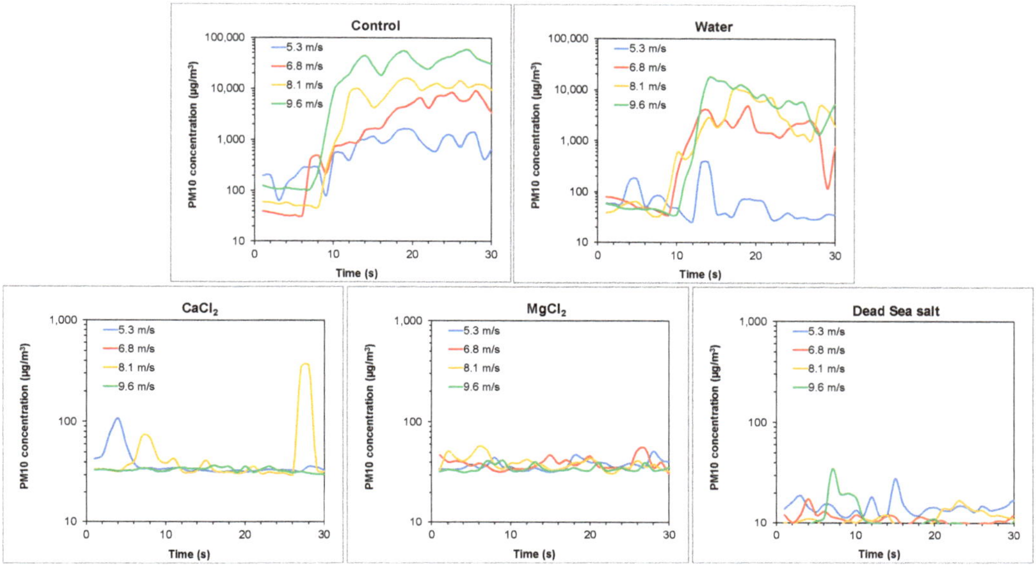

Figure 4. Wind-driven PM_{10} emissions from the Ze'elim soil two weeks following treatment with stabilizers. Note the differences in the values of the Y-axis between Control and Water to the brines.

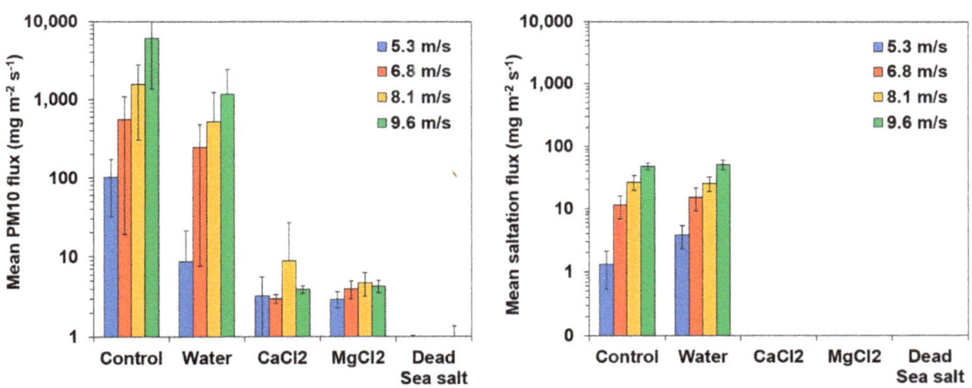

Figure 5. Winddriven PM_{10} and saltation fluxes from the Ze'elim soil treated with stabilizers.

To quantitatively evaluate the impact of the treatments on the resuspension of PM_{10} and saltating particles, suppression efficiencies were calculated (Tables 6 and 7). Treating the Ze'elim soil with brine solutions resulted in effective stabilization, as shown by significantly reduced fluxes compared to the control and high resuspension suppression efficiencies of >97% (Figure 5, Tables 6 and 7) for all experimental conditions. The impact of the brine solutions was directly related to the salt treatment, as slightly reduced PM_{10} fluxes and unchanged saltation fluxes were observed in soils misted with tap water only.

Table 6. Suppression efficiencies of wind-driven PM_{10} emission from the Ze'elim soil treated with stabilizers.

Wind Velocity/ Treatment	5.3 m/s	6.8 m/s	8.1 m/s	9.6 m/s
Tap Water	92%	56%	66%	81%
$CaCl_2$	97%	99%	99%	100%
$MgCl_2$	97%	99%	100%	100%
Dead Sea salt	100%	100%	100%	100%

Table 7. Suppression efficiencies of wind-driven saltating particle emission from the Ze'elim soil treated with stabilizers.

Wind Velocity/ Treatment	5.3 m/s	6.8 m/s	8.1 m/s	9.6 m/s
Tap Water	0%	0%	4%	0%
$CaCl_2$	100%	100%	100%	100%
$MgCl_2$	100%	100%	100%	100%
Dead Sea salt	100%	100%	100%	100%

While all salt solutions efficiencies may be operationally relevant, interestingly, the most effective suppression effect on overall resuspension was achieved by the Dead Sea salt treatment, yielding 100% suppression efficiency over all wind velocities tested. For the prepared calcium and magnesium salt solutions, the efficiencies were less for lower wind speeds.

To evaluate the durability of the stabilization technique, re-testing was performed four weeks following the wind tunnel experiments described above (six weeks from the day of treatment). These time points were chosen because while operations may start immediately, they may continue over several weeks, so it is necessary to study the longer-term effectiveness. Re-testing resuspension of PM_{10} concentrations from the Ze'elim soil is presented in Figure 6. Treatment with all three stabilizers resulted in low average PM_{10} concentrations similar to background levels (~20 $\mu g/m^3$). Resuspension levels of saltating particles were undetected (no particles were collected). These results demonstrated that treating the Ze'elim soil with brine solutions resulted in effective stabilization six weeks post-treatment.

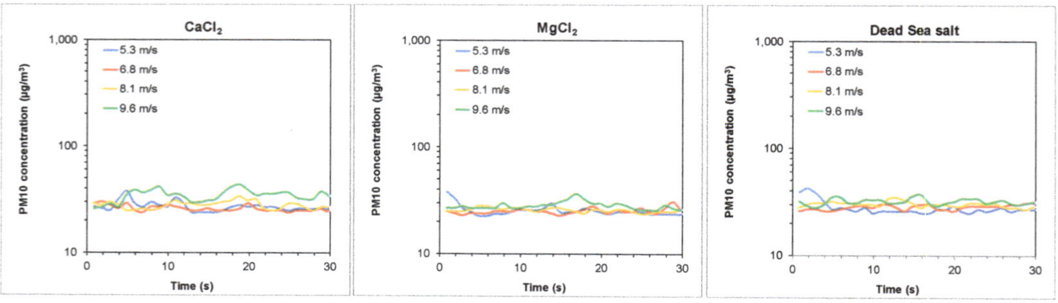

Figure 6. Wind-driven PM_{10} emissions from the Ze'elim soil six weeks following treatment with stabilizers.

3.3. *Effectiveness of Brine Stabilizers on Resuspension Suppression from the Yamin Soil*

Yamin soil was subjected to treatments and resuspension testing similar to those performed on the Ze'elim soil. Soils were treated with different stabilizers, left to dry for two weeks, and then tested for particle emission in the wind tunnel. PM_{10} concentrations recorded during the experiment, representing wind-induced dust emission from the Yamin soil following different treatments, are presented in Figure 7.

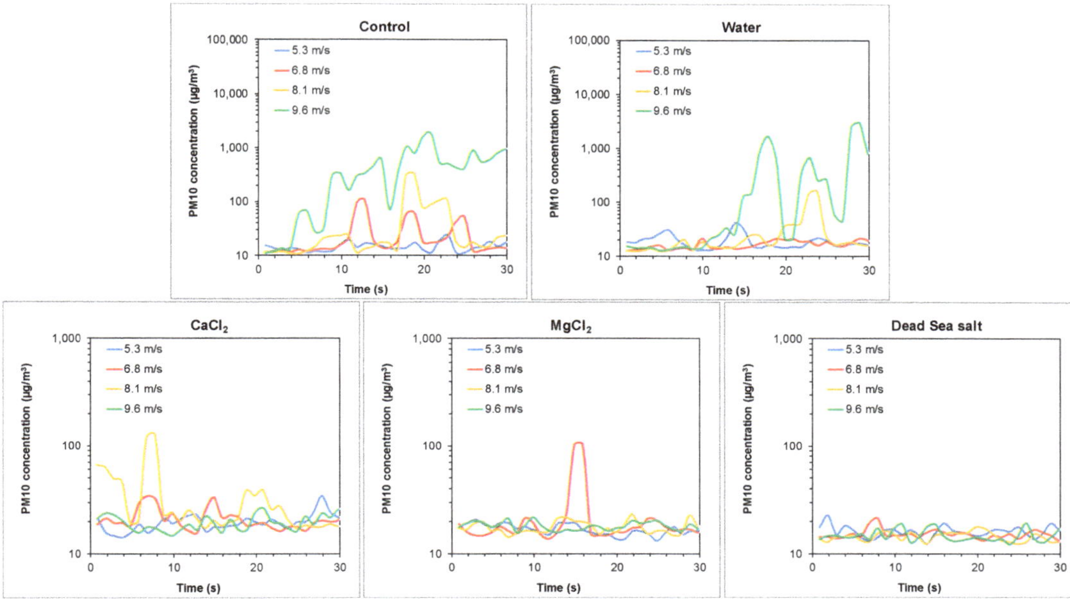

Figure 7. Wind-driven PM_{10} emissions from the Yamin soil two weeks following treatment with stabilizers. Note the differences in the values of the Y-axis between Control and Water to the brines.

As shown for the Ze'elim soil, higher wind velocities resulted in higher PM_{10} resuspension levels from the untreated Yamin soil (control). In contrast, significantly lower (by at least an order of magnitude) resuspension levels were observed from this soil when compared with the Ze'elim soil, as demonstrated by lower PM_{10} concentrations recorded under identical conditions (Figures 4 and 7, control).

As shown in Figure 7, extremely low resuspension levels were detected in brine treated soils, demonstrating that the soils were effectively stabilized following the treatments. The Dead Sea salt treatment was the most effective dust suppressor for the Yamin soil, similar to the results obtained for the Ze'elim soil.

Figure 8 show the mean PM_{10} fluxes and the mean saltation fluxes from the Yamin soil under different treatments, tested under four wind velocities. As shown for the Ze'elim soil, the resuspension fluxes of saltating particles from the Yamin soil were significantly lower (by at least an order of magnitude) than dust particles, supporting that PM_{10} are the major resuspension contributors under untreated conditions.

Tables 8 and 9 present the calculated suppression efficiencies of PM_{10} and saltating particle resuspension from the Yamin soils. Suppression efficiencies could not be calculated under the lowest wind velocity because PM_{10} measurements were low (around background levels), and no saltating particles could be collected and measured (noted NA). Treating the soil with brine solutions resulted in effective stabilization, as shown by significantly reduced fluxes compared to the control, along with resuspension suppression efficiencies (>94%). The most effective suppression effects on overall resuspension were achieved by the $MgCl_2$ and Dead Sea salt treatments, yielding 100% suppression efficiency over all wind velocities tested.

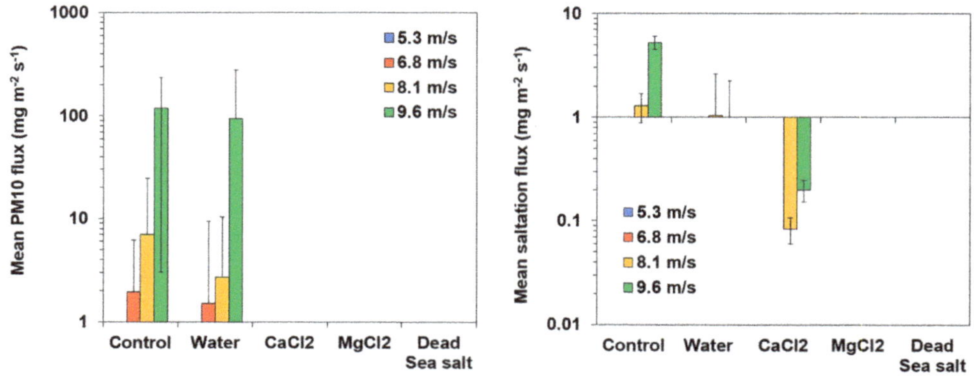

Figure 8. Wind-driven PM_{10} and saltation fluxes from the Yamin soil treated with stabilizers.

Table 8. Suppression efficiencies of wind-driven PM_{10} emission from the Yamin soil treated with stabilizers.

Wind Velocity/ Treatment	5.3 m/s	6.8 m/s	8.1 m/s	9.6 m/s
Tap Water	NA [1]	22%	61%	21%
$CaCl_2$	NA [1]	100%	100%	100%
$MgCl_2$	NA [1]	100%	100%	100%
Dead Sea salt	NA [1]	100%	100%	100%

[1] Not available (NA) means values could not be calculated because the mean mass of flux of the control sample was zero.

Table 9. Suppression efficiencies of wind-driven saltating particle emission from the Yamin soil treated with stabilizers.

Wind Velocity/ Treatment	5.3 m/s	6.8 m/s	8.1 m/s	9.6 m/s
Tap Water	NA [1]	NA [1]	19%	81%
$CaCl_2$	NA [1]	NA [1]	94%	96%
$MgCl_2$	NA [1]	NA [1]	100%	100%
Dead Sea salt	NA [1]	NA [1]	100%	100%

[1] Not available (NA) means values could not be calculated because the mean mass of flux of the control sample was zero.

Analogous to the Ze'elim soil, the durability of the stabilization technique was evaluated on the Yamin soil by retesting resuspension from the treated trays following four additional weeks. Figure 9 present the PM_{10} concentrations recorded during the wind tunnel experiment from the Yamin soil. Treatment with all three stabilizers resulted in average PM_{10} concentrations similar to background levels (~20 μg/m^3). Resuspension levels of saltating particles were undetected (no particles were collected). These results demonstrated that treating the Yamin soil with brine solutions resulted in effective stabilization even six weeks post-treatment.

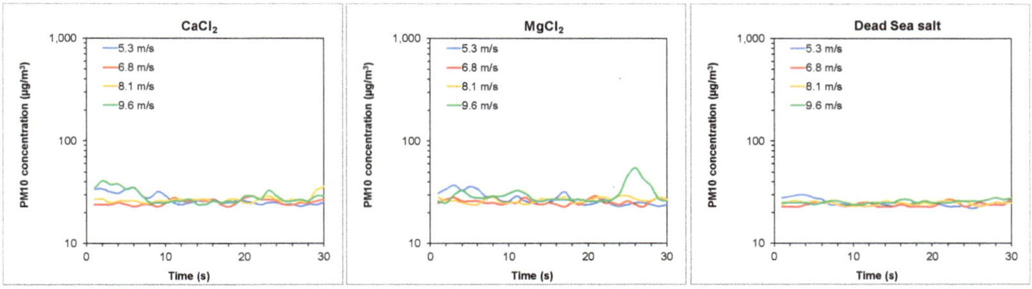

Figure 9. Wind-driven PM_{10} emissions from the Yamin soil six weeks following treatment with stabilizers.

4. Discussion

Treating the two soils with salt/brine solutions resulted in reduced particle resuspension, as shown by extremely low PM_{10} fluxes (equivalent to background levels) and high resuspension suppression efficiencies (>94%). The impact of the brine solutions was directly related to the salt treatment rather than the wetting of the soils since similar particle resuspension fluxes were obtained from untreated soils or soils sprayed with tap water only. Brine solutions are, therefore, effective stabilizers, leading to reduced resuspension of soil particles. These results are consistent with previous work performed by Katra et al. [3], which tested the impact of diverse dust control products of synthetic and organic polymers (Lignin, Resin, Bitumen, PVA, Brine) on unpaved roads. The authors showed that some products significantly reduced dust emission from quarry roads, especially when using magnesium chloride (Brine).

All three salt\brine solutions tested in this study function primarily by helping cement small particles into larger ones that are more difficult to resuspend [2]. Their capability to enhance the cohesion of smaller particles is expected to vary with the composition of the salt solution, as well as the specific particles involved. Aiding in this cohesion is the fact that salts such as $CaCl_2$ and $MgCl_2$ are hygroscopic, so when they dry out after being applied (usually by spraying an aqueous solution), some water may be present, which helps enhance cohesion [2]. The effectiveness of the stabilizers is expected to occur immediately after the applied solutions dry, which in desert climates is expected to not take long, as shown in this work. While all salt solutions have operational relevance, the most effective stabilizer was the Dead Sea salt solution, yielding 100% resuspension suppression efficiency of PM_{10} and saltating particles over all wind velocities tested. The motivation to test the Dead Sea salt solution as a stabilizer was it being an easily available, natural resource of salts. Saltwater from the Dead Sea can be derived directly from the sea or procured locally. $MgCl_2$ and $CaCl_2$ were also highly effective but slightly less effective than the Dead Sea salt in limiting PM_{10} resuspension from the Ze'elim soil (>97%). $CaCl_2$ was also slightly less effective in limiting the resuspension of saltating particles from the Yamin plateau (>94%). The Dead Sea solution is expected to contain other substances, such as specific ions and humic substances that help retain hydration, which may enhance the cohesion of small particles.

Significantly lower resuspension levels were observed from the Yamin soil when compared with the Ze'elim soil (>10-fold difference), indicated by lower PM_{10} concentrations recorded under identical conditions (Figures 6 and 9). This may result from differences in the cohesiveness of the soil particles between the two soil types, rather than the content of the PM_{10} in the soil (Table 5), which is significantly higher in the Yamin than the Ze'elim soil (28 wt% and 7 wt%, respectively). It demonstrates the role of sand transport in dust-PM_{10} emission from sandy soils [14].

Resuspension fluxes of saltating particles from the two soils were >10-fold lower than dust particles, demonstrating that PM_{10} are the major resuspension contributors under natural conditions. This result confirms that dust emission is expected to cause the major spread of the contamination in the case of an emergency event in the Negev desert, highlighting the importance of limiting resuspension of contaminated dust. Treating the soils with brine solutions resulted in effective stabilization six weeks post-treatment, supporting that this technique can effectively limit resuspension of contaminated soil after an emergency event for a prolonged duration, allowing sufficient time for decision making and management of further actions. This is particularly important in desert environments where continued drying could otherwise lead to increased resuspension.

Our results highlight the importance of considering the soil properties at a specific site when considering the impacts and mitigation of resuspension. The two soils in this study have characteristics that contribute to their ability to be resuspended, e.g., small organic content and low moisture content. Therefore, they may be considered "worst cases", such that the results may also be applicable to many other types of soils for which resuspension may be inherent less favored.

While the salt solutions appear to increase the cohesiveness of small particles and thus reduce wind-induced resuspension, complex mechanisms appear to govern the disintegration of the cohesive/cemented particles and their subsequent resuspension. Therefore, to validate the applicability of stabilization techniques, it is essential to test the impact of stabilizers in specific situations which induce different types of physical stresses other than wind. Two operationally relevant cases are the movement of vehicles and foot traffic. EPA investigated simulated vehicle and foot traffic in controlled laboratory studies [15]. Together, the results of the present study, along with the EPA study, suggest the relevancy and urgency of testing stabilization techniques on a larger scale area under natural environmental conditions.

Author Contributions: Conceptualization, H.R.-A. and A.S.; methodology, H.R.-A., A.S., I.K., T.S., S.S., J.A. and M.M.; writing—H.R.-A. and A.S.; writing—review and editing, I.K. and M.M.; funding acquisition, A.S. and T.S. All authors have read and agreed to the published version of the manuscript.

Funding: This research was funded by The Nuclear Research Centre Negev, Beer-Sheva, Israel.

Institutional Review Board Statement: Not applicable.

Informed Consent Statement: Not applicable.

Data Availability Statement: Data are contained within the article.

Conflicts of Interest: The authors declare no conflict of interest.

Disclaimer: Mention of trade names, products, or services does not convey approval, endorsement or recommendation by any agency. The research presented was not performed or funded by EPA and was not subject to EPA's quality system requirements. The views expressed in this article are those of the author(s) and do not necessarily represent the views or the policies of the U.S. Environmental Protection Agency.

References

1. Argonne National Laboratory. *Municipal and Commercial Equipment for Radiological Response and Recovery in an Urban Environment: State of Science, Research Needs, and Evaluation of Implementation towards Critical Infrastructure Resilience*; ANL/NE-17/37; Argonne National Laboratory: Lemont, IL, USA, 2018.
2. Environmental Protection Agency. *Wide Area Stabilization of Radiological Particulate Contamination*; EPA/600/R-16/067; U.S. Environmental Protection Agency: Washington, DC, USA, 2017. Available online: https://cfpub.epa.gov/si/si_public_file_download.cfm?p_download_id=530543&Lab=NHSRC (accessed on 25 January 2022).
3. Katra, I. Comparison of Diverse Dust Control Products in Wind-Induced Dust Emission from Unpaved Roads. *Appl. Sci.* **2019**, *9*, 5204. [CrossRef]
4. Shao, Y.; Raupach, M.R.; Findlater, P.A. Effect of saltation bombardment on the entrainment of dust by wind. *J. Geophys. Res. Atmos.* **1993**, *98*, 12719–12726. [CrossRef]
5. Swet, N.; Katra, I. Reduction in soil aggregation in response to dust emission processes. *Geomorphology* **2016**, *268*, 177–183. [CrossRef]

6. Environmental Protection Agency. *Evaluation of Hydrogel Technologies for the Decontamination of 137Cs from Building Material Surfaces*; EPA/600/R-17/035; U.S. Environmental Protection Agency: Washington, DC, USA, 2017. Available online: https://cfpub.epa.gov/si/si_public_file_download.cfm?p_download_id=530777&Lab=NHSRC (accessed on 25 January 2022).
7. Environmental Protection Agency. *Technical Report for the Demonstration of Wide Area Radiological Decontamination and Mitigation Technologies for Building Structures and Vehicles*; EPA/600/R-16/019; U.S. Environmental Protection Agency: Washington, DC, USA, 2016. Available online: https://cfpub.epa.gov/si/si_public_file_download.cfm?p_download_id=529008&Lab=NHSRC (accessed on 25 January 2022).
8. Bustos, M.; Cordo, O.; Girardi, P.; Pereyra, M. Evaluation of the Use of Magnesium Chloride for Surface Stabilization and Dust Control on Unpaved Roads. *Transp. Res. Rec.* **2015**, *2473*, 13–22. [CrossRef]
9. Gonzalez, A.; Aitken, D.; Heitzer, C.; Lopez, C.; Gonzalez, M. Reducing mine water use in arid areas through the use of a byproduct road dust suppressant. *J. Clean. Prod.* **2019**, *230*, 46–54. [CrossRef]
10. Ben-Hur, M.; Cohen, R.; Danon, M.; Nachshon, U.; Katra, I. Evaluation of groundwater salinization risk following application of anti-dust emission solutions on unpaved roads in arid and semiarid regions. *App. Sci.* **2021**, *11*, 1771. [CrossRef]
11. Katra, I.; Laor, S.; Swet, N.; Kushmaro, A.; Ben-Dov, A. Shifting cyanobacterial diversity in response to agricultural soils associated with dust emission. *Land Degrad Dev.* **2017**, *28*, 878–886. [CrossRef]
12. Edri, A.; Dody, A.; Tanner, S.; Swet, N.; Katra, I. Variations in dust-related PM_{10} emission from an arid land due to surface composition and topsoil disturbance. *Arab J. Geosci* **2016**, *9*, 607. [CrossRef]
13. Katra, I.; Elperin, T.; Fominykh, A.; Krasovitov, B.; Yizhaq, H. Modeling of particulate matter transport in atmospheric boundary layer following dust emission from source areas. *Aeolian Res.* **2016**, *20*, 147–156. [CrossRef]
14. Swet, N.; Elperin, T.; Kok, J.; Martin, R.; Yizhaq, H.; Katra, I. Can active sands generate dust particles by wind-induced processes? *Earth Planet. Sci. Lett.* **2019**, *506*, 371–380. [CrossRef]
15. Magnuson, M.; Stilman, T.; Serre, S.; Archer, J.; James, R.; Xia, X.; Lawrence, M.; Raveh-Amit, H.; Sharon, A. Stabilization/Containment of Radiological Particle Contamination to Enhance First Responder, Early Phase Worker, and Public Safety—Part 2. *Appl. Sci.* 2022, *submitted*.

applied sciences

MDPI

Article

Part 2: Stabilization/Containment of Radiological Particle Contamination to Enhance First Responder, Early Phase Worker, and Public Safety

Matthew Magnuson [1,*], Terry Stilman [2], Shannon Serre [3], John Archer [1], Ryan James [4], Xiaoyan Xia [4], Mitchell Lawrence [4], Erin Tamargo [5], Hadas Raveh-Amit [6] and Avi Sharon [7]

1 EPA Office of Research and Development, Homeland Security Materials and Management Division, 26 W Martin Luther King Dr., Cincinnati, OH 45268, USA; archer.john@epa.gov
2 EPA Region 4, 61 Forsyth St., SW, Atlanta, GA 30303, USA; stilman.terry@epa.gov
3 EPA Office of Land and Emergency Management, Consequence Management Advisory Division, WJC-N, Washington, DC 20002, USA; serre.shannon@epa.gov
4 Battelle Memorial Institute, 505 King Ave., Columbus, OH 43201, USA; jamesr@battelle.org (R.J.); xia@battelle.org (X.X.); lawrencem1@battelle.org (M.L.)
5 Irregular Warfare Technical Support Directorate, Alexandria, VA 22350, USA; erin.tamargo@iwtsd.gov
6 Department of Chemistry, Nuclear Research Centre Negev, P.O. Box 9001, Beer Sheva 8419000, Israel; hadasa@nrcn.gov.il
7 Environmental Research Unit, Nuclear Research Centre Negev, P.O. Box 9001, Beer Sheva 8419000, Israel; avish@nrcn.gov.il
* Correspondence: magnuson.matthew@epa.gov

Abstract: The application of stabilization technologies to a radiologically contaminated surface has the potential for reducing the spread of contamination and, as a result, decreasing worker exposure to radiation. Three stabilization technologies, calcium chloride ($CaCl_2$), flame retardant Phos-Chek® MVP-Fx, and Soil$_2$O^{TM} were investigated to evaluate their ability to reduce the resuspension and tracking of radiological contamination during response activities such as vehicle and foot traffic. Concrete pavers, asphalt pavers, and sandy soil walking paths were used as test surfaces, along with simulated fallout material (SFM) tagged with radiostrontium (Sr-85) applied as the contaminant. Radiological activities were measured using gamma spectrometry before and after simulated vehicle operation and foot traffic experiments, conducted with each stabilization technology and without application as a nonstabilized control. These measurements were acquired separately for each combination of surface and vehicle/foot traffic experiment. The resulting data describes the extent of SFM removed from each surface onto the tires or boots, the extent of SFM transferred to adjacent surfaces, and the residual SFM remaining on the tires or boots after each experiment. The type of surface and response worker actions influenced the stabilization results. For instance, when walked over, less than 2% of particles were removed from nonstabilized concrete, 4% from asphalt, and 40% of the particles were removed from the sand surface. By contrast, for vehicle experiments, ~40% of particles were again removed from the sand, but 7% and 15% from concrete and asphalt, respectively. In most cases, the stabilization technologies did provide improved stabilization. The improvement was related to the type of surface, worker actions, and stabilizer; a statistical analysis of these variables is presented. Overall, the results suggest an ability to utilize these technologies during the planning and implementation of response activities involving foot and vehicle traffic. In addition, resuspension of aerosolizable range SFM was monitored during walking path foot traffic experiments, and all stabilizing agents decreased the measured radioactivity, with the Soil$_2$O^{TM} decrease being 3 fold, whereas the $CaCl_2$ and Phos-Chek MVP-Fx surfaces generated no detectable radioactivity. Overall, these results suggest that the stabilization technologies decrease the availability of particles respirable by response workers under these conditions.

Keywords: stabilization; containment; radiological contamination; cesium; strontium

Citation: Magnuson, M.; Stilman, T.; Serre, S.; Archer, J.; James, R.; Xia, X.; Lawrence, M.; Tamargo, E.; Raveh-Amit, H.; Sharon, A. Part 2: Stabilization/Containment of Radiological Particle Contamination to Enhance First Responder, Early Phase Worker, and Public Safety. *Appl. Sci.* **2022**, *12*, 3861. https://doi.org/10.3390/app12083861

Academic Editor: Itzhak Katra

Received: 9 February 2022
Accepted: 9 April 2022
Published: 11 April 2022

1. Introduction

Discharge of a radiological dispersal device (RDD) or an accidental radiological release may cause contamination over a wide area. During response and recovery, emergency responders, recovery workers, and the general public may be exposed to radioactivity due to direct contact and inhalation. Direct contact results from "hands-on" activities with contaminated material or exposure to unshielded radiation sources. Inhalation exposure could originate from dust generated by the release incident or during demolition or decontamination work [1], as well as from the resuspension of such dust during work activities. The risk of tracking contamination via vehicle or foot traffic is of significant concern. Tracking can lead to contaminant spread which enlarges the contaminated area, potentially increasing the time and expense of decontamination efforts.

Control of the contamination source area to prevent resuspension and tracking could both reduce radioactivity exposure and reduce the spread of contamination. Hence, there is a need for technologies and methodologies to limit exposure, reduce resuspension, and limit the tracking of radioactive materials. Such technologies have been widely investigated for nonradioactive materials due to the importance of controlling urban particulate matter and its associated human health effects. These technologies broadly fall into two categories: physical removal and stabilization [1–4]. Physical removal approaches include sweeping, washing, wiping, and vacuuming. Stabilization technologies prevent particles from spreading by resuspension or tracking, and such technologies are routinely used for dust control in industries such as road construction, and at mining sites.

Technologies for preventing secondary transport of soluble and particulate radiological contamination from roadways, roadside vegetation, and adjacent soils have recently been reviewed by Saito [5]. Although the focus of the review was preventing resuspension associated with decontamination activities, not necessarily resuspension arising directly from vehicle and foot traffic, it suggests that widely available and easily applicable stabilization technologies have the potential to minimize worker exposure by reducing the spread of contamination. These technologies may also minimize the resuspension of radioactive particles from surfaces, thus reducing the time and resources needed for additional decontamination operations [5]. Technologies for minimizing resuspension are an active research area because of keen interest in controlling dust and other particulate matter from anthropogenic sources due to its potential impact on human health [3,6–12].

In a previous study, the U.S. Environmental Protection Agency (EPA) assembled stakeholders with expertise in radiological stabilization and performed a downselection exercise to prioritize experimental testing of stabilization technologies (e.g., water, fire retardants, gels, foams, and clays) based on literature searches and the personal experience of the stakeholders [13]. Twenty-four technologies were identified, and, based on stakeholder ranking, a fire retardant (Phos-Chek MVP-F) and two dust suppression technologies (Soil$_2$O and $CaCl_2$) were downselected to evaluate the efficacy of particle stabilization during vehicle and foot traffic. This evaluation determined that cesium (Cs)-137 quantitatively bound to both Phos-Chek MVP-F and Soil$_2$O, while $CaCl_2$ application increased sorption of aqueous Cs-137 onto the surface matrix, Arizona road dust. During driving activities, the transfer of particles from treated surfaces was least for Phos-Chek MVP-F fire retardant, followed by Soil$_2$O. The transfer of particles was greatest for surfaces treated with $CaCl_2$. An evaluation of the impacts on decontamination processes, waste generation, and the environment following stabilization suggested that Phos-Chek MVP-F, $CaCl_2$, and Soil$_2$O demonstrated the feasibility of using these materials, traditionally used for other purposes, for radiological stabilization [13]. A second EPA project evaluated the same technologies qualitatively, and the results indicated that the Phos-Chek MVP-F was more effective at particle stabilization than Soil$_2$O and $CaCl_2$ [14].

A recent collaborative effort between the EPA and the Nuclear Research Centre Negev (NRCN) studied wind-induced resuspension. It highlighted the need to validate stabilization technologies for the physical stresses from specific resuspension processes other than wind because these physical stresses lead to the disintegration of stabilized particles and, ultimately, resuspension [15]. This paper builds on that collaboration and includes the Irregular Warfare Technical Support Directorate [16]. This paper uses pilot-scale experiments with radiolabeled simulated fallout material (SFM) to investigate stabilization efficacy against specific stresses related to vehicle and foot traffic on surfaces of interest for resuspension and tracking, especially during urban response and recovery. The tests are designed to describe the extent of particle removal from each surface onto the tires of vehicles or the boots of responders, the extent of SFM transferred to adjacent surfaces, and the residual SFM remaining on the tires or boots after each experiment. Other tests investigate the resuspension of aerosolizable range particles potentially subject to inhalation during foot traffic.

2. Materials and Methods

2.1. Description of Stabilization Materials

$CaCl_2$ (CESCO Solutions Inc., Bellingham, WA, USA), Phos-Chek MVP-Fx (Perimeter Solutions, Rancho Cucamonga, CA, USA), and $Soil_2O$ (Geltech Solution, Jupiter, FL, USA) were investigated in this study. The Supplemental Information includes information sheets that summarize operationally important aspects of these materials. Briefly, $CaCl_2$ (Table S1) is a hygroscopic material attracting moisture from the air that works to increase the moisture level on surface particles, decreasing particle resuspension and transfer. $CaCl_2$ is applied as a solution, which, upon drying, creates a weak cementation effect by binding fine particulates together. This moisture binds fine aggregate particles to decrease particle resuspension and transfer. $CaCl_2$ is extensively used for dust control during construction projects and on unpaved roads, as it is easy to use and widely available. Phos-Chek MVP-Fx (Table S2) is a gum-thickened, medium viscosity fire retardant that provides accurate aerial drops for wildland fire control in forest, bush, or grassland. With added colorant, Phos-Chek MVP-Fx is a high visibility powder concentrate that can readily be mixed with water. $Soil_2O$ (Table S3) is a dust control agent that can effectively suppress particulate matter from entering the air. $Soil_2O$ is a copolymer intended to maintain the moisture level in the soil by preventing evaporation, thus decreasing the propensity of particles to be available for dust formation.

2.2. Experimental Procedure

2.2.1. Preparation of Simulated Fallout Material (SFM) Containing Radiostrontim (Sr-85)

Arizona test dust with a particle size ranging from a few micrometers (μm) to a few hundred μm (Powder Technology Inc., Arden Hills, MN, USA) was selected to be used in the evaluation testing since the particles in selected particle size ranges were susceptible to resuspension and tracking. To prepare the mixture of particles, equal amounts of ultrafine test dust (ISO 12103-1, A1) and coarse test dust (ISO 12103-1, A4) were mixed. To make radioactive Sr-85 contaminated particles, an aqueous solution that contained 500 microcuries (μCi) Sr-85 (strontium chloride in 0.5 M HCl, Eckert & Ziegler Isotope Products, Valencia, CA) was added to ~250 g (g) of substrate particles using a sprayer (BS-3 3 oz Locking Personal Sprayer, Sprayco, Liwnia, MI, USA). The particles were well-mixed by manually shaking the container side to side for approximately 10 s and then manually rotating the container for approximately 30 s. This process was repeated until the required volume of aqueous Sr-85 was added to the particles. After being allowed to dry overnight in a well-ventilated exhausted box in a fume hood (T: 22.0 ± 0.7 °C; RH: 24.9 ± 0.5%), the spiked substrate particles were mixed again for approximately 10 min to ensure a homogenous mixture. Then, the dry particles were transferred to saltshakers (Tablecraft Products Company, 3-oz glass with stainless steel top, UNSPSC# 52152013), used for application. For simulated vehicle experiments and walking experiments on concrete and asphalt surfaces, ~3 g of particles were transferred from each saltshaker to contaminate one testing surface. For walking experiments on the soil walking path surface, ~40 g of particles were transferred from each saltshaker for one experiment.

To characterize the particle size of the contaminated particles and confirm the lack of particle clumping during contamination, an identical particle mixture was prepared, and one aliquot was wetted with non-radiological aqueous solvent and then dried using a procedure similar to the procedure for the preparation of radiological particles. The particle size was determined for both aliquots using a standard test method for particle-size analysis of soils, ASTM D422-63. The analytical results of the particle sizes are listed in Table 1, and the wetted and nonwetted samples produced similar results.

Table 1. Particle size distribution of wetted and nonwetted simulated fallout material.

Particle Size Analysis	Wetted	Nonwetted
Clay (%)	24.00	28.00
Silt (%)	59.90	56.70
Sand (%)	16.10	15.30
Organic Matter (%)	0.43	0.25
Fine Gravel (%)	0	0
Sand Fractions (%)		
1 mm	0.10	0.00
0.25 mm	1.10	0.00
0.15 mm	0.60	0.20
0.05 mm	14.30	15.10

2.2.2. Preparation of Surfaces for Testing

The contaminated fallout material was applied to three surfaces for the simulated vehicle-rolling and foot-traffic experiments. These surfaces included asphalt pavers (6″ × 12″ × 1.75″, Asphalt Products, Hanover, PA, USA), concrete pavers (12″ × 12″ × 1.75″, Pewter Square Concrete Step Stone, The Home Depot, Atlanta, GA, USA), and sand (Garick premium play sand, Garick LLC, Cleveland, OH, USA) loosely placed on vinyl fabric (80 Vinyl Military Spec Fabric, Herculite Products Inc., Emigsville, PA, USA) covering the floor of a tent. All the pavers were purchased in bulk and used in as-received condition. The soil bags were opened in the laboratory a few days before use; thus, the soil was equilibrated to laboratory conditions (T: 21.9 ± 0.7 °C and RH: 54 ± 7%).

The Garick sand utilized was selected as follows: Two (2) Israeli sandy soil samples, including "Ze'elim" and "Rotem", were received and characterized. Two (2) local soil samples, including Ohio mulch (Ohio Mulch, Columbus, OH, USA) and the Garick sand with similar particle sizes, were characterized and compared at the same time. The test results are listed in Table 2.

Table 2. Particle Size Distribution of Different Soils.

Particle Size Analysis	Israel Soil Samples		Experimental Bulk Soil	
	Ze'elim	Rotem	Ohio Mulch	Garick
Clay (%)	4	2	1.5	2
Silt (%)	4	1.2	1.6	0.8
Sand (%)	92	96.8	74.4	97.2
Organic Matter (%)	0.08	0	0.87	0.05
Fine Gravel (%)	0	0	22.5	0
Sand Fractions (%)				
1 mm	0.1	1	12.6	1.4
0.25 mm	28.9	87.3	59.6	74
0.15 mm	10.6	5.8	1.4	12.3
0.05 mm	52.4	2.7	0.08	9.5
Texture Class	Sand	Sand	Sand	Sand

Garick sand was selected for the study because of its particle sizes in the sand fraction, similar to Rotem particle sizes. While this sand is technically a "soil", the material will be referred to throughout this document as a "sand surface" because the surface is primarily sand and also because one of the stabilization products is called "Soil$_2$O" which tends to cause some linguistic confusion.

2.2.3. Preparation of Stabilization Technologies

Based on manufacturers' instructions and previous studies [13,14], aqueous solutions were prepared for each stabilization technology for the dry runs, and the concentrations were finalized for actual testing. The preparation procedure for each stabilization technology is summarized below. The deionized (DI) water was from a Barnstead DI water system (D12681, Barnstead International, Dubuque, IA, USA).

$CaCl_2$ solution was prepared by weighing 100 g of $CaCl_2$ pellets and transferring them to a 2 L volumetric flask. The flask was partially filled with DI water and mixed until solids were completely dissolved. Then, the flask was filled to the volume of 1600 mL with DI water and mixed again until the solution was homogenous. Soil$_2$O (5 g) was weighed and transferred to a 2 L volumetric flask. The flask was partially filled with DI water and mixed until the powder was dissolved. Then, the flask was filled to a volume of 1600 mL with DI water and mixed again until the solution was homogenous. Phos-Chek MVP-Fx (110 g) was weighed and transferred to a 1 L volumetric flask. The flask was filled partially with DI water and mixed until solids were completely wet, and then the flask was filled to the 1 L volume with DI water and mixed well.

2.2.4. Radiological Contamination and Stabilization of Surfaces

In an actual fallout event, the material loading level would vary greatly depending on the height of a possible explosion, ground characteristics below a possible explosion, distance from the release point, meteorological conditions, and ventilation of residences or offices. Previous fallout decontamination research [17,18], mostly outdoor, has used surface densities of approximately 20 mg/cm^2, so a similar density was used as the surface loading. To apply contaminated particles to the testing surface, approximately 3 g of particles were measured and transferred into a saltshaker and rotated to mix well. One shaker was emptied onto each target area, where foot or tire contact was made of the simulated traffic surface (12.5 cm by 15 cm) corresponding to 16 mg/cm^2 and 5 to 6 µCi of Sr-85. For walking experiments on the sand walking path surface, 40 g of contaminated particles was applied to each target area (45 cm by 45 cm), corresponding to 20 mg/cm^2 and 80 µCi of Sr-85. The contaminated concrete, asphalt, and sand walking path surfaces are shown in Figure 1.

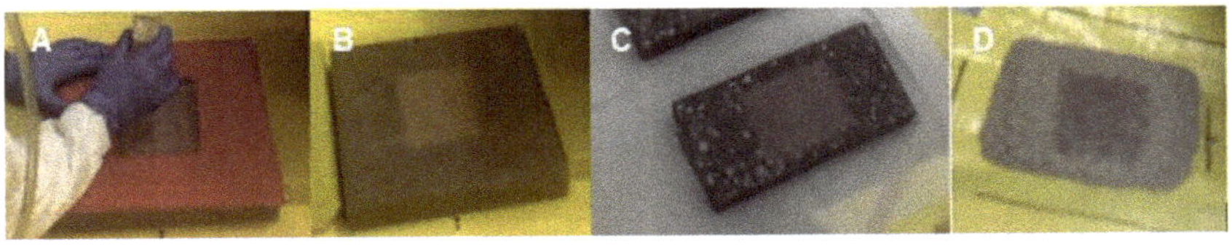

Figure 1. Application of simulated fallout material (SFM) to tested surfaces. (**A**) Use of stencil to define target area; (**B**) Concrete; (**C**) Asphalt; (**D**) Sand.

2.2.5. Measurement of Sr-85 Activity

The measurement of gamma radiation of Sr-85 from the testing surfaces was performed using a sodium iodide (NaI) spectrometer (Canberra InSpector 1000, Canberra Industries, Inc., Meriden, CT, USA) with different customized stand frames (Figure 2) made of Plexiglass® to ensure a constant distance (1.27 cm) between the detector and the measured surfaces. After applying Sr-85 contaminated particles to the testing surface or following the application of stabilization technologies on the contaminated surfaces, the center of the contaminated area was measured for the activity of Sr-85. To track the transfer of Sr-85 as the result of the movement of simulated vehicles and walking, the contaminated/stabilized surfaces, transferred surfaces that were previously uncontaminated surfaces, and the tires/boots were measured for the activity of Sr-85 after rolling simulated vehicle/walking across the surfaces. The instrument performance testing of Canberra InSpector 1000 was conducted daily and met the acceptance criteria. Specifically, the instrument's accuracy was monitored through daily performance checks, which included analysis of a sealed sample of 1 μCi Sr-85. The sealed sample was analyzed in the same geometry with respect to the probe with a 100-s acquisition time. The instrument performance was considered consistent if the relative percent difference (RPD) of the measured activity and the theoretical value of Sr-85 prepared due to decay were within 10% throughout the testing. Based on the half-life of Sr-85 (64.8 days), the theoretical values were calculated on each experiment day. Across the two months of testing, the RPD ranged from -2.4% to 2.5%, indicating the consistency of instrument performance.

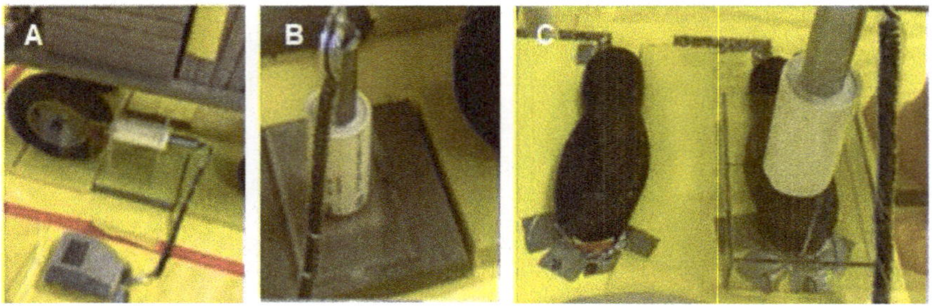

Figure 2. Activity Measurement Using Frames Customized to (**A**) Tires; (**B**) Surfaces; and (**C**) Boots.

2.2.6. Evaluation of Stabilization Technologies

The stabilization technologies were applied to the surfaces studied as described below, and the exact application varied with the type of experiment and the specific surfaces involved. As an example, Figure 3 shows the stabilization technologies for treated surfaces before and after drying. Figure 3 is intended to highlight the appearance of the technologies, whereas later figures illustrate the steps of evaluating the technologies.

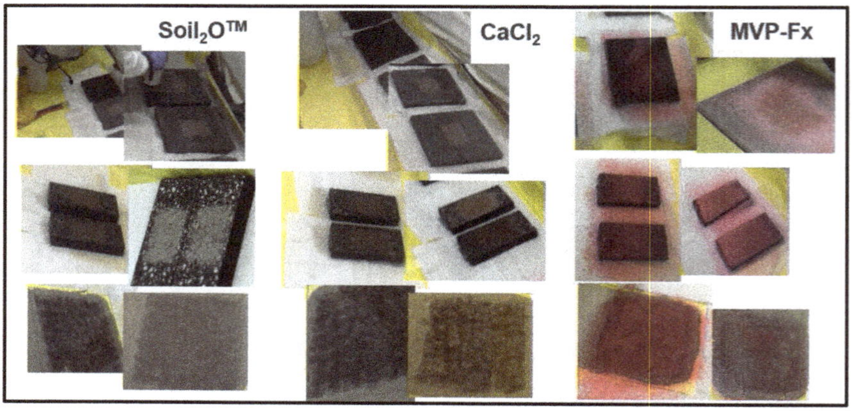

Figure 3. Surfaces Treated with Stabilization Technologies (Left: Wet; Right: Dry).

Simulated vehicle experiments and walking experiments were conducted on the three surfaces treated with three stabilization technologies on each surface. Quality control (QC) samples included background and nonstabilized controls (NSCs). All the radiological stabilization technology testing experiments were performed in a containment tent (Dual Chamber Tent, LANCS Industries, Kirkland, WA, (15 ft long × 8 ft wide × 7 ft high) as shown in Figure 4, which was located in a laboratory. The temperature and relative humidity (RH) were continuously monitored in the tent, and the data were stored in a HOBO data logger. Throughout stabilization technology evaluation experiments, the average temperature and RH in the tent were 21.9 ± 0.7 °C and 53.5% ± 6.6%, respectively.

Figure 4. Radiological Containment Tent.

The radiological testing technicians wore personal protective equipment (PPE) that included protective coveralls, gloves, shoe covers, boots, powered air-purifying respirators, and a personal air sampler (AirChek 52, SKC Inc., 863 Valley View Road, Eighty-Four, PA, USA). The AirChek 52 personal air sampler collected particulates on a glass fiber filter during each experiment. The filters were analyzed for Sr-85 activity to monitor for possible inhalational exposure to the radiological testing technician (all results were below background levels). The tent was connected to a high-efficiency particle air filtration system, and two air sampling systems (LV-10, F&J Specialty Products Inc., Ocala, FL, USA) were operated at 3 cubic feet per minute (cfm) to collect air samples on 47 mL round quartz fiber filters within the radiological containment tent during experiments. As shown in Figure 5, an additional air sampling system (HV-1, F&J Specialty Products, Inc., Ocala, FL, USA) operating at approximately 5 cfm was used during walking experiments on the sand walking path surfaces at low height (~30 cm above the contaminated sand surface) to collect air samples and characterize particle resuspension of aerosolizable range particles. The sampling system utilized 47 mm FP-XM glass fiber filters (F&J Specialty Products, Inc., Ocala, FL, USA).

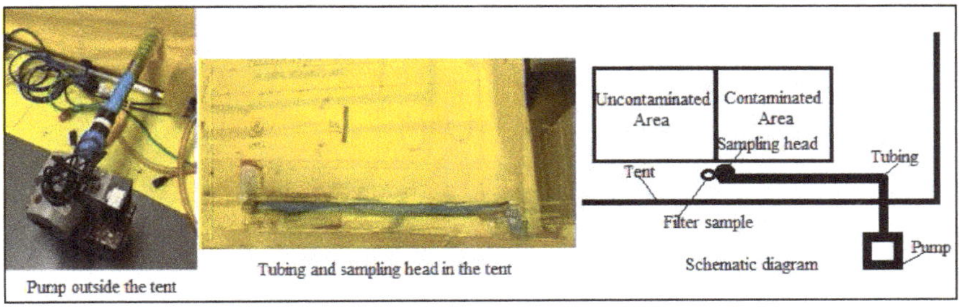

Figure 5. Schematic Diagram of High-Volume Sampler and Filter Sample Setup (**left**); with pictures of the pump (**right**) and tubing and sampling head (**middle**) to show a relationship to the containment tent.

2.2.7. Simulated Vehicle Experiments

For simulated vehicle experiments, approximately 3 g of Sr-85 contaminated particles was applied to each testing surface with a surface area of 12.5 cm × 15 cm (using a template to maintain repeatable particle coverage). Four (4) contaminated surfaces without the application of any stabilization technologies were prepared for each testing surface as NSCs. Uncontaminated surfaces served as blanks for background measurements. After the surface was contaminated with Sr-85, a stabilization technology solution was sprayed onto the contaminated surfaces (Figure 6A) and allowed to dry in the tent (drying time ranged from overnight to 3 days, as the surfaces did not all seem to dry completely overnight). $CaCl_2$ and $Soil_2O$ solutions were sprayed using a two-gallon multipurpose pump-up Scotts sprayer with the cone spray setting (Item 190498, https://www.scotts.com, accessed on 9 April 2022). Phos-Chek MVP-Fx solution was sprayed using a 0.5-gallon multipurpose handheld pump-up sprayer (Item 56167, http://www.harborfreight.com, accessed on 9 April 2022) because the other sprayer would quickly clog while using Phos-Chek MVP-Fx. Four (4) contaminated surfaces with each stabilization technology were prepared for each testing surface material. A small wagon (Model # H-2547; deck size: 24″ × 48″, load capacity: 3000 lb; wheel type: 16″ pneumatic; https://www.uline.com/, accessed on 9 April 2022) fitted with pneumatic tires with tread comparable to applicable vehicle tires (Figure 6B) was loaded with 800 lb of concrete pavers that generated a contact pressure similar to applicable vehicles with downforce of 2000 lb. The evaluation testing for one testing surface material was set up under four tires simultaneously (Figure 6C). The wagon was manually pulled to move across contaminated/stabilized surfaces, as shown in Video S1 in Supplemental Information. The radiological activities of the contaminated/stabilized surfaces were measured before and after simulated vehicle rolling.

Figure 6. Sprayers Used in Study (**A**), Wagon Used to Simulate Applicable Vehicle (**B**), and Simulated Vehicle Tire (**C**).

The contaminated surfaces were replaced with uncontaminated surfaces, and the wagon was pushed backward across the uncontaminated transfer surfaces. The radiological activities of the previously uncontaminated and post-traffic contaminant transfer surfaces were measured to evaluate the magnitude of Sr-85 transfer. The radiological activities of Sr-85 from the tires having contacted the center of the contaminated/stabilized surfaces were measured to quantify the residual of Sr-85 contaminated particles on the tire. All measurements were done using an NaI spectrometer with different customized stand frames made of Plexiglass® to ensure a constant distance (1.27 cm) between the detector and the measured surfaces. For testing experiments on sand surfaces, both contaminated/stabilized and transferred surfaces were prepared by filling a 25 cm × 25 cm template on a piece of Herculite® plastic sheeting (Emigsville, PA, USA) with equal amounts of premeasured sand. Then, the sand surfaces were contaminated/stabilized the same way as the concrete and asphalt pavers. The same rolling procedures were applied to sand surfaces, and the corresponding measurements were performed.

2.2.8. Straight-Line Walking Experiments

For foot traffic experiments, similar testing was performed with technicians wearing rubber boots (Item: Dunlog 86020, https://www.dunlopboots.com/, accessed on 9 April 2022) while walking across the testing surfaces. As with the simulated vehicle experiments, approximately 3 g of Sr-85 contaminated particles was applied to each testing surface with a surface area of 12.5 cm × 15 cm using a template. Four (4) contaminated surfaces were prepared for each testing surface as experiment controls. For stabilization technology testing experiments, the testing surfaces were dosed first with Sr-85 contaminated particles. After the surface was contaminated, a stabilization technology solution was sprayed onto the contaminated surfaces using a sprayer and allowed to dry (drying time ranged from overnight to 3 days, as the surfaces did not all dry completely overnight). $CaCl_2$ and $Soil_2O$ solutions were also sprayed using a 2 gallon, multipurpose Scott's sprayer with the cone spray setting. Phos-Chek MVP-Fx solution was sprayed using the same 0.5 gallon, multipurpose handheld sprayer as described above. Four (4) contaminated surfaces treated with each stabilization technology were prepared for each testing surface. The Sr-85 contaminated surfaces treated with/without stabilization technologies were tested first by walking across alternately placed contaminated surfaces, with one footstep on each surface, and subsequently onto previously uncontaminated surfaces (Figure 7 and Video S2). The boots were taken off and placed on a customized boot stand. The contaminated surfaces were measured for activity before and after walking. The activity of previously uncontaminated surfaces and Sr-85 transferred to the boots was measured.

Figure 7. Straight-line Walking Experiment.

2.2.9. Circuit Path Walking Experiments

For the sand circuit walking path surfaces, the walking path included a contaminated sand surface (45 cm × 45 cm) connected with an uncontaminated surface (1st step and 2nd step including transfer surfaces, 45 cm × 90 cm) and non-sand surface (3rd step to 8th step and Start/Stop point). Figure 8 shows the post-walking condition of the control experiment and the walking path schematic diagram, and Video S3 shows an example experiment. To prepare the sand walking path, a premeasured amount of approximately 11 kg of loose sand was evenly distributed onto the defined area, including the contaminated area and the 1st step and 2nd step area. Approximately 40 g of Sr-85 contaminated particles were applied to the contaminated area of the walking path. To stabilize the contaminated surfaces, $CaCl_2$, Phos-Chek MVP-Fx, and $Soil_2O$ solutions were sprayed using sprayers, as mentioned previously. One NSC experiment was contaminated with Sr-85 particles but was not sprayed with any stabilization technology. The other three experiments were performed similarly, but the contaminated surfaces were treated separately with the three stabilization technologies. The experiments were conducted to take the 1st step onto one-half of the 45 cm section of contaminated sand (with or without stabilization technologies). The next two steps were onto two (2) 45 cm sections of previously uncontaminated sand, followed by the remainder of the loop on the previously clean Herculite surface. Each time the walker stepped onto the contaminated surface, they alternated which foot they started with to keep the potential transfer as similar as possible. The walking path experiment was conducted for 5 continuous minutes (in the same direction) for approximately 36 passes.

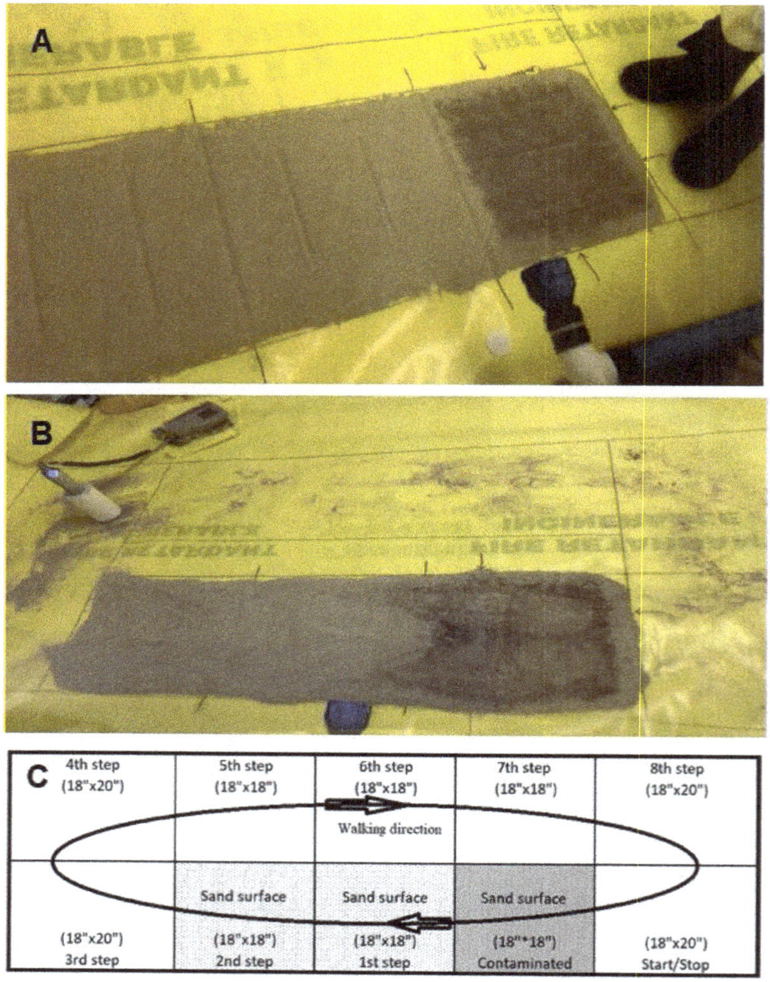

Figure 8. Appearance of Circuit Walking Path (**A**) Pre- and Post- (**B**) Walking for (**C**) Schematic Diagram for Circuit Walking Path Experiment. Panel (**B**) also depicts the use of the radiation detector for this type of experiment.

The contaminated/stabilized surfaces were measured for activity before and after walking. The transferred surfaces and boots were also measured for the activity of Sr-85 to estimate the amount of transfer. All the measurements were accomplished by using an NaI spectrometer with customized stands made of Plexiglass® to ensure a constant distance (1/2″) and geometry between the detector and the measured surfaces (Figure 9). The areas of each step where particles accumulated along the non-sand portion of the walking path were also measured for the activity of Sr-85 to further evaluate the transfer. A filter sample was placed on the floor at the edge of the contaminated soil surface to monitor the spread of Sr-85 contaminated particles, and an air sample with the sampling head deployed about 30 cm above the edge of the contaminated area was collected from each experiment using the same type of filters (47 mm FP-XM glass fiber filter, F&J Specialty Products, Inc., Ocala, FL, USA) to measure the resuspension of Sr-85 contaminated particles.

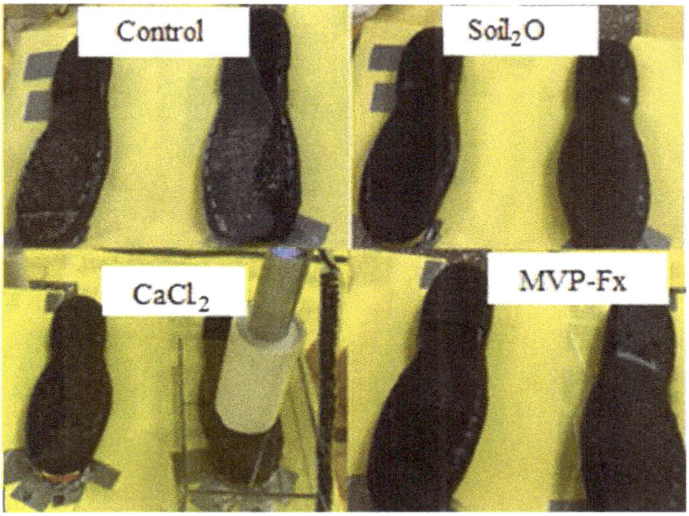

Figure 9. Customized stand for activity measurements from boots, shown for $CaCl_2$. Also visible is residue on boots from contaminated asphalt without stabilization technology, indicated as the control, and with the labeled stabilization technologies. The use of the radiation detector is depicted for the $CaCl_2$ stabilizer; its use was similar for the other stabilizers and the nonstabilized control.

2.2.10. Calculation of Percent Removal, Percent Transfer, and Percent Residual

The efficacy of stabilization technologies was evaluated through the calculation of percent removal (%R) after the application of stabilization technologies on different testing surfaces. The efficacy was also evaluated by calculating the percent transfer (%T) of activity from contaminated/stabilized surfaces to previously uncontaminated surfaces. These metrics were calculated using the following equations:

$$\%R = (1 - A_r/A_o) \times 100\% \quad (1)$$

$$\%T = A_t/A_o \times 100\% \quad (2)$$

where A_o is the activity from the contaminated/stabilized surface before vehicle-rolling or foot traffic, A_r is the activity from the contaminated/stabilized surface after vehicle rolling or foot traffic, and A_t is the activity from the previously uncontaminated surface after foot traffic or vehicle rolling experiments. The results were extrapolated by averaging to reflect the transfer of the whole surface.

The percent residual (%Res) remaining on the tires and boots after each experiment was calculated using the %T equation based on the residual activity of Sr-85 contaminated particles on tires/boots as the final activity and the activity of Sr-85 contaminated/stabilized testing surfaces as the original activity:

$$\%Res = A_{Res}/A_0 \times 100\% \quad (3)$$

where A_0 is the activity from the contaminated/stabilized surface before vehicle-rolling or foot traffic, and A_{Res} is the residual activity from the tires/boots. All the calculations were completed using the activity data collected from the experiments. Since each experiment lasted approximately 1.5 to 3 h, the data set for each experiment was used directly in the above calculations without any correction for radioactive decay. An independent group *t*-test at a 95% confidence interval was used to investigate differences in percent removal, transfer, and residual on tires/boots among different stabilization technologies.

3. Results and Discussion

The efficacy of the stabilization technologies selected was measured for each testing surface during simulated vehicle and foot traffic experiments. The %R of activity (from Sr-85 tagged particles) from the contaminated surface, %T of activity to a previously uncontaminated surface, and %Res activity remaining on the tire or boot after the vehicle and foot traffic experiment were calculated using the equations mentioned previously. All activities in the calculations were background-corrected.

3.1. Simulated Vehicle Experiments

The average %R, average %T, average %Res, and their corresponding standard deviation (SD) for each stabilization technology and testing surface are shown in Table 3. Four (4) replicates were conducted per stabilization technology and testing surface.

Table 3. Efficacy of stabilization technologies during simulated vehicle experiments, expressed as average ± standard deviation.

Test Surface	Stabilizaiton Technology	Initial CPS (Average)	Removal (%R)	Transferred (%T)	Residual (%R)
Asphalt	Control	1497	15 ± 1	8 ± 1	9 ± 1
	$CaCl_2$	1089	5 ± 4	1 ± 1	4 ± 2
	MVP-Fx	1292	2 ± 1	0.2 ± 0.6	0.5 ± 0.4
	$Soil_2O$	1210	10 ± 5	5 ± 2	9 ± 7
Concrete	Control	1773	7 ± 3	5 ± 2	3 ± 1
	$CaCl_2$	1219	0.9 ± 0.7	1 ± 1	0.8 ± 0.9
	MVP-Fx	1346	3 ± 1	1 ± 1	2 ± 1
	$Soil_2O$	1382	5 ± 1	4 ± 1	2 ± 1
Sand	Control	1363	41 ± 8	9 ± 3	20 ± 2
	CaCl2	1303	17 ± 4	3 ± 2	5 ± 3
	MVP-Fx	1312	16 ± 6	3 ± 1	2 ± 1
	$Soil_2O$	1325	25 ± 2	6 ± 2	11 ± 4

3.1.1. Percent Removal from Contaminated Surfaces

Figure 10 shows the average %R of Sr-85 contaminated particles from the NSCs and the surfaces stabilized with $CaCl_2$, Phos-Chek MVP-Fx, and $Soil_2O$ treated pavers. Table 4 lists the *p*-values from the *t*-tests performed. The *p*-values indicate the probability that the groupings of data were from the same population, i.e., statistically similar at the 95% level.

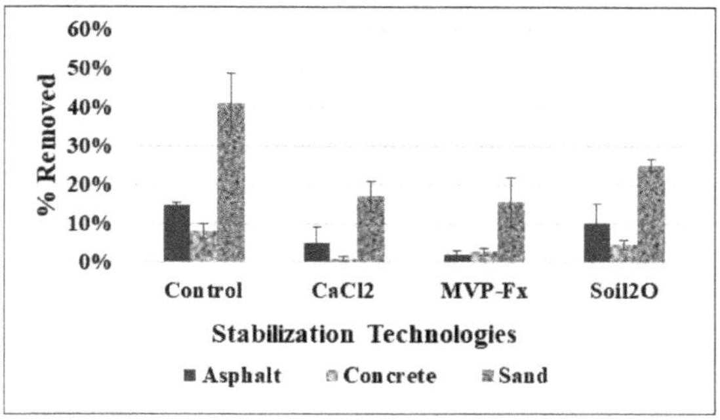

Figure 10. Percent Removal of Sr-85 Contaminated Particles for Vehicle Experiments.

Table 4. Statistical results (*p*-values) for simulated vehicle experiments percent removal. Gray shading indicates a significant difference ($p < 0.05$).

Surface	Stab. Tech.	$CaCl_2$	MVP-Fx	$Soil_2O$
Asphalt	Control	0.0172	<0.0001	0.1764
	$CaCl_2$		0.2535	0.1655
	MVP-Fx			0.0475
Concrete	Control	0.0255	0.0394	0.2081
	$CaCl_2$		0.0133	0.0010
	MVP-Fx			0.0322
Sand	Control	0.0017	0.0023	0.0251
	$CaCl_2$		0.6876	0.0101
	MVP-Fx			0.0256

Generally, the potential for removal, indicated by the control %R, was highest for the sand surface, followed by asphalt and concrete. Interpreting the efficacy values in the figure in combination with the *p*-values reveals the following for specific surfaces.

Asphalt surfaces. $CaCl_2$ (5 ± 4 %R) and MVP-Fx (2 ± 1 %R) resulted in significantly decreased %R compared to the NSC (15 ± 1 %R), while $Soil_2O$ (10 ± 5 %R) did not. MVP-Fx (2 ± 1 %R) demonstrated significantly lower %R than $Soil_2O$ (10 ± 5 %R), but no significant difference from $CaCl_2$ (5 ± 4 %R). $CaCl_2$ (5 ± 4 %R) and $Soil_2O$ (10 ± 5 %R) generated %R values that were not significantly different from one another.

Concrete surfaces. $CaCl_2$ (0.9 ± 0.7 %R) and MVP-Fx (3 ± 1 %R) resulted in significantly decreased %R compared to the NSC (8 ± 2 %R), while $Soil_2O$ (5 ± 1 %R) did not. $CaCl_2$ (0.9 ± 0.7 %R) demonstrated significantly lower %R than MVP-Fx (3 ± 1 %R) and $Soil_2O$ (5 ± 1 %R). MVP-Fx (3 ± 1 %R) demonstrated significantly lower %R than $Soil_2O$ (5 ± 1 %R), but significantly higher than $CaCl_2$ (0.9 ± 0.7 %R). $Soil_2O$ (5 ± 1 %R) demonstrated %R values that were significantly higher than $CaCl_2$ (0.9 ± 0.7 %R) and MVP-Fx (3 ± 1 %R).

Sand surfaces. $CaCl_2$ (17 ± 4 %R), MVP-Fx (16 ± 6 %R), and $Soil_2O$ (25 ± 2 %R) resulted in significantly decreased %R compared to the NSC (41 ± 8 %R). $CaCl_2$ (17 ± 4 %R) and MVP-Fx (16 ± 6 %R) generated %R values that were not significantly different from one another. $Soil_2O$ (25 ± 2 %R) generated %R values that were significantly higher than $CaCl_2$ (17 ± 4 %R) and MVP-Fx (16 ± 6 %R).

3.1.2. Percent Transfer to Uncontaminated Surfaces

Figure 11 shows the average percent transfer of Sr-85-contaminated particles from the NSCs and from the surfaces that were stabilized with $CaCl_2$, Phos-Chek MVP-Fx, and $Soil_2O$ treated pavers to pavers that were previously uncontaminated. Table 5 shows the *p*-values from the *t*-tests performed on the data. The *p*-values indicate the probability that the groupings of data were from the same population, i.e., statistically similar.

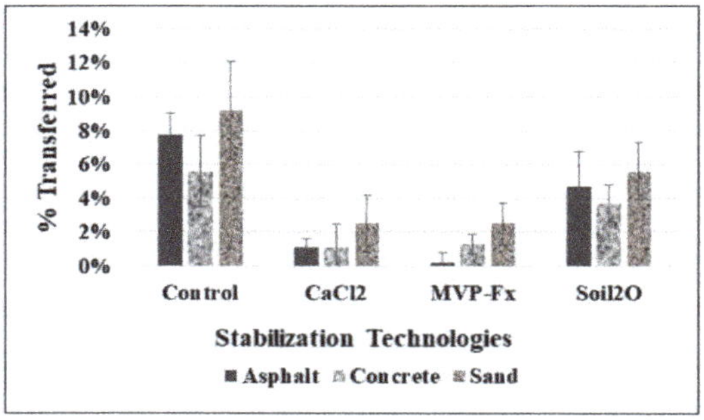

Figure 11. Percent Transfer of Sr-85 Contaminated Particles for Vehicle Experiments.

Table 5. Statistical results (*p*-values) for simulated vehicle experiments percent transfer. Gray shading indicates a significant difference ($p < 0.05$).

Surfaces	Stab. Tech.	$CaCl_2$	MVP-Fx	$Soil_2O$
Asphalt	Control	<0.0001	<0.0001	0.0444
	$CaCl_2$		0.0802	0.0160
	MVP-Fx			0.0065
Concrete	Control	0.0139	0.0072	0.2285
	$CaCl_2$		0.8527	0.0281
	MVP-Fx			0.0073
Sand	Control	0.0070	0.0055	0.0759
	$CaCl_2$		0.9631	0.0411
	MVP-Fx			0.0277

In general, the potential for transfer, indicated by the control %T, was similar for the three surfaces, as the average %T values ranged from approximately 6% to 9%. Interpreting the efficacy values in the figure in combination with the *p*-values from the table reveals the following for specific surfaces.

Asphalt. $CaCl_2$ (1 $\pm$ 1 %T), MVP-Fx (0.2 $\pm$ 0.6 %T), and $Soil_2O$ (5 $\pm$ 2 %T) resulted in significantly decreased %T compared to the NSC (8 $\pm$ 1 %T); the statistical power of the $CaCl_2$ and MVP-Fx ($p < 0.0001$) were much higher than $Soil_2O$ ($p = 0.04$). $CaCl_2$ (1 $\pm$ 1 %T) and MVP-Fx (0.2 $\pm$ 0.6 %T) generated %T values not significantly different from one another, and both produced %T values that were significantly less than $Soil_2O$ (5 $\pm$ 2 %T).

Concrete. $CaCl_2$ (1 $\pm$ 1 %T) and MVP-Fx (1 $\pm$ 1 %T) resulted in significantly decreased %R compared to the NSC (6 $\pm$ 2 %T), while $Soil_2O$ (4 $\pm$ 1 %T) did not. $CaCl_2$ (1 $\pm$ 1 %T) and MVP-Fx (1 $\pm$ 1 %T) generated %T values that were not significantly different from one another, and both produced %T values that were significantly less than $Soil_2O$ (4 $\pm$ 1 %T).

Sand. $CaCl_2$ (3 $\pm$ 2 %T) and MVP-Fx (3 $\pm$ 1 %T) resulted in significantly decreased %R compared to the NSC (9 $\pm$ 3 %T), while $Soil_2O$ (6 $\pm$ 2 %T) did not. $CaCl_2$ (3 $\pm$ 2 %T) and MVP-Fx (3 $\pm$ 1 %T) generated %T values that were not significantly different from one another, and both produced %T values that were significantly less than $Soil_2O$ (6 $\pm$ 2 %T).

3.1.3. Percent Residual Activity on Simulated Vehicle Tires

Figure 12 shows the average percent residual of Sr-85 contaminated particles that remained on simulated vehicle tires after each experiment, including the NSCs and surfaces stabilized with $CaCl_2$, Phos-Chek MVP-Fx, and $Soil_2O$. Table 6 shows the *p*-values from the *t*-tests performed on the data. The *p*-values indicate the probability that the groupings of data were from the same population, i.e., statistically similar.

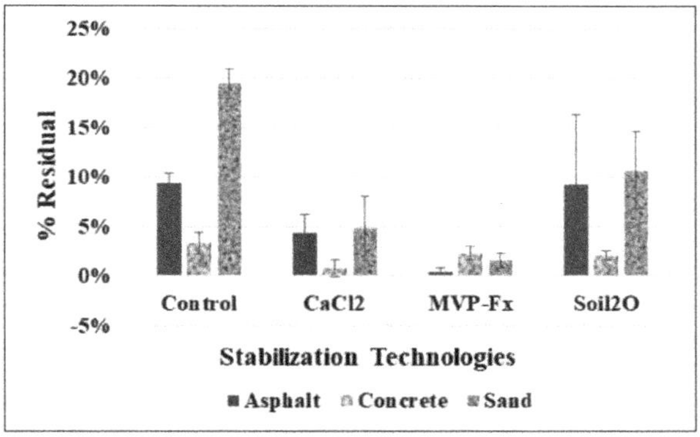

Figure 12. Percent Residual Activity of Sr-85 Contaminated Particles for Vehicle Experiments.

Table 6. Statistical results (*p*-values) for simulated vehicle experiments percent residual. Gray shading indicates a significant difference ($p < 0.05$).

Surfaces	Stab. Tech.	$CaCl_2$	MVP-Fx	$Soil_2O$
Asphalt	Control	0.0036	<0.0001	0.9709
	$CaCl_2$		0.0267	0.2282
	MVP-Fx			0.0907
Concrete	Control	0.0089	0.1401	0.0743
	$CaCl_2$		0.0388	0.0404
	MVP-Fx			0.7378
Sand	Control	0.0002	<0.0001	0.0056
	$CaCl_2$		0.1503	0.0601
	MVP-Fx			0.0169

The potential for residual, indicated by the control %Res, was highest for the sand surface, followed by asphalt and concrete. Interpreting the efficacy values in the figure in combination with the *p*-values from the table reveals the following for specific surfaces.

Asphalt. $CaCl_2$ (4 ± 2 %Res) and MVP-Fx (0.5 ± 0.4 %Res) resulted in significantly decreased %Res compared to the NSC (9 ± 1 %T), while $Soil_2O$ (9 ± 7 %Res) did not. MVP-Fx (0.5 ± 0.4 %Res) demonstrated significantly lower %R than $CaCl_2$ (4 ± 2 %Res), and both of those produced less %Res than $Soil_2O$ (9 ± 7 %Res). However, the difference between MVP-Fx and $Soil_2O$ was not statistically significant (likely due to the relatively high variability of the $Soil_2O$ result).

Concrete. Only $CaCl_2$ (0.8 ± 0.9 %Res) resulted in significantly decreased %Res compared to the NSC (3 ± 1 %Res), while MVP-Fx (2 ± 1 %Res) and $Soil_2O$ (2 ± 1 %Res) did not. $CaCl_2$ (0.8 ± 0.9 %Res) resulted in significantly lower %Res compared to both MVP-Fx (2 ± 0.7 %Res) and $Soil_2O$ (2 ± 0.5 %Res), which were not statistically different.

Sand. $CaCl_2$ (5 ± 3 %Res), MVP-Fx (2 ± 1 %Res), and $Soil_2O$ (11 ± 4 %Res) resulted in significantly decreased %Res compared to the NSC (20 ± 2 %Res). MVP-Fx (2 ± 1 %Res) demonstrated significantly lower %Res than $Soil_2O$ (11 ± 4 %Res) but was not significantly different from $CaCl_2$ (5 ± 3 %Res). $CaCl_2$ (5 ± 3 %Res) and $Soil_2O$ (11 ± 4 %Res) generated %Res that were not significantly different from one another.

3.1.4. Simulated Vehicle Experimental Observations and Conclusions

Clearly, the sand surface provided the largest potential for transfer based on the sand having the largest %R and %Res during the NSC experiments. Overall, in most cases, the stabilization technologies provided improved stabilization compared to the specific NSC results, with $CaCl_2$ and MVP-Fx generating lower average %R values and %T values in most cases. Both $CaCl_2$ and MVP-Fx reduced the %R by at least a factor of 2 for all three surfaces tested, while $Soil_2O$ had a milder impact, demonstrating a factor of 0.4. However, there were a few instances in which one or more of the three stabilization technologies did not demonstrate a statistically significant improvement at the 95% confidence level. It is a limitation imposed by the radiological containment enclosure that vehicle speeds and vehicles representative of what would be used on a response location could not be studied. Airflow caused by the movement of vehicles and by the exhaust and fans coming from within the vehicle is extremely difficult to be simulated within radiological containment.

For data interpretation purposes, the %R and %T results are more reliable than the %Res results because of the nature of the measurement. Specifically, the %R and %T measurements were made by locating the detector probe in pre-marked positions on pavers with the same geometry with respect to the detector probe. For the %Res measurements, the detector probe was held near the tire, which had a curved contour both vertically and horizontally. In addition, the tires were measured at the center location of the expected tire contamination when the tire was oriented perpendicularly to the floor. Therefore, occasionally some particles, estimated < 10% in most cases, were observed to have fallen off the tire before the measurement was made.

3.2. Simulated Foot Traffic Experiments

Simulated foot traffic (i.e., walking) experiments were performed wearing boots with both straight-line walking on contaminated pavers and circuit walking on a path. Contaminated/stabilized pavers were measured for Sr-85 activity before and after walking to determine the %R. The transfer surfaces were measured for the activity of Sr-85 to evaluate the %T, and the boots were measured to determine the %Res on the boots after the experiment. The average %R, %T, %Res, and corresponding SD for each stabilization technology and testing surface are provided in Table 7. Four (4) replicates were conducted per stabilization technology and testing surface.

Table 7. Efficacy of stabilization technologies for simulated foot traffic experiments, expressed as average ± standard deviation.

Test Surface	Stabilizaiton Technology	Initial CPS (Average)	Removal (%R)	Transferred (%T)	Residual (%R)
Asphalt	Control	1918	4 ± 1	1 ± 0.6	2 ± 0.3
	$CaCl_2$	1283	1 ± 1	ND	0.8 ± 0.7
	MVP-Fx	1671	ND	ND	ND
	$Soil_2O$	1688	2 ± 1	0.2 ± 0.2	0.2 ± 0.2
Concrete	Control	1869	2 ± 1	1 ± 0.1	1 ± 0.4
	$CaCl_2$	1794	1 ± 1	ND	ND
	MVP-Fx	1685	3 ± 1	0.9 ± 0.2	0.4 ± 0.2
	$Soil_2O$	1575	2 ± 1	1 ± 0.2	0.6 ± 0.1
Sand	Control	3401	40 ± 4	8 ± 2	3.0 ± 0.4
	$CaCl_2$	2683	38.0 ± 0.4	0.6 ± 0.6	3 ± 1
	MVP-Fx	2544	14.0 ± 0.2	2.0 ± 0.2	0 ± 0
	$Soil_2O$	2968	24.0 ± 0.9	0.8 ± 0.2	2 ± 0

3.2.1. Percent Removal from Contaminated Surfaces for Straight-Line Walking

Figure 13 shows the average %R of Sr-85 contaminated particles from the NSCs and the surfaces stabilized with $CaCl_2$, Phos-Chek MVP-Fx, and $Soil_2O$ treated pavers. Table 8 shows the *p*-values from the *t*-tests performed on the data. The *p*-values indicate the probability that the groupings of data were from the same population, i.e., statistically similar.

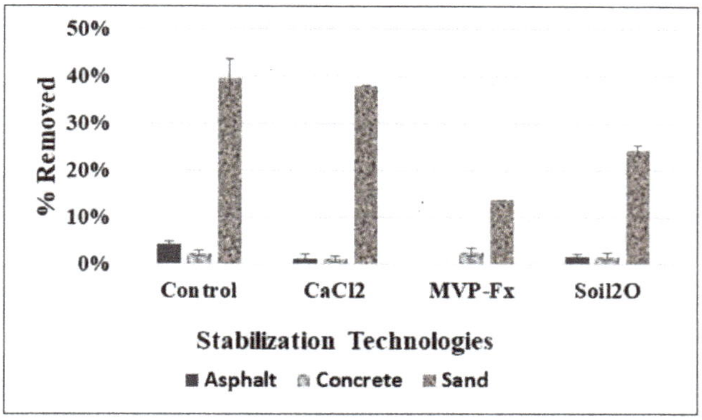

Figure 13. Percent Removal of Sr-85 Contaminated Particles for Straight-line Walking Experiments.

Table 8. Statistical results (p-values) for foot traffic experiment percent removal. Gray shading indicates a significant difference ($p < 0.05$).

Surfaces	Stab. Tech.	$CaCl_2$	MVP-Fx	$Soil_2O$
Asphalt	Control	0.0024	<0.0001	0.0009
	$CaCl_2$		0.0366	0.5490
	MVP-Fx			0.0027
Concrete	Control	0.0429	0.5701	0.2801
	$CaCl_2$		0.0230	0.3127
	MVP-Fx			0.1435

The potential for %R appears to be highest for the sand surfaces, followed by asphalt and concrete, as indicated by the control %R values. However, the sand results were determined from the 5 min continuous walking experiment, so the results are not exactly comparable. Interpreting the efficacy values in the figure in combination with the p-values from the table reveals the following for specific surfaces.

Asphalt. $CaCl_2$ (1 ± 1 %R), $Soil_2O$ (2 ± 1 %R) and MVP-Fx (Not Detected [ND]) resulted in significantly decreased %R compared to the NSC (4 ± 1 %R). MVP-Fx (ND) demonstrated significantly lower %R than $Soil_2O$ (2 ± 1 %R) and $CaCl_2$ (1 ± 1 %R). $CaCl_2$ (1 ± 1 %R) and $Soil_2O$ (2 ± 1 %R) had %R values not significantly different from one another.

Concrete. Only $CaCl_2$ (1 ± 1 %R) resulted in significantly decreased %R compared to the NSC (2 ± 1 %R) while MVP-Fx (3 ± 1 %R) and $Soil_2O$ (2 ± 1 %R) did not. $CaCl_2$ (1 ± 1 %R) demonstrated significantly lower %R than MVP-Fx (3 ± 1 %R). $CaCl_2$ (1 ± 1 %R) did not demonstrate significantly different %R from $Soil_2O$ (2 ± 1 %R). MVP-Fx (3 ± 1 %R) did not demonstrate significantly different %R from $Soil_2O$ (2 ± 1 %R).

Sand. Note: There was a single experiment for each stabilization technology, and the SD was based on the right and left sides of the path [N = 2], so no p-values were calculated. MVP-Fx (14 ± 0.2 %R) and $Soil_2O$ (24 ± 0.9 %R) resulted in decreased %R compared to the NSC (40 ± 4 %R), while $CaCl_2$ (38 ± 0.4 %R) did not. MVP-Fx (14 ± 0.2 %R) generated the lowest %R.

3.2.2. Percent Transfer to Uncontaminated Surfaces for Straight-Line Walking

Figure 14 shows the average %T of Sr-85 contaminated particles from the NSCs and from the surfaces stabilized with $CaCl_2$, Phos-Chek MVP-Fx, and $Soil_2O$ treated pavers to pavers that were previously uncontaminated. Table 9 shows the p-values from the t-tests performed on the data. The p-values indicate the probability that the groupings of data were from the same population, i.e., statistically similar.

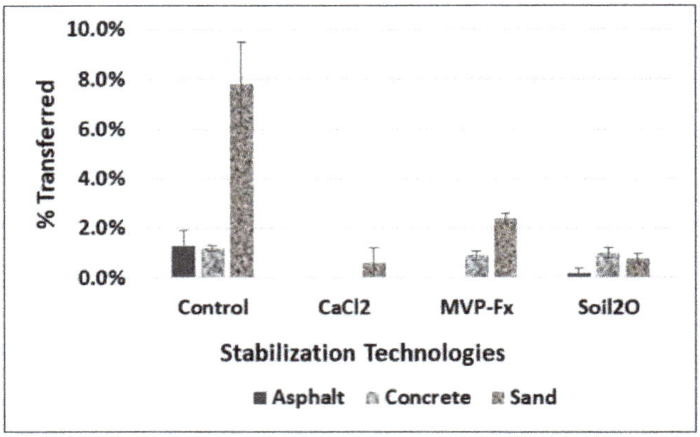

Figure 14. Percent Transfer of Sr-85 Contaminated Particles for Straight-line Walking Experiments.

Table 9. Statistical results (p-values) for foot traffic experiments percent transfer. Gray shading indicates a significant difference ($p < 0.05$).

Surfaces	Stab. Tech.	$CaCl_2$	MVP-Fx	$Soil_2O$
Asphalt	Control	0.0035	0.0121	0.0108
	$CaCl_2$		0.7594	0.0422
	MVP-Fx			0.0010
Concrete	Control	<0.0001	0.0407	0.1677
	$CaCl_2$		<0.0001	<0.0001
	MVP-Fx			0.4502

The potential for %T appeared to be highest for the sand surface, as evidenced by the control %T being more than four times larger than the other %T values. However, the sand results were determined from the 5 min continuous walking experiment, so the results are not exactly comparable. Interpreting the efficacy values in the figure in combination with the p-values from the table reveals the following for specific surfaces.

Asphalt. $CaCl_2$ (ND %T), MVP-Fx (ND %T), and $Soil_2O$ (0.2 ± 0.2 %T) resulted in significantly decreased %T compared to the NSC (1 ± 1 %T). $CaCl_2$ (ND %T) and MVP-Fx (ND %T) generated %T values that were not significantly different from one another, and both produced %T values that were significantly less than $Soil_2O$ (0.2 ± 0.2 %T).

Concrete. $CaCl_2$ (ND %T) resulted in significantly decreased %T compared to the NSC (1 ± 0.1 %T), while MVP-Fx (0.9 ± 0.2 %T) and $Soil_2O$ (1 ± 0.2 %T) did not. $Soil_2O$ (1 ± 0.2 %T) and MVP-Fx (0.9 ± 0.2 %T) generated %T values that were not significantly different from one another, and both produced %T values that were significantly higher than $CaCl_2$ (ND %T).

Sand. Note: There was a single experiment for each stabilization technology, and the SD was based on the right and left sides of the path [N = 2], so no p-values were calculated. $CaCl_2$ (0.6 ± 0.6 %T), MVP-Fx (2 ± 0.2 %T), and $Soil_2O$ (0.8 ± 0.2 %T) resulted in decreased %T compared to the NSC (8 ± 2%T).

3.3. Circuit Walking Path Results

For each stabilization technology, the average percent transfer of each step is provided in Table 10 and Figure 15. See Figure 8 for the NSC results for transferred sand and experimental schematic. The first step was always non-detect as the walker was always stepping from the noncontaminated area, after 8 steps on the non-sand surface, onto that area. The second step was taken from the contaminated area to the previously uncontaminated sand, so this area was expected to have had the highest degree of transfer. However, for the NSC and $CaCl_2$, the fourth step (i.e., the footstep from the second step area) had the highest transfer. The $CaCl_2$ stabilized sand seemed to form a crust that bound to the boot upon the first step, similar to the NSC as suggested by the $CaCl_2$ %R, but the crust did not release from the boot until a step was made onto the non-sand surface. The third step area was expected to be relatively lower in transfer because the step was coming from the first step area. The measurable radioactivity was likely a product of kicking from the second step area or release from the boot upon stepping onto a different surface type. The radiological activities measured in the second step area, fourth step area, and sixth step area were generally higher, and more evident in the NSC than in the third, fifth, and seventh step area due to the more direct transfer of Sr-85 contaminated particles from the boot contact with those particles during every lap. The transfer of particles into the odd number step areas was due mostly to particles that had remained on the boot from at least the previous lap, or before.

Table 10. Percent transfer of walking experiments to sand and non-sand.

Step Number	Control	$Soil_2O$	$CaCl_2$	MVP-Fx
1	ND [1]	ND	ND	ND
2	8%	4%	0.6%	1%
3	3%	0.8%	3%	0.8%
4	10%	2%	6%	0.9%
5	1%	0.2%	NC	1%
6	4%	ND	0.3%	ND
7	ND	2%	ND	ND
8	NC [2]	ND	2%	ND

[1] ND = Not detected. [2] NC = Not collected because no particles were deposited.

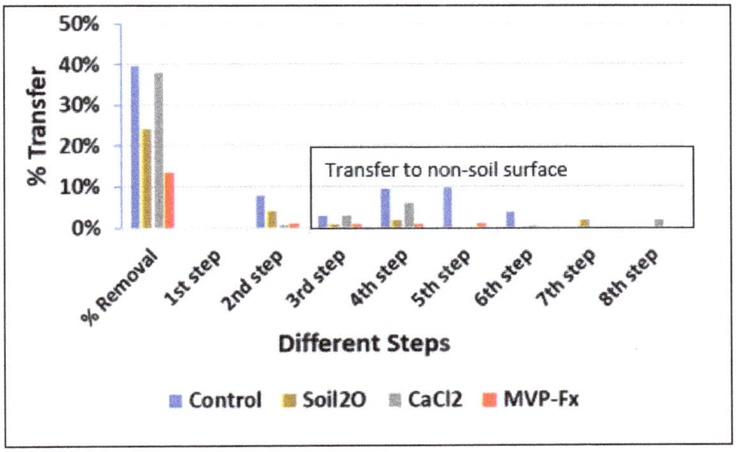

Figure 15. Percent Transfer for Walking Experiment on the Sand Walking Path.

3.3.1. Percent Residual Activity on Boots from Circuit Walking

Figure 16 shows the average %Res of Sr-85 contaminated particles that remained on boots after each experiment, including the NSCs and surfaces stabilized with $CaCl_2$, Phos-Chek MVP-Fx, and $Soil_2O$. Table 11 shows the p-values from the *t*-tests performed on the data. The *p*-values indicate the probability that the groupings of data were from the same population, i.e., statistically similar.

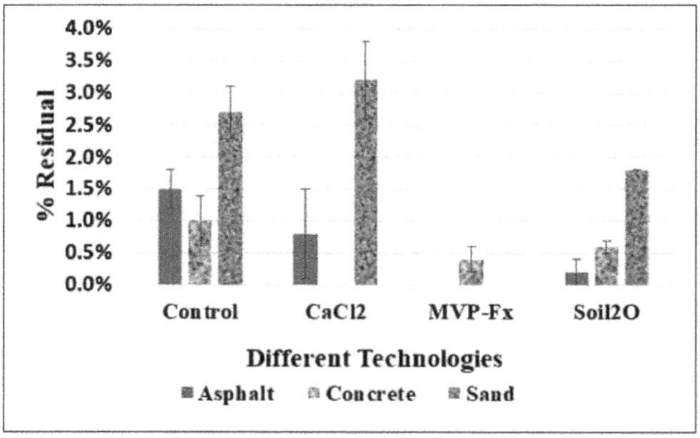

Figure 16. Percent Residual of Sr-85 Contaminated Particles from Foot Traffic Experiments.

Table 11. Statistical results (p-values) for foot traffic experiments percent residual. Gray shading indicates a significant difference ($p < 0.05$).

Surfaces	Stab. Tech.	$CaCl_2$	MVP-Fx	$Soil_2O$
Asphalt	Control	0.1143	<0.0001	0.0004
	$CaCl_2$		0.0481	0.1662
	MVP-Fx			0.0023
Concrete	Control	0.0120	0.0553	0.1460
	$CaCl_2$		0.0799	0.0157
	MVP-Fx			0.1769

The potential for residual appeared to be highest for the sand surface, as indicated by the control %Res. However, the sand results were determined from the 5 min continuous walking experiment, so the results are not exactly comparable. Interpreting the efficacy values in the figure in combination with the p-values from the table reveals the following for specific surfaces.

Asphalt. $Soil_2O$ (0.2 ± 0.2 %Res) and MVP-Fx (ND %Res) resulted in significantly decreased %Res compared to the NSC (2 ± 0.3 %T), while $CaCl_2$ (0.8 ± 0.7 %Res) did not. MVP-Fx (ND %Res) demonstrated significantly lower %Res than $CaCl_2$ (0.8 ± 0.7 %Res) and $Soil_2O$ (0.2 ± 0.2 %Res).

Concrete. Only $CaCl_2$ (ND %Res) resulted in significantly decreased %Res compared to the NSC (1 ± 0.4 %Res), while MVP-Fx (0.4 ± 0.2 %Res) and $Soil_2O$ (0.6 ± 0.1 %Res) did not, although MVP-Fx was close to being significantly different ($p = 0.056$). $CaCl_2$ (ND %Res) resulted in significantly lower %Res compared to $Soil_2O$ (0.6 ± 0.1 %Res).

Sand. Note: There was a single experiment for each stabilization technology, and the SD was based on the right and left sides of the path [N = 2], so no p-values were calculated. MVP-Fx (ND %Res) and $Soil_2O$ (2 ± 0 %Res) resulted in decreased %Res compared to the NSC (3 ± 0.4 %Res). $CaCl_2$ (3 ± 0.6 %Res) generated higher average %Res than NSC (3 ± 0.4 %Res).

3.3.2. Resuspension during Circuit Walking

A filter sample was deployed near the edge of the transition area (from Sr-85 contaminated sand surface to uncontaminated loose sand area) for each walking experiment on the sand walking path to monitor the spread of Sr-85 contaminated particles adjacent to the contaminated area. During the experiments, there were no visible particles transferred (kicked or dragged, etc.) over to the filter, and the measurements for each experiment were extremely close to the background measurements. The NSC sample produced a non-detect result, while each stabilization technology experiment produced less than 4 counts per second (cps), indicating a minimal presence of radiological material.

For the air sampler (5 cfm flow) set up approximately 30 cm above the ground near the contaminated area to collect an air sample for each sand walking path experiment, radiological activity was detected only from the air samples collected during the NSC experiment (2.8×10^{-10} μCi/mL) and the $Soil_2O$ stabilization technology (9.7×10^{-11} μCi/mL). These values are two to three orders of magnitude less than the derived air concentration (DAC) value (6×10^{-7} μCi/mL for Sr-85 particles potentially retained for years after exposure) listed in the US regulations for standards for radiation protection [19]. The DAC is defined as the concentration of a given radionuclide in the air which, if breathed for a working year of 2000 h under conditions of light work (1.2 cubic meters of air per hour), results in an intake of one annual limit on intake. The DAC values are presented for information only because they are intended to control chronic occupational exposure and are calculated based on a semi-infinite cloud of uniform concentration. This is not necessarily the exposure these experiments are intended to simulate, so the relationship between DAC and the experimental results may differ in an actual incident.

Particles containing approximately 80 μCi were placed on the surface prior to the walking experiment. For the NSC experiment, approximately 2×10^{-4} μCi was measured on the air sample, and for the $Soil_2O$ experiment, approximately 7×10^{-5} μCi was measured on the air sample. The ratio of measured activity in the air to the activity available on the ground was 2.5×10^{-6} and 9×10^{-7} for the NSC and $Soil_2O$ experiments, respectively. Albeit slight, the $Soil_2O$ experiments indicated a decreased availability of particles for inhalation. Neither air sample collected during the $CaCl_2$ and Phos-Chek MVP-Fx experiments generated detectable radiological activity, indicating they also functioned to decrease the availability of particles. During each simulated vehicle and foot traffic experiments (and all the experimental preparatory work, including particle contamination), two air samples (3 cfm flow) were collected in the general work area of the containment tent. The average concentration of these samples was 4.9×10^{-13} μCi/mL, approximately six orders of magnitude less than the DAC.

3.3.3. Foot Traffic Experimental Observations and Conclusions

Similar to the vehicle experiments, the individual results can be viewed in the context of the statistical result summaries above. For the asphalt and concrete experiments, less than 5% of the particles were removed from the NSC experiments, whereas during the walking path experiment, almost 40% of the particles were removed. In most cases, the stabilization technologies provided improved stabilization compared to the surface-type-specific NSC experiment results. However, there were a few instances in which one or more of the three stabilization technologies did not demonstrate statistically significant improvement or had high removals even with the stabilizer. This often occurred for sand which had the highest NSC, suggesting sand is a less stable surface to begin with, which is arguably intuitive when compared to more solid surfaces like concrete and asphalt. This also tended to occur for the $CaCl_2$, which tended to form a crust on the boot, increasing the removal from the surface of radioactivity contained in the crust.

Also, similar to the simulated vehicle experiments, the %R and %T results are more reliable than the %Res results because of the nature of the measurement. The %R and %T measurements were made by locating the detector probe in pre-marked positions on pavers with the same geometry with respect to the detector probe. For the %Res measurements, the detector probe was held near the boots, which had an uneven contour, and the sand would bind to somewhat different locations depending on the experiment.

4. Conclusions

Removal percentages, with or without stabilizers, were higher from the soil than from asphalt or concrete. Therefore, stabilization of contaminated soils surfaces can represent an important goal for stabilizer application. The data and statistical analysis presented can aid in selecting appropriate stabilizers because they indicate the potential for stabilization techniques to enable operations in contaminated areas. In addition, the process of performing the studies revealed numerous operational details and observations for each of the stabilization technologies pertaining to application, surface material interactions, health and safety concerns, and estimated cost. These are summarized in Tables S1–S3 for each of the three stabilizers.

Related to the selection of the stabilizer at a particular site, this study also suggests which of the several measures of contaminant spread may be most useful to determine from future studies, whether on the laboratory, pilot, or field scale. Namely, in general, percent removal from surface (%R) and percent transferred from contaminated to an uncontaminated surface (%T) seemed more useful in performing statistical comparisons between experiments. The value of the percent residue left on the tire or boot (%Res) was often limited by the ability to reliably measure the radioactivity on the tire or boot, often for practical reasons.

Many conclusions from the specific types of experiments are discussed above, but further refinement of this information would result from field-scale studies, which better represent all the forces that stabilized particles are subjected to. For instance, based on the logistical constraints for walking experiments in the contamination enclosure tent, it was not possible to extensively vary operationally important variables like worker mass, walking speed, stride length, boot design and size, and other operator-specific variables that could affect the physical stresses that walking imparts upon the stabilized radioactivity and its ultimate resuspension. Likewise, for the vehicle experiments, it was impossible to achieve realistic vehicle speeds in the radiological containment enclosure. Other variables could include tire tread pattern and acceleration rate, which may be even more important as electric, high torque vehicles, especially private ones, become more common. Electric vehicles may result in higher resuspension due to the higher mass of their batteries [20,21]. Further, differently shaped vehicles will have different aerodynamics and the resulting potential for air disturbance. Airflow caused by the movement of vehicles and by the exhaust and fans issuing from within the vehicle is extremely difficult to be simulated within a radiological containment tent.

Testing additional stabilization agents for resuspension resulting from foot and vehicle traffic is warranted. For instance, in a study of wind-driven suspension, different saltwater compositions yielded somewhat different results, with some results dependent on the nature of the physical stresses on the stabilized particles [15]. Interestingly, this dependency did not exist for one saltwater mixture, namely that from an aqueous solution from the Dead Sea. Such dependencies can introduce uncertainty into operational decisions. Thus, in devising studies related to foot and vehicle traffic resuspension, achieving the desirable goal of seeking stabilizing technologies independent of the surface seems promising.

Supplementary Materials: The following supporting information can be downloaded at: https://www.mdpi.com/article/10.3390/app12083861/s1, Table S1: $CaCl_2$ Operational Information, Table S2: Phos-Chek MVP-Fx Operational Information, Table S3: Soil$_2$O Operational Information, Video S1: Simulated Vehicle Experiment Example, Video S2: Straight-line Walking Experiment Example, Video S3: Circuit Walking Path Experiment Example.

Author Contributions: Conceptualization, M.M., T.S., S.S., J.A., E.T., H.R.-A. and A.S.; methodology, M.M., T.S., S.S., J.A., H.R.-A., A.S. and R.J.; software, X.X.; investigation, R.J., X.X. and M.L.; data curation, X.X.; writing—original draft preparation, M.M., R.J. and X.X.; writing—review and editing, M.M., T.S., S.S., J.A., H.R.-A. and A.S. All authors have read and agreed to the published version of the manuscript.

Funding: This research was funded by IWTSD via contract to Battelle Memorial Institute.

Institutional Review Board Statement: Not applicable.

Informed Consent Statement: Not applicable.

Data Availability Statement: All data is found within the tables of this manuscript.

Conflicts of Interest: The authors declare no conflict of interest.

Disclaimer: Financial support by IWTSD does not constitute an express or implied endorsement of the results or conclusions of the project by either IWTSD or the Department of Defense. This document has undergone review in accordance with U.S. Environmental Protection Agency (EPA) policy prior to publication. Note that approval for publication does not signify that the contents necessarily reflect the views of the Agency. Mention of trade names, products, or services does not convey EPA approval, endorsement, or recommendation.

References

1. Kaminski, M.D.; Lee, S.D.; Magnuson, M. Wide-area decontamination in an urban environment after radiological dispersion: A review and perspectives. *J. Hazard. Mater.* **2016**, *305*, 67–86. [CrossRef] [PubMed]
2. Amato, F.; Querol, X.; Johansson, C.; Nagl, C.; Alastuey, A. A review on the effectiveness of street sweeping, washing and dust suppressants as urban PM control methods. *Sci. Total Environ.* **2010**, *408*, 3070–3084. [CrossRef]
3. Gulia, S.; Goyal, P.; Goyal, S.; Kumar, R. Re-suspension of road dust: Contribution, assessment and control through dust suppressants—A review. *Int. J. Environ. Sci. Technol.* **2019**, *16*, 1717–1728. [CrossRef]
4. Norman, M.; Johansson, C. Studies of some measures to reduce road dust emissions from paved roads in Scandinavia. *Atmos. Environ.* **2006**, *40*, 6154–6164. [CrossRef]
5. Saito, H.; Sutton, M.; Zhao, P.; Lee, S.D.; Magnuson, M. Review of technologies for preventing secondary transport of soluble and particulate radiological contamination from roadways, roadside vegetation, and adjacent soils. *Environ. Adv.* **2020**, *1*, 100003. [CrossRef]
6. Xu, L.; Pei, Z. Preparation and optimization of a novel dust suppressant for construction sites. *J. Mater. Civ. Eng.* **2017**, *29*, 04017051. [CrossRef]
7. Zhan, Q.; Qian, C.; Yi, H. Microbial-induced mineralization and cementation of fugitive dust and engineering application. *Constr. Build. Mater.* **2016**, *121*, 437–444. [CrossRef]
8. Zhang, B.; Wang, Y.; Zhao, X.; Cao, L.; Tong, R. Effectiveness of road dust suppressants: Insights from particulate matter-related health damage. *Environ. Geochem. Health* **2021**, *43*, 4139–4162. [CrossRef] [PubMed]
9. Parvej, S.; Naik, D.L.; Sajid, H.U.; Kiran, R.; Huang, Y.; Thanki, N. Fugitive Dust Suppression in Unpaved Roads: State of the Art Research Review. *Sustainability* **2021**, *13*, 2399. [CrossRef]
10. Tsogt, B.; Oh, S.-Y. Preparations and application of dust suppressants from biomass-based materials. *J. Air Waste Manag. Assoc.* **2021**, *71*, 1386–1396. [CrossRef] [PubMed]
11. Stallworth, A.M.; Chase, E.H.; McDevitt, B.; Marak, K.K.; Freedman, M.A.; Wilson, R.T.; Burgos, W.D.; Warner, N.R. Efficacy of oil and gas produced water as a dust suppressant. *Sci. Total Environ.* **2021**, *799*, 149347. [CrossRef] [PubMed]
12. Li, S.; Zhao, B.; Lin, H.; Shuang, H.; Kong, X.; Yang, E. Review and prospects of surfactant-enhanced spray dust suppression: Mechanisms and effectiveness. *Process Saf. Environ. Prot.* **2021**, *154*, 410–424. [CrossRef]
13. USEPA. Wide Area Stabilization of Radiological Particulate Contamination. EPA/600/R-616/067. 2017. Available online: https://cfpub.epa.gov/si/si_public_record_report.cfm?Lab=NHSRC&dirEntryId=335602 (accessed on 9 April 2022).
14. USEPA. Technical Report for the Demonstration of Wide Area Radiological Decontamination and Mitigation Technologies for Building Structures and Vehicles. EPA/600/R-616/019. 2016. Available online: https://cfpub.epa.gov/si/si_public_record_report.cfm?Lab=NHSRC&dirEntryId=312072 (accessed on 9 April 2022).
15. Raveh-Amit, H.; Sharon, A.; Katra, I.; Stilman, T.; Serre, S.; Archer, J.; Magnuson, M. Limiting Wind-Induced Resuspension of Radioactively Contaminated Particles to Enhance First Responder, Early Phase Worker and Public Safety—Part 1. *Appl. Sci.* **2022**, *12*, 2463. [CrossRef]
16. IWTSD. Irregular Warfare Technical Support Directorate. 2022. Available online: https://www.cttso.gov (accessed on 9 April 2022).
17. Clark, D.E.; Cobbin, W.C. *Removal Effectiveness of Simulated Dry Fallout from Paved Areas by Motorized and Vacuumized Street Sweepers*; Naval Radiological Defense Lab: San Francisco, CA, USA, 1963. Available online: https://apps.dtic.mil/sti/pdfs/AD0456495.pdf (accessed on 9 April 2022).
18. Wiltshire, L.L.; Owen, L.W. *Three Tests of Firehosing Technique and Equipment for the Removal of Fallout from Asphalt Streets and Roofing Materials*; Naval Radiological Defense Lab: San Francisco, CA, USA, 1966. Available online: https://apps.dtic.mil/sti/pdfs/AD0640491.pdf (accessed on 9 April 2022).
19. NRC. Appendix B to Part 20—Annual Limits on Intake (ALIs) and Derived Air Concentrations (DACs) of Radionuclides for Occupational Exposure; Effluent Concentrations; Concentrations for Release to Sewerage. 2021. Available online: https://www.nrc.gov/reading-rm/doc-collections/cfr/part020/part020-appb.html (accessed on 9 April 2022).
20. Alam, M.S.; Hyde, B.; Duffy, P.; McNabola, A. Analysing the Co-Benefits of transport fleet and fuel policies in reducing $PM_{2.5}$ and CO_2 emissions. *J. Clean. Prod.* **2018**, *172*, 623–634. [CrossRef]
21. Piscitello, A.; Bianco, C.; Casasso, A.; Sethi, R. Non-exhaust traffic emissions: Sources, characterization, and mitigation measures. *Sci. Total Environ.* **2021**, *766*, 144440. [CrossRef] [PubMed]

applied sciences

MDPI

Article

Effectiveness of Food Processing By-Products as Dust Suppressants for Exposed Mine Soils: Results from Laboratory Experiments and Field Trials

Justus Freer *, Maximilian Lübeck, Johannes L. Sieger, Bernd G. Lottermoser and Marius Braun

Institute of Mineral Resources Engineering, RWTH Aachen University, 52062 Aachen, Germany
* Correspondence: freer@mre.rwth-aachen.de

Abstract: In this study, the effectiveness of biodegradable food processing by-products (chicory vinasses, corn steep liquor, decantation syrup, and palatinose molasses) as dust suppressants on mine soils has been precisely quantified using controlled laboratory experiments and field trials. Laboratory experiments using a wind tunnel indicate that rainfall intensity and repetitive wetting and drying cycles affect the by-products' effectiveness. In addition, field trials conducted using soil plots at an open-pit lignite mine (Germany) demonstrate that the tested biomaterials can effectively reduce dust emissions under field conditions, despite the fact that rainfall led to the leaching of the applied biomaterials, decreasing the additives' concentrations on the soil surface and impairing the materials' effectiveness to suppress wind erosion. Thus, food processing by-products may be used for short-term dust mitigation at mine sites and represent environmentally benign alternatives to dust control chemicals detrimental to the environment.

Keywords: dust suppressants; dust control; food industry; wastes; by-products; field testing; leaching test

Citation: Freer, J.; Lübeck, M.; Sieger, J.L.; Lottermoser, B.G.; Braun, M. Effectiveness of Food Processing By-Products as Dust Suppressants for Exposed Mine Soils: Results from Laboratory Experiments and Field Trials. *Appl. Sci.* **2022**, *12*, 11551. https://doi.org/10.3390/app122211551

Academic Editor: Itzhak Katra

Received: 10 October 2022
Accepted: 9 November 2022
Published: 14 November 2022

1. Introduction

Dust suppressants effectively mitigate particulate matter emissions from exposed mine soils by changing the physical properties of soil surfaces [1]. Their effectiveness and durability are decisively impacted by environmental factors such as precipitation, radiation or the presence of microorganisms. For example, heavy rainfall can significantly impair the performance of ligninsulfonates, which are widely applied as non-traditional dust suppressants, due to their water solubility [2]. Exposure to ultraviolet radiation causes photo-oxidation of polymer-based reagents, where polymer chains cleave into smaller segments, resulting in deterioration of the mechanical properties [3]. Biogenic amendments (e.g., biopolymers or molasses), in turn, can be naturally decomposed by microorganisms. Carbohydrates and proteins are recognized as food sources and are split into oligomers, dimers, and monomers by microorganisms' intra- and extracellular enzymes [4]. Hence, field trials are imperative to evaluate the actual performance of dust suppressants and to enable an appropriate product selection for applications at mine and mineral processing sites. A variety of field studies (e.g., [5–7]) have examined the effectiveness of dust suppressants on heavily trafficked surfaces such as unpaved roads. However, there is a lack of experimental research investigating the effects of amendments on reducing wind-blown dust emissions from exposed mine soils under field conditions [8].

Currently applied and investigated dust suppressants (e.g., electrochemical products, magnesium chloride, ligninsulfonates, biopolymers) have limitations and disadvantages, such as high product prices, detrimental effects on the environment, insolubility in cold water, or yielding viscous solutions at low concentrations [9–12]. Food processing wastes and by-products may pose sustainable alternatives to established dust suppressants. They are usually inexpensive, readily available and generated in high volumes [13–15]. Moreover, many of these materials are biodegradable and contain carbohydrates and proteins, which provide techno-functional properties to bind single soil particles [14]. A study has recently shown that wastes and by-products from the food industry can enhance the unconfined compressive strength of soils, which indicates an improved wind erosion resistance and thus an enhanced dust control effectiveness [16]. Freer et al. [16] concluded that among the biomaterials studied, chicory vinasses, corn steep liquor, decantation syrup, and palatinose molasses have the highest potential to be applied as dust suppressants. In a follow-up study, Freer et al. [17] conducted laboratory-based wind tunnel tests and demonstrated that these biomaterials could indeed effectively mitigate wind-induced soil losses. Even though these results are promising for the actual application of wastes and by-products from the food industry as dust suppressants on mine soils, the study did not consider abiotic (e.g., radiation, water) and biotic factors (e.g., bacteria), which may diminish the effectiveness of reagents to reduce dust emissions [5]. Rainfall, in particular, may be a controlling factor on the amendments' effectiveness because water-soluble reagents can be leached from the surfaces of treated soils and substrates.

This study aimed to evaluate the effectiveness of four biodegradable by-products from the food industry (chicory vinasses, corn steep liquor, decantation syrup, and palatinose molasses) in reducing dust emissions from mine soils. The effectiveness of dust suppression from soils is defined herein as the degree to which by-products from the food industry successfully reduce wind erosion from a selected sandy mine soil as measured with and without dust control. Laboratory experiments were performed to study the effects of rainfall intensities and wetting–drying cycles on the wind erosion resistance of treated soil surfaces. In addition, field trials were conducted to establish the effectiveness of applied biomaterials under actual field conditions that are indicative of possible commercial-scale application and practice. For this study, a low-cost and low-maintenance experimental setup was applied that is feasible for remote and difficult-to-access areas and allows various dust control products to be tested. Thus, this research extends our knowledge of the application of biodegradable dust suppressants on mine soils and supports the implementation of sustainable dust control measures at mine and mineral processing sites.

2. Materials and Methods

2.1. Mine Soil

Test substrate used in this study was sampled from a sand and gravel pit in North-Rhine Westphalia, Germany. According to DIN EN ISO 14688-1:2018-05 [18], the mine soil can be classified as medium-coarse sand (D_{50} = 0.65 mm). X-ray diffraction analysis indicated that the sand is dominated by quartz, with minor amounts of orthoclase, microcline and clinochlore. Detailed material characterization of the test substrate can be found in Freer et al. [16].

2.2. By-Products from the Food Industry

Four by-products from the food industry were investigated as dust suppressants. Each reagent is biodegradable and readily soluble in water. Further information on the tested biomaterials can be taken from Freer et al. [17].

2.2.1. Chicory Vinasses

Chicory vinasses is a viscous liquid obtained as a by-product during inulin and oligofructose production from chicory. Its dry matter (DM) content is 63.3%, crude protein being the main component (45–51% of DM). Samples of chicory vinasses were acquired from BENEO-Orafti S.A., Tienen, Belgium [16].

2.2.2. Corn Steep Liquor

Corn steep liquor is a by-product of the steeping process of corn and has a DM content of 27.0%. DM components are crude protein (36–49% of DM), starch, reducing sugars and crude fiber. Cargill Deutschland GmbH, Düsseldorf, Germany supplied samples of corn steep liquor [16].

2.2.3. Decantation Syrup

Decantation syrup is obtained as a viscous liquid during the production of candy sugar and has a DM content of 72.0%. Its DM consists mainly of sucrose, besides small amounts of fructose, glucose, and oligosaccharides. BENEO-Orafti S.A. provided samples of decanter syrup [16].

2.2.4. Palatinose Molasses

Palatinose molasses is the concentrated liquid fraction from the isomaltulose production from sugar beets and has a DM content of 68.2%. DM components are isomaltulose and trehalulose (>50% of DM) and sucrose, isomaltose, fructose and glucose (28–42% of DM). Samples of palatinose molasses were acquired from BENEO-Orafti S.A. [16].

2.3. *Laboratory Experiments*

2.3.1. Materials

Selected biomaterials were dissolved in pure water to their specific concentration using a magnetic stirrer to prepare homogeneous solutions (Figure 1). Chosen reagent concentrations are shown in Table 1. It should be noted that the reagent concentration is referred to as the dry matter content of the additive. The specific concentration of each additive and its application rate were selected based on Freer et al. [17] who identified the optimum parameters, whereby a further increase in concentration and application rate has an insignificant effect on the biomaterial's effectiveness to reduce dust emissions.

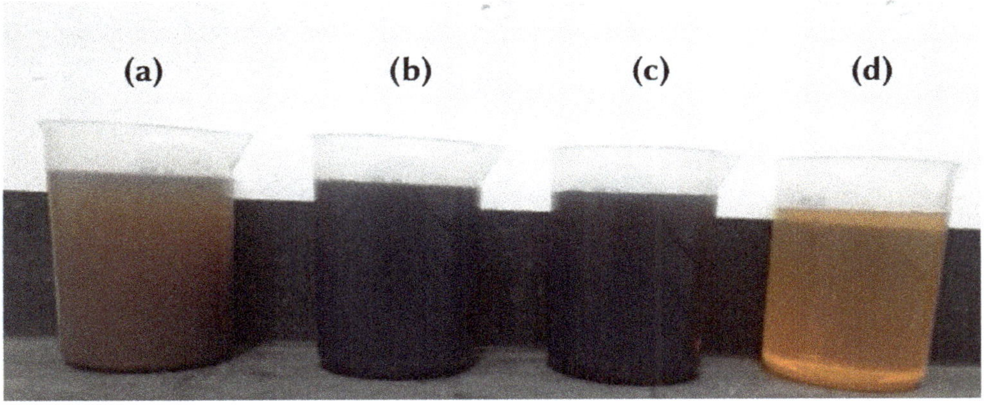

Figure 1. Prepared biomaterial solutions: (**a**) chicory vinasses, (**b**) corn steep liquor, (**c**) decantation syrup, and (**d**) palatinose molasses.

Table 1. Applied additive concentration and application rate of each biomaterial.

Biomaterial	Concentration (%)	Application Rate (L/m^2)
Chicory vinasses	10.0	1.5
Corn steep liquor	5.0	0.75
Decantation syrup	6.0	1.0
Palatinose molasses	6.0	1.0

Three soil replicates were prepared for each wind erosion test to evaluate the precision of measured soil losses, while dry, untreated soil was selected for control samples to determine the applied biomaterials' quantitative dust control effectiveness. The prepared solutions were then sprayed on soil surfaces according to their specific application rate (Table 1). In order to ensure consistent solution application covering the entire soil surfaces and to prevent any disturbance of soil surfaces, the solutions were carefully applied with a hand-held trigger sprayer. As other studies have stated (e.g., [19]), the effective performance of dust suppressants requires careful and uniform product application. Prior to wind tunnel experimentation, prepared soils were air-dried for seven days at ambient laboratory temperature.

2.3.2. Experimental Methodology

Wind erosion tests were performed using a laboratory wind tunnel at the Institute of Mineral Resources Engineering, RWTH Aachen University, Germany (Figure 2). This facility has already been used for evaluating the effectiveness of biogenic dust suppressants [17]. Sample trays were placed centrally in the test section and exposed to a laminar airflow of 13.57 m/s for 120 s.

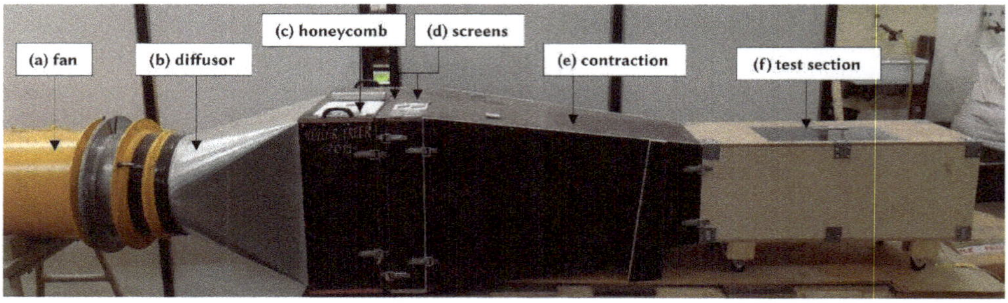

Figure 2. Setup of laboratory experiments. Constructed wind tunnel includes different components to produce a laminar airflow: (**a**) fan, (**b**) diffuser, (**c**) honeycomb slits, (**d**) two screens, (**e**) contraction cone, and (**f**) test section. A dGAL4-30/30 axial fan (Korfmann Lufttechnik, Witten, Germany) with a diameter of 400 mm driven by a 1.5 kW electric motor was used to generate a constant wind speed of 13.57 m/s.

In subsequent laboratory experiments, treated soil surfaces were investigated to find whether different rainfall intensities and repetitive wetting-drying cycles impact the dust control effectiveness of the investigated biomaterials. Four different rainfall intensities were tested: (a) 0.25 L/m^2, (b) 1 L/m^2, (c) 2 L/m^2, and (d) 4 L/m^2. For this, pure water was applied on the soil surfaces using a hand-held trigger sprayer. The first application of water was carried out seven days after applying biomaterials. Then, individual samples were moistened with their chosen rainfall intensities (wet–dry cycle 1) and stored in a room at ambient temperature. After seven days of drying, the specimens were tested again in the wind tunnel and then wetted once again (wet–dry cycle 2). The same procedure was used for the following three wet–dry cycles (3–5). The sample tray was weighed before and after each measurement using a PES 4200-2M (Kern & Sohn, Balingen-Frommern, Germany) with a resolution of 0.01 g. Untreated mine soil was selected as a control to calculate the dust control effectiveness for each experimental setup. Quantitative dust control effectiveness is therefore defined as the ratio of total weight loss between treated and untreated mine soil after four wet–dry cycles. Statistical analysis of measured soil losses included calculating arithmetic mean and standard deviation values.

2.4. Field Trials

2.4.1. Test Site and Materials

A field test site was established at an open-pit lignite mine in North Rhine-Westphalia, Germany (Figure 3c), where field experimentations were conducted between July and August 2020. Initially, the test site was cleared of vegetation and topsoil. Then, 18 test plots were constructed using wooden frames (length 700 mm, width 400 mm, height 56 mm) and geotextile fabric, which was attached to the bottom surface of each frame to separate the test substrate from the underlying soil and to allow for water drainage (Figure 3a).

Then, each test plot was filled with mine soil up to the top of each wooden frame. Each soil surface was then levelled, and finally, soil surfaces were treated with selected additives. The reagent solutions' preparation and application on the substrate surface were identical to the procedure described in Section 2.3.1. Untreated mine soil and soil only treated with water were set up as a control group to establish the dust control effectiveness of selected biomaterials.

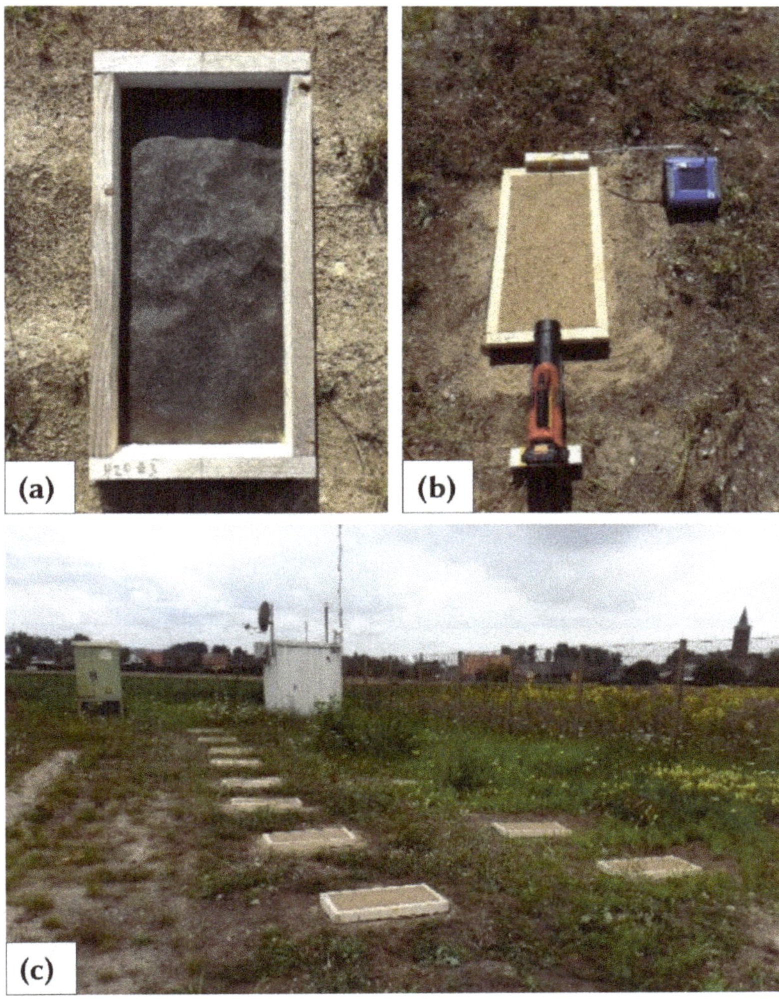

Figure 3. Images of field test site: (**a**) construction design of individual test plot showing wooden frame and geotextile base, (**b**) experimental setup of a single test plot with electric blower, isokinetic sampling probe and dust monitor DustTrak DRX 8533 (TSI, Shoreview, USA), and (**c**) entire test site with installed soil plots and nearby weather station.

Three measurement series were carried out on each sampling day to determine the average background particulate matter concentration at the test site (cf. Table A4, Appendix A). Meteorological data were obtained from a nearby weather station for the duration of the study (12:00, 17 July 2020–12:00, 14 August 2020). Daily mean data included temperature (°C), solar radiation (W/m^2), humidity (%) and precipitation (L/m^2). A summary of the recorded meteorological data is given in Table A1, Appendix A. The lowest daily mean temperature for the study period was 16.04 °C, with the highest daily mean temperature of 28.04 °C. In total, a precipitation amount of 39.44 L/m^2 was measured. The highest rainfalls were recorded at 9, 23, 26 and 27 days after soil treatment with 4.8 L/m^2, 4.25 L/m^2, 17.48 L/m^2, and 8.84 L/m^2, respectively.

2.4.2. Experimental Methodology

To simulate wind erosion events, a commercially available electric blower was repetitively applied to each treated soil plot surface, and dust measurements were carried out at particular times over a period of 28 days. At each test plot, suspended dust concentrations were measured 3, 7, 14, 21 and 28 days after substrate treatment, whereby $PM_{2.5}$, PM_{10}, and total suspended particle (TSP) concentrations were determined in the air (i.e., mg per m^3), using a DustTrak DRX 8533 from TSI with a log interval of 1 s. The DustTrak DRX 8533 records aerosol concentrations in a range of 0.001–150 mg/m^3 (±0.1% of reading). Isokinetic sampling was applied to avoid biased measurements of particulate concentration, using a 90-degree bent tubing with a streamlined inlet. A 2 mm inlet diameter was selected to match the probe inlet velocity to the generated airflow velocity. The isokinetic probe was placed centrally opposite the applied wind direction at the height of 450 mm above the wooden frame (Figure 3b). The duration of each measurement was 60 s, whereby the test plot surface was subjected to a continuous airstream from the electric blower for 45 s, which generated a constant wind speed of 14.65 m/s (i.e., 52 km/h). The wind speed was measured using an anemometer positioned at the test plot end. However, it should be noted that such blowers generate a turbulent flow and do not meet the aerodynamic criteria of portable field wind tunnels, simulating the atmospheric flow causing natural wind erosion [20]. Consequently, the measured dust emissions do not equate to dust emissions caused by natural wind erosion. Measured particulate matter concentrations were initially analyzed for the presence of outliers and extreme values using box-plot graphics. Implausible or illogical values (e.g., negative values) were removed from the data set before the mean concentration of TSP, PM_{10}, and $PM_{2.5}$ were calculated. In addition, photographic images were taken before and after each measurement to document changes in the soil plot surfaces.

3. Results

3.1. Laboratory Testing

3.1.1. Effect of Rainfall Intensities on Wind-Induced Soil Losses

The wind-induced soil losses of samples treated with different biomaterial solutions and wetted with four varying rainfall intensities are shown in Figure 4. Each bar represents the calculated mean from triplicates and indicates the total weight loss (TWL) after four wet–dry cycles. Prior to the first wet–dry cycle, soil treated with chicory vinasses, corn steep liquor, decantation syrup, and palatinose molasses showed only negligible weight losses of 0.09 g (SD = 0.02), 0.12 g (SD = 0.16), 0.04 g (SD = 0.01), and 0.03 g (SD = 0.01), respectively. From Figure 4, it can be seen that wetting of samples increased mass losses at any investigated rainfall intensity. Samples treated with chicory vinasses can be considered the most susceptible to wind erosion, showing the highest mass losses at a rainfall intensity of 0.25 L/m^2 with a TWL of 63.04 g. Samples treated with corn steep liquor, decantation syrup, and palatinose molasses were less prone to wetting and peaked in mass losses at a rainfall intensity of 0.25 L/m^2 with a TWL of 4.38 g, 2.79 g, and 13.52 g, respectively. By comparison, untreated soil has a TWL of 125.28 g (SD = 19.34) after four measuring cycles.

Mass losses of each biomaterial-treated soil wetted with higher rainfall intensities (1–4 L/m^2) were significantly lower than those moistened with only 0.25 L/m^2. With increasing rainfall intensity, the TWL of soil treated with chicory vinasses decreased to 5.35 g (1 L/m^2), 0.6 g (2 L/m^2), and 2.12 g (4 L/m^2), respectively. Corn steep liquor behaved similarly and showed a reduced TWL of 1.24 g (1 L/m^2), 0.7 g (2 L/m^2), and 2.38 g (4 L/m^2). By contrast, all soil samples treated with decantation syrup and palatinose molasses had a TWL of <1 g. A summary of the laboratory test data is shown in Table A2, Appendix A.

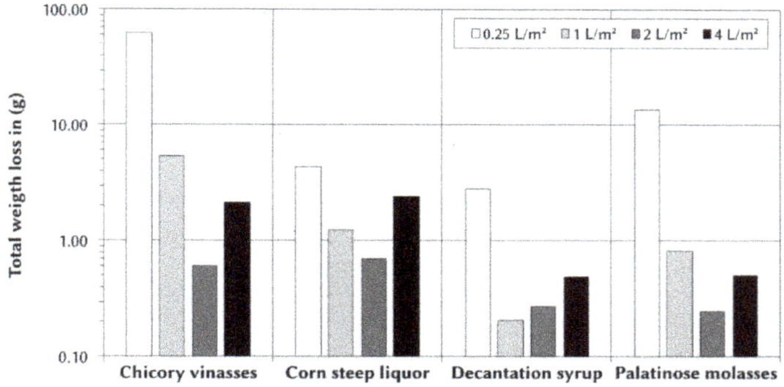

Figure 4. Wind-induced total weight loss (g) of treated mine soils wetted with four different rainfall intensities after four wet–dry cycles (mean values, n = 3). By comparison, untreated soil showed a total weight loss of 125.28 g after four measuring cycles.

3.1.2. Effect of Wet–Dry Cycles on Wind-Induced Soil Losses

Figure 5 depicts the effect of wet–dry cycles on wind-induced soil losses of biomaterial-treated sand wetted with different rainfall intensities. Each bar represents the calculated mean of three replicates. It is apparent from the diagrams (Figure 5a–d) that (i) the soil losses varied for each wet–dry cycle and rainfall intensity, and (ii) the biomaterials behaved differently. Chicory vinasses and corn steep liquor showed similar characteristics at a rainfall intensity of 0.25 L/m^2. The lowest soil losses occurred after the first wet–dry cycle, while a considerable increase was observed after the fourth. Samples treated with decantation syrup behaved slightly differently, whereby the soil losses achieved their minimum at wet–dry cycle 3 and its peak at wet–dry cycle 4. By contrast, the weight losses of palatinose molasses-treated soil peaked at wet–dry cycle 1, then decreased and subsequently increased at wet–dry cycle 4 (to a similar level to wet–dry cycle 1). At higher rainfall intensities (1–4 L/m^2), the mass loss data followed no clear trend. Either the soil losses per wet–dry cycle did not differ significantly, or there were isolated outliers. In part, samples showed almost no soil losses with increasing wet–dry cycles. A summary of the soil loss data after each wet–dry cycle is given in Table A3, Appendix A.

3.2. Field Trials

3.2.1. Measured Dust Concentrations

Figure 6 illustrates the concentrations of dust particle sizes ($PM_{2.5}$, PM_{10}, TSP) measured for soils treated with different biomaterial solutions. Each bar represents the calculated mean value from triplicate analyses. The data demonstrate that mine soils treated with any of the biomaterials showed a significant reduction of dust emissions after 3 days. With increasing exposure time, dust emissions increased steadily and peaked 21 days after the solutions were applied. By comparison, the data of water-treated and untreated mine soil show a different trend. The highest dust emissions were recorded on day 3 and continued to decline subsequently. In some cases, dust emissions of the control group (i.e., water-treated and untreated mine soil) achieved lower PM concentrations than treated soil samples. For example, untreated mine soil emitted less dust emissions from day 21 onwards than all test plots prepared with biomaterials.

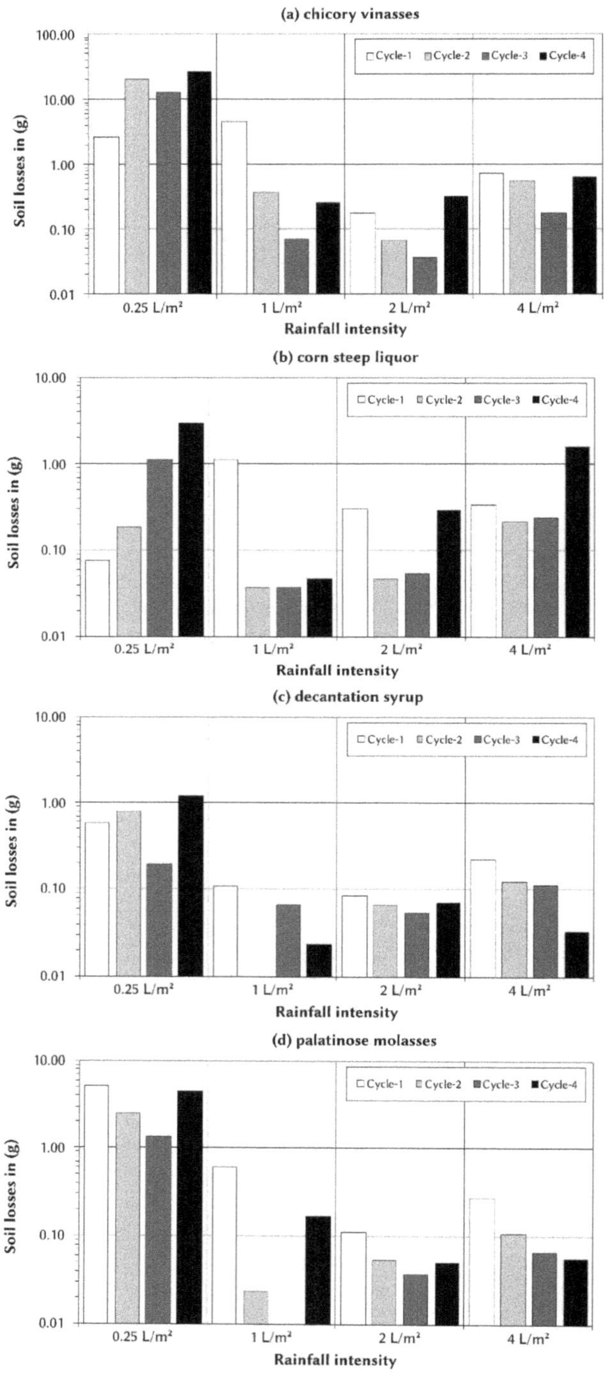

Figure 5. Effect of wet–dry cycles on wind-induced soil losses (g) from mine soils treated with different food processing by-products (mean values, n = 3): (**a**) chicory vinasses, (**b**) corn steep liquor, (**c**) decantation syrup, and (**d**) palatinose molasses.

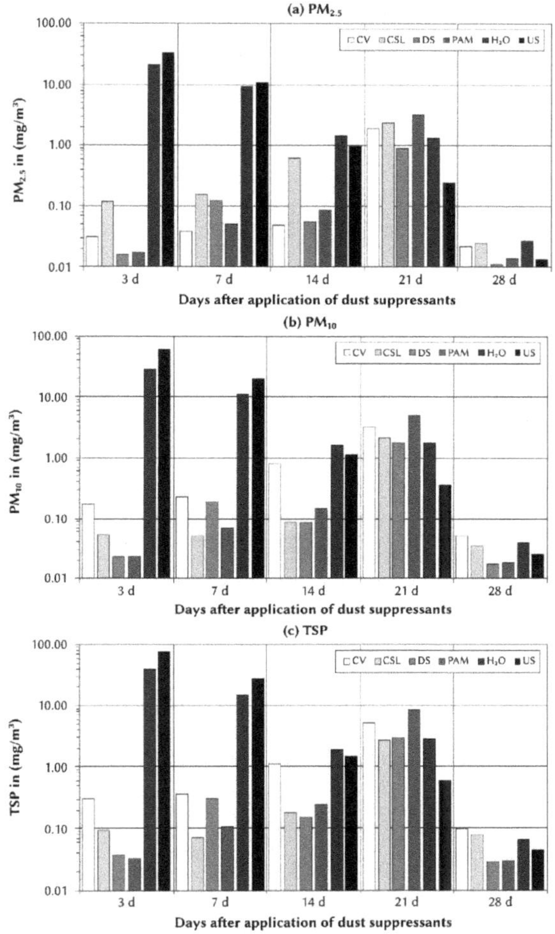

Figure 6. Wind-induced dust emissions from treated and untreated mine soils (mean values, n = 3): (**a**) $PM_{2.5}$, (**b**) PM_{10}, and (**c**) TSP. Dust emissions were measured 3, 7, 14, 21, and 28 days after application of dust suppressants. Abbreviations: CV = chicory vinasses, CSL = corn steep liquor, DS = decantation syrup, PAM = palatinose molasses, H_2O = water-treated soil, US = untreated soil.

The peak values of biomaterial-treated plots were significantly lower than those of the control group (i.e., water-treated and untreated soil). For untreated soil, PM_{10} emissions amounted to 59.026 mg/m^3 ($PM_{2.5}$: 32.078 mg/m^3, TSP: 77.075 mg/m^3). Soil treated with pure water achieved a PM_{10} concentration of 27.649 mg/m^3 ($PM_{2.5}$: 20.981 mg/m^3, TSP: 39.176 mg/m^3). By contrast, applications of palatinose molasses and corn steep liquor produced a decrease of PM_{10} emissions to 5.003 mg/m^3 ($PM_{2.5}$: 3.152 mg/m^3, TSP: 8.482 mg/m^3), and 3.184 mg/m^3 ($PM_{2.5}$: 2.351 mg/m^3, TSP: 5.169 mg/m^3), respectively. In comparison, chicory vinasses and decantation syrup reduced PM_{10} emissions to 2.148 mg/m^3 ($PM_{2.5}$: 1.863 mg/m^3, TSP: 2.697 mg/m^3), and 1.756 mg/m^3 ($PM_{2.5}$: 0.900 mg/m^3, TSP: 2.896 mg/m^3), respectively. Among the investigated biomaterials, decantation syrup demonstrated the highest effect on reducing $PM_{2.5}$ and PM_{10} emissions after 21 days, while chicory vinasses was the most effective additive in suppressing TSP emissions. The lowest dust emissions were measured for all soil plots, including the control group (i.e., water-treated and untreated mine soil), 28 days after soil treatment.

A summary of the mean dust particle concentrations in air as detected during the field trial from variously treated mine soils is shown in Table A4, Appendix A.

3.2.2. Visual Inspection of Soil Surfaces

Figure 7 illustrates the typical surface characteristics of eroded soils as observed during the field trials. Surfaces of the control group (i.e., water-treated and untreated mine soil) and the treated soil samples changed differently over time. Soil surfaces prepared with food processing by-products showed minor indications of wind erosion (i.e., disturbance of soil surfaces, removal of smaller soil particles) after 14 days and 21 days, respectively, without any significant changes in horizontal soil thickness (Figure 7b). By contrast, plots with water-treated and untreated soil formed a conical-shaped depression that expanded slightly in depth over time (Figure 7a). The highest soil losses were observed at the beginning of the study. Almost all test plots indicated wildlife tracks that can be attributed primarily to birds. One of the water-treated plots had to be removed from the field trial after seven days because it was extensively damaged. On day 28, all specimens appeared to have a high moisture content, and their surfaces showed visible traces of raindrops and thus were exposed to obvious rainfall events.

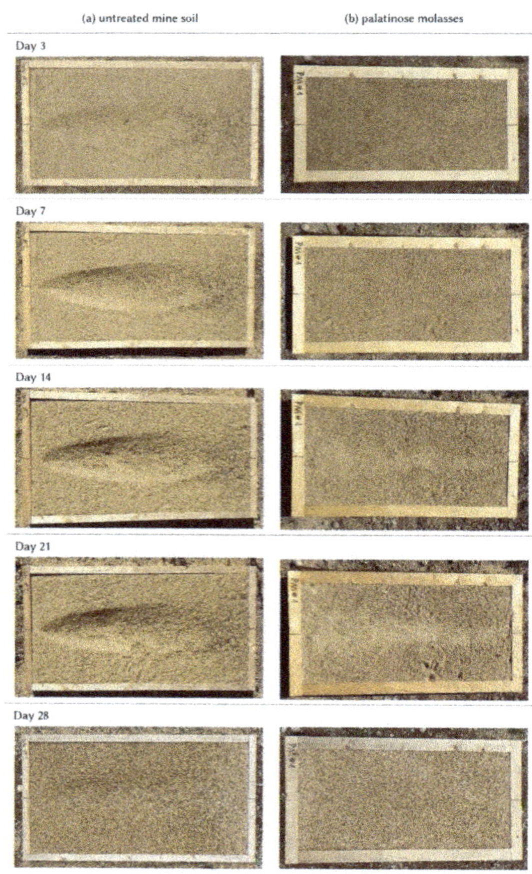

Figure 7. Photographic images of soil surfaces observed on day 3, 7, 14, 21, and 28 after the dust suppressants were applied: (**a**) untreated mine soil, and (**b**) mine soil treated with palatinose molasses. The size of each plot is 0.7 m × 0.4 m.

4. Discussion

4.1. Laboratory Experiments

Results of this study demonstrate that exposure of treated mine soils to different rainfall intensities influenced the amount of soil loss. In particular, increasing rainfall intensities from 0.25 to 4 L/m^2 led to decreasing soil losses (Figure 4). Possible explanations for this reduced wind erosion might be that greater amounts of applied water led to greater particle cohesion in treated mine soils, or that the infiltrated water had not completely evaporated. As shown in several studies (e.g., [21–25]), higher moisture content improves soil cohesion and thus the resistance to wind erosion. The increase in soil cohesion can be attributed to adsorptive and capillary forces [25,26]. Regardless, it appears that more water retained in the soil samples improved the inter-particle bonding. This effect might be intensified by the food processing by-products as they consist of carbohydrates and proteins, which can augment the water retention capacity of soils and retard moisture loss from soils [27,28].

By contrast, exposure of treated samples to repeated wet–dry cycles resulted in increased soil losses and thus reduced the effectiveness of tested food processing by-products to suppress dust emissions (Figure 5). It appears that water applied at frequent intervals leached the water-soluble amendments from the substrates to such an extent that it reduced the cohesion between soil particles. This created soil surfaces that were less resistant to aerodynamic lift forces and saltation bombardment, resulting in higher soil losses. In fact, saltating sand grains are mainly responsible for aeolian dust entrainment [29–31]. Further analysis of the soil loss data (Table A2, Appendix A) reveals differences between the food processing by-products in their susceptibility to leaching. While chicory vinasses showed the highest soil losses, palatinose molasses, corn steep liquor and decantation syrup were less prone to leaching. Such different responses to wetting and rainfall intensities may be explained by the composition of the tested additives. The chosen biomaterials comprise mixtures of carbohydrates and proteins, which have different affinities to water molecules. Mono- and disaccharides provide multiple hydroxyl groups that form hydrogen bonds with water [32]. Structural features such as branching structure, charged groups and, molecular weight affect the solubility of polysaccharides, which can promote or hinder intra- and intermolecular association [33]. Protein solubility is primarily influenced by the hydrophobic and hydrophilic amino acid residues exposed to water [34,35]. However, it is not possible to explain why some biomaterials are more water-soluble than others because the exact material compositions are not accessible.

4.2. Field Trials

Field experiments are subject to interferences that need to be considered when interpreting and comparing the effectiveness of dust suppressants. The test site is located close to an open-pit mine and to other sources of PM (e.g., unpaved roads), which could have led to dust deposition between measurements. However, it can be expected that such an additional dust input would have impacted all soil plots in a similar manner. Observed physical impairments of soil plot surfaces caused by wildlife may also be a potential source of error (Figure 7b). Disturbances affect the erodibility of soil surfaces and thus the generation of dust emissions [36,37]. Consequently, results should be interpreted cautiously as to which amendments were most effective in suppressing dust emissions.

Conducted field trials demonstrate that investigated food processing by-products can significantly reduce wind erosion and dust emissions ($PM_{2.5}$, PM_{10}, and TSP) from coarse mine soils (Figure 6). The observed decrease in dust emissions is likely attributed to carbohydrates and proteins contained in the tested amendments, which agglomerate soil particles through adhesive and cohesion forces. The aforementioned carbohydrates and proteins likely bind to soil particles upon evaporation of applied solutions. The resulting improvement in inter-particle bonding enhances the soil's mechanical properties, creating a surface with a higher erosion threshold [16].

Field trial data also demonstrate that dust emissions from soil plots treated with food processing by-products progressively increased with time. Such a decrease in dust suppression effectiveness probably resulted from reactions at the soil surface, which weakened the bonding between soil particles and thus reduced the resistance to wind erosion. It is assumed that rainfall during the study period (Table A1, Appendix A) resulted in the leaching of the amendments from the soils since they are water-soluble. However, visual inspections of the treated soil plots show intact surfaces unlike that of the untreated soil (Figure 7). Hence, natural rainfall did not impact on the integrity of treated soil surfaces, allowing amendments to bind surficial soil particles and prevent their erosion.

By contrast to the treated soil plots, dust emissions from the untreated mine soil and mine soil treated only with water constantly decreased with time (Figure 6). These trends can be explained by the observed changes in surface textures (Figure 7a,b) and the experimental setup. Near-surface wind velocity is crucial for the dynamics of wind erosion [38]. Soil particles are detached and transported when the aerodynamic drag and lift forces exceed the gravitational and inter-particle forces [39]. Moreover, dust particles are entrained by the impact of saltating sand grains [29]. During the study period, erosion of the control group plots formed conical-shaped depressions and changes in horizontal soil thickness (Figure 7a). In parallel, the position and orientation of the electric blower did not change. Hence, it is assumed that the wind velocity at the soil surface decreased over time, which reduced the forces available to erode soil and consequently dust emissions from water-treated and untreated soil plots.

At the end of the field trial (i.e., after 28 days), the lowest dust concentrations in air were measured for all soil plots, including the control group. From the meteorological data (Table A1, Appendix A), it can be concluded that previous rainfall (i.e., 27.7 L/m^2 between days 25–27) significantly increased the moisture content of the soil. Soil moisture directly contributes to soil cohesion and substantially affects the resistance to wind erosion [21]. It is generally recognized that an increase in inter-particle forces is induced by the absorbed water film and capillary forces [26]. However, Ravi et al. [40] state that capillary forces are mainly responsible for increasing cohesion forces in sandy soils, and water adsorption around the soil particles can be neglected when the soil is relatively wet.

Only a few studies have conducted field experiments simulating wind erosion on exposed mine soils treated with biodegradable reagents. Hence, only limited field trial data on dust suppressants exist to compare the study results. Chang et al. [41] have already stated a lack of studies evaluating the durability and reliability of biopolymers for geotechnical applications (i.e., dust control, soil stabilization, erosion control) under field conditions. For example, Park et al. [42] tested two biocompatible polymers, polyethylene glycol and poloxamer, to suppress wind-blown dust emissions from a bauxite tailings substrate. Their experiments showed that polyethylene glycol and poloxamer solutions could reduce PM_{10} emissions by up to 44% and 66%, respectively, compared to water-treated samples. Katra [8] demonstrated similar performances for calcium lignosulfonate and tree resin on bedding substrate used for unpaved roads in a calcareous quarry. Conducted wind tunnel experiments indicated a 63% and 53% reduction in PM_{10} emissions from lignosulfonate and tree resin solutions, respectively, compared to untreated samples [8]. However, only limited conclusions can be drawn due to partially substantial distinctions in experimental design, substrate type, climatic conditions, and other factors. Regardless, the chosen food processing by-products displayed a distinct ability to suppress PM emissions in the field (Figure 6).

4.3. *Possible Applications of Food Processing By-Products as Dust Suppressants at Mine Sites*

4.3.1. Technical Aspects

The present study demonstrates that food processing by-products can considerably suppress dust emissions from sandy mine soils (Figure 6). However, the results also demonstrate that the tested biomaterials are more likely to be used only for temporary or short-term dust control practices due to their susceptibility to be leached from soil surfaces upon rainfall events and possible degradation in the environment. Factors that may contribute to the degradation of biomaterials include solar radiation, heat, abiotic oxidation, biotic reactions of microorganisms and saltating soil particles' impact force [1,5,43–45]. Repetitive application of the amendments to mine soils may counteract their diminishing dust control effectiveness over time.

Considering their water-solubility, tested amendments could either be sprayed on substrate surfaces or incorporated into mine soils, which is usually applied on traffic areas [1]. This method involves higher single application costs but extends the service life of treatments [46,47]. However, it appears challenging to exploit this durability benefit due to the amendments' susceptibility to leaching and biodegradability.

The fact that the tested reagents could likely be applied using existing application equipment (e.g., water trucks with spray bars) already available at many mine sites is promising. The biomaterials readily dissolve in water and have not notably impaired the rheology of the aqueous solution. By contrast, thermal-gelling biopolymers, such as gellan gum or agar gum, require heated water to be dissolved or applied, respectively, and yield highly viscous solutions even at low concentrations [11,12]. As Chang et al. [41] already stated, any application equipment may need to be modified or developed considering the rheological properties of biopolymers.

The tested biomaterials contain carbohydrates and proteins, which promote seed germination and plant growth and serve as sources of energy for soil bacteria [48,49]. Therefore, an application of food processing by-products can provide nutrients to plants while simultaneously stabilizing the soil surface until the vegetation cover has been developed. Hence, the use of food processing by-products in mine closure and mine site rehabilitation activities might also support revegetation efforts of mined lands.

4.3.2. Financial Considerations

The application of food processing by-products as dust suppressants appears to have financial benefits compared to commercially available biopolymers. By-products from the food industry are usually inexpensive, generated in high volumes, and readily available [13–15]. Chicory vinasses, for example, is priced at less than 100 USD/t (data according to the supplier). By contrast, the biopolymer xanthan gum, extensively investigated for soil stabilization and wind erosion control in previous studies, ranges in price from 1500 to 4000 USD/t [41,50]. However, growing demand for biopolymers and the associated increase in production volume are expected to reduce biopolymer prices in the future [41,50,51].

Specific emphasis should be paid to equipment operating costs, which can frequently escalate the total costs of dust suppression [52]. Here, significant cost drivers include the suppressants' durability and their related re-application intervals required to sustain a sufficient level of dust control [46]. Compared to biopolymers, tested biomaterials will likely need shorter re-application intervals because rainfall may leach them from surface substrates more easily. Some biopolymers have thermal-gelation properties, such as gellan gum or agar gum, which only dissolve entirely in hot water [53]. Likewise, proteins (e.g., casein) with lower hydrophilicity probably provide higher resistance to rainfall and subsequent leaching [54]. However, a reliable cost-effectiveness assessment requires further research, considering biomaterials durability, application intervals and application rates. One possible approach to progress this knowledge would be the economic evaluation model developed by Thompson and Visser [46], which enables the costing of the establishment, application and maintenance activities associated with the use of dust suppressants.

4.3.3. Environmental Issues and Implications

Even though this study shows encouraging results for applying food processing by-products as dust suppressants, potential adverse effects and issues associated with the future application on mine soils need to be considered. Rainfall events could leach the tested biomaterials through soils to receiving ground and surface waters, adding an organic load to aquatic ecosystems [9,55]. Organic matter decreases dissolved oxygen concentration in freshwater bodies due to bacteria and other microorganisms' growth, resulting in stresses on aquatic life [56]. Also, carbohydrates and proteins in the tested reagents could attract animals and insects or cause problems with mould growth on the treated surface or inside the application equipment [7,57].

If such amendments are applied to tailings surfaces, further problems could arise. Recycled water from tailings storage facilities may contain appreciable concentrations of dust suppressant residues, which could adversely affect mineral processing methods such as flotation due to reactions with other reagents [42]. Park et al. [42] also remarked that polymers could impair tailings ponds stability because of augmented water retention and associated changes in substrate properties (e.g., pore pressure).

4.4. Outlook

Any comparison of field trial data on dust suppressants is hampered by the fact that different studies pursue individual experimental designs and measurement methods. Hence, there is a clear need for standardized simulated dust suppression tests that provide information that can be used to predict dust emissions from soils. Such a method would be used to determine probable dust dispersion and develop cumulative dust emission and suppression data to support mine permit application requirements. Those methods could also be tools to generate data used to design and implement best management practices and treatment processes needed by mining operations to meet compliance requirements.

In the mining industry context, a standard testing methodology should be feasible in remote and relatively inaccessible areas (e.g., tailings sites), easy to implement and allow for comparative sampling of various dust control products with minor financial and personnel requirements. Accordingly, experimental setups with field boundary layer wind tunnels, usually used to quantify dust emissions from soils, are not expedient because they are difficult to transport, labor-intensive and pose substantial methodological challenges [58,59].

By contrast, methodologies adopting soil plots eroded with commercial electric blowers do pose suitable alternatives, even though measured dust emissions do not reflect dust emissions caused by natural wind erosion since electric blowers do not simulate the atmospheric flow. Those experiments can establish the effectiveness of dust suppressants over different test intervals at locations where they are intended to be applied, indicative of possible commercial-scale application and practice [59]. Moreover, such field trials are easy to conduct and enable cost-effective and low-maintenance testing of various soil treatments with replicates. Future steps that need to be taken to establish a standardized methodology are (i) harmonization of experimental parameters (e.g., applied wind speed, exposure time to simulated wind erosion) and sample preparation, and (ii) definition of requirements on electric blowers (e.g., outlet diameter), dust monitors and test site. It is also recommended to compare measurements of the proposed methodology with the portable in situ wind erosion lab (PI-SWERL) concept. The PI-SWERL is a cost-effective method to assess the potential of soil surfaces to generate dust emissions and correlates with traditional wind tunnel tests [60].

5. Conclusions

This study investigated the effectiveness of four biodegradable by-products from the food industry in suppressing fugitive dust emissions from mine soils. The following main conclusions can be drawn:

1. Carbohydrate- and protein-rich by-products from food processing reduce PM emissions from sandy mine soils as tested in the laboratory and field trials, indicating their potential to be applied as dust suppressants at mine and mineral processing sites.
2. Rainfall impairs the effectiveness of tested biomaterials due to progressive leaching and loss of the amendments from treated soils. While rain may decrease the additives' concentration in surface layers, treated soil surfaces remain physically intact.

3. Soil loss data from laboratory tests indicate that biomaterials' effectiveness decreases with repetitive wet–dry cycles. Exposure of treated mine soils to increasing rainfall intensities led to decreasing soil losses, possibly due to enhanced inter-particle bonding, incomplete evaporation of infiltrated water or the augmented water retention capacity of the soil.
4. There is a clear need for standardized simulated dust suppression tests.

In conclusion, this study demonstrates that food processing by-products can be used for short-term dust mitigation at mine and mineral processing sites. Their application as dust suppressants would reduce dust control expenses and diminish chemical products detrimental to the environment.

Author Contributions: Conceptualization, J.F., J.L.S. and M.B.; methodology, J.F. and J.L.S.; formal analysis, J.F.; investigation, J.F. and M.L.; data curation, J.F.; writing—original draft preparation, J.F.; writing—review and editing, J.F., J.L.S. and B.G.L.; visualization, J.F.; project administration, J.F. All authors have read and agreed to the published version of the manuscript.

Funding: This research received no external funding.

Institutional Review Board Statement: Not applicable.

Informed Consent Statement: Not applicable.

Data Availability Statement: The data that support the findings of this study are available from the corresponding author, upon reasonable request.

Acknowledgments: The authors would like to thank the anonymous reviewers for their helpful comments and suggestions.

Conflicts of Interest: The authors declare no conflict of interest.

Appendix A

Table A1. Meteorological data for the study site and study period (12:00, 17 July 2020–12:00, 14 August 2020).

Days after Application	Date	Temperature (°C)			Humidity (%)	Global Radiation (W/m^2)	Precipitation (L/m^2)
		Min	Max	Mean			
0	17.07.20	18.61	21.84	20.10	68.48	217.51	0.01
1	18.07.20	12.90	27.56	21.21	65.89	292.44	0.29
2	19.07.20	15.21	28.30	22.24	59.82	278.16	0.08
3	20.07.20	14.33	22.48	17.78	68.82	135.69	0.00
4	21.07.20	10.04	21.78	16.04	62.25	306.56	0.05
5	22.07.20	9.80	22.06	16.37	62.31	293.56	0.00
6	23.07.20	9.35	25.89	18.69	56.23	310.89	0.18
7	24.07.20	17.29	24.84	20.46	58.70	202.74	0.02
8	25.07.20	14.94	25.30	21.05	64.87	109.04	0.21
9	26.07.20	16.04	23.25	19.53	72.65	220.62	4.80
10	27.07.20	12.87	30.25	21.40	58.40	175.06	0.74
11	28.07.20	16.46	23.63	20.45	57.46	222.27	0.01
12	29.07.20	13.19	23.39	18.29	59.13	267.38	0.04
13	30.07.20	10.60	28.22	20.07	55.16	298.25	0.04
14	31.07.20	15.15	35.56	25.78	42.94	309.94	0.06
15	01.08.20	21.46	30.95	25.43	55.01	205.48	0.67
16	02.08.20	16.92	23.92	20.33	63.76	191.84	0.01
17	03.08.20	10.99	23.27	17.73	61.83	202.76	0.09
18	04.08.20	12.98	23.33	17.79	60.71	279.16	0.02
19	05.08.20	12.39	30.37	21.79	44.47	300.26	0.03
20	06.08.20	17.17	31.67	24.82	39.82	290.72	0.02
21	07.08.20	16.91	33.98	26.52	41.35	293.58	0.07
22	08.08.20	17.54	35.84	28.04	42.61	282.54	0.03

Table A1. *Cont.*

Days after Application	Date	Temperature (°C)			Humidity (%)	Global Radiation (W/m²)	Precipitation (L/m²)
		Min	Max	Mean			
23	09.08.20	21.27	33.20	25.08	58.56	114.01	4.25
24	10.08.20	19.09	33.44	26.74	58.13	242.12	0.02
25	11.08.20	21.63	34.42	26.65	57.93	179.42	1.38
26	12.08.20	19.86	31.68	23.68	75.36	182.34	17.48
27	13.08.20	20.05	31.98	24.84	71.71	196.22	8.84
28	14.08.20	18.62	25.79	21.84	74.67	96.42	0.00

Table A2. Total weight loss (g) of mine soils treated with different food processing by-products and rainfall intensities (arithmetic mean with standard deviation, n = 3). The quantitative dust control effectiveness represents the ratio of total weight loss between treated and untreated mine soil. Untreated mine soil showed a total weight loss of 125.28 g after four measuring cycles.

Rainfall Intensity	Total Weight Loss (g)		Dust Control Effectiveness (%)
	M	SD	
	Chicory vinasses		
0.25 L/m²	63.04	31.29	49.68
1 L/m²	5.35	2.72	95.73
2 L/m²	0.60	0.09	99.52
4 L/m²	2.12	0.40	98.31
	Corn steep liquor		
0.25 L/m²	4.38	1.12	96.50
1 L/m²	1.24	0.27	99.01
2 L/m²	0.70	0.23	99.44
4 L/m²	2.38	0.96	98.10
	Decantation syrup		
0.25 L/m²	2.79	0.79	97.78
1 L/m²	0.21	0.08	99.84
2 L/m²	0.27	0.05	99.78
4 L/m²	0.49	0.22	99.61
	Palatinose Molasses		
0.25 L/m²	13.52	3.27	89.21
1 L/m²	0.81	0.20	99.35
2 L/m²	0.25	0.06	99.80
4 L/m²	0.50	0.11	99.60

Note. M = mean, SD = standard deviation.

Table A3. Summary of soil loss (g) from treated mine soils after each wet–dry cycle at different rainfall intensities (arithmetic mean, n = 3).

Rainfall Intensity	Soil Loss (g)			
	WDC-1	WDC-2	WDC-3	WDC-4
		Chicory vinasses		
0.25 L/m²	2.68	20.37	13.03	26.97
1 L/m²	4.65	0.38	0.07	0.25
2 L/m²	0.18	0.07	0.04	0.32
4 L/m²	0.74	0.56	0.17	0.65
		Corn steep liquor		
0.25 L/m²	0.08	0.19	1.14	2.98
1 L/m²	1.12	0.04	0.04	0.05
2 L/m²	0.31	0.05	0.05	0.29
4 L/m²	0.34	0.21	0.24	1.58

Table A3. *Cont.*

Rainfall Intensity	Soil Loss (g)			
	WDC-1	WDC-2	WDC-3	WDC-4
		Decantation syrup		
0.25 L/m^2	0.58	0.80	0.19	1.21
1 L/m^2	0.11	0.01	0.07	0.02
2 L/m^2	0.08	0.07	0.05	0.07
4 L/m^2	0.22	0.12	0.11	0.03
		Palatinose molasses		
0.25 L/m^2	5.18	2.52	1.34	4.48
1 L/m^2	0.61	0.02	0.01	0.17
2 L/m^2	0.11	0.05	0.04	0.05
4 L/m^2	0.27	0.11	0.07	0.06

Note. WDC = wet–dry cycle.

Table A4. Mean concentrations of dust particles ($PM_{2.5}$, PM_{10}, TSP) in air from mine soils treated with various food processing by-products. Dust emissions were measured 3, 7, 14, 21, and 28 days after application of dust suppressants.

	Concentrations of Dust Particles (mg/m^3)														
	$PM_{2.5}$	PM_{10}	TSP	$PM_{2.5}$	PM_{10}	TSP	$PM_{2.5}$	PM_{10}	TSP	$PM_{2.5}$	PM_{10}	TSP	$PM_{2.5}$	PM_{10}	TSP
	Day 3			Day 7			Day 14			Day 21			Day 28		
							Chicory vinasses								
M	0.031	0.053	0.093	0.039	0.049	0.070	0.048	0.088	0.180	1.863	2.148	2.697	0.022	0.034	0.078
SD	0.065	0.094	0.171	0.039	0.045	0.068	0.154	0.306	0.725	3.274	3.519	3.933	0.055	0.076	0.194
							Corn steep liquor								
M	0.118	0.174	0.303	0.154	0.234	0.374	0.617	0.796	1.123	2.351	3.184	5.169	0.025	0.051	0.096
SD	0.340	0.438	0.697	0.749	1.042	1.586	0.943	1.135	1.522	6.043	7.872	12.140	0.048	0.109	0.232
							Decantation syrup								
M	0.016	0.023	0.036	0.124	0.191	0.301	0.055	0.085	0.150	0.900	1.756	2.896	0.011	0.017	0.028
SD	0.024	0.040	0.083	0.223	0.632	1.364	0.152	0.232	0.495	5.047	9.799	15.472	0.008	0.015	0.041
							Palatinose molasses								
M	0.017	0.023	0.032	0.050	0.067	0.104	0.088	0.144	0.247	3.152	5.003	8.483	0.014	0.018	0.030
SD	0.043	0.057	0.085	0.050	0.075	0.152	0.138	0.204	0.354	11.445	14.409	22.296	0.026	0.028	0.037
							Pure water								
M	20.981	27.649	39.176	9.197	11.297	15.114	1.441	1.623	1.905	1.335	1.756	2.849	0.027	0.039	0.064
SD	20.422	24.394	32.320	9.938	11.206	13.960	2.926	3.170	3.554	2.241	2.747	4.265	0.044	0.054	0.098
							Untreated soil								
M	32.078	59.026	77.075	10.926	19.476	28.218	0.980	1.142	1.461	0.242	0.360	0.603	0.013	0.024	0.043
SD	31.216	52.371	58.956	8.332	14.301	20.837	1.224	1.326	1.561	0.433	0.717	1.256	0.020	0.033	0.063
							Background load								
M	0.008	0.009	0.010	0.016	0.019	0.023	0.019	0.025	0.032	0.018	0.021	0.026	0.007	0.009	0.014
SD	0.003	0.006	0.010	0.067	0.068	0.080	0.005	0.010	0.026	0.003	0.006	0.014	0.002	0.006	0.026

Note. M = mean, SD = standard deviation.

References

1. Piechota, T.; van Ee, J.; Batista, J.; Stave, K.; James, D. *Potential Environmental Impacts of Dust Suppressants: Avoiding Another Times Beach*; University of Nevada: Las Vegas, NV, USA, 2004.
2. Bolander, P.; Yamada, A. *Dust Palliative Selection and Application Guide*; San Dimas Technology & Development Center: San Dimas, CA, USA, 1999.
3. Yousif, E.; Haddad, R. Photodegradation and photostabilization of polymers, especially polystyrene: Review. *Springerplus* **2013**, *2*, 398. [CrossRef] [PubMed]

4. Nair, N.R.; Sekhar, V.C.; Nampoothiri, K.M.; Pandey, A. Biodegradation of biopolymers. In *Current Developments in Biotechnology and Bioengineering*; Pandey, A., Negi, S., Soccol, C.R., Eds.; Elsevier: Amsterdam, The Netherlands, 2017; pp. 739–755. ISBN 978-0-444-63662-1.
5. Edvardsson, K.; Gustafsson, A.; Magnusson, R. Dust suppressants efficiency study: In situ measurements of dust generation on gravel roads. *Int. J. Pavement Eng.* **2012**, *13*, 11–31. [CrossRef]
6. Monlux, S.; Mitchell, M. Chloride stabilization of unpaved road aggregate surfacing. *Transp. Res. Rec.* **2007**, *1989–2*, 50–58. [CrossRef]
7. Gillies, J.A.; Watson, J.G.; Rogers, C.F.; DuBois, D.; Chow, J.C.; Langston, R.; Sweet, J. Long-term efficiencies of dust suppressants to reduce PM10 emissions from unpaved roads. *J. Air Waste Manag.* **1999**, *49*, 3–16. [CrossRef] [PubMed]
8. Katra, I. Comparison of diverse dust control products in wind-induced dust emission from unpaved roads. *Appl. Sci.* **2019**, *9*, 5204. [CrossRef]
9. Chang, I.; Im, J.; Cho, G.-C. Geotechnical engineering behaviours of gellan gum biopolymer treated sand. *Can. Geotech. J.* **2016**, *53*, 1658–1670. [CrossRef]
10. Mudgil, D.; Barak, S.; Khatkar, B.S. Guar gum: Processing, properties and food applications—A Review. *J. Food Sci. Technol.* **2014**, *51*, 409–418. [CrossRef]
11. Hataf, N.; Ghadir, P.; Ranjbar, N. Investigation of soil stabilization using chitosan biopolymer. *J. Clean. Prod.* **2018**, *170*, 1493–1500. [CrossRef]
12. Kunz, B.K.; Little, E.E.; Barandino, V.L. Aquatic toxicity of chemical road dust suppressants to freshwater organisms. *Arch. Environ. Contam. Toxicol.* **2021**, *82*, 294–305. [CrossRef]
13. Mahro, B.; Gaida, B.; Schüttmann, I.; Zorn, H. Survey of the amount and use of biogenic residues of the German food and biotech industry. *Chem. Ing. Tech.* **2015**, *87*, 537–542. [CrossRef]
14. Anal, A.K. *Food Processing by-Products and their Utilization*, 1st ed.; Wiley: Chichester, UK, 2017; ISBN 9781118432921.
15. Jõgi, K.; Bhat, R. Valorization of food processing wastes and by-products for bioplastic production. *Sustain. Chem. Pharm.* **2020**, *18*, 100326. [CrossRef]
16. Freer, J.; Bucher, P.G.; Braun, M.; Lottermoser, B.G. Food processing by-products and wastes as potential dust suppressants at mine sites: Results from unconfined compressive strength testing. *J. Air Waste Manag.* **2022**, *72*, 1012–1026. [CrossRef]
17. Freer, J.; Sieger, J.L.; Lottermoser, B.G.; Braun, M. Effectiveness of Food Processing by-Products in Suppressing Wind-Induced Dust Emissions from Mine Soils: Results from Laboratory Wind Tunnel Experiments. 2022. Available online: https://ssrn.com/abstract=4220674 (accessed on 27 September 2022).
18. German Institute for Standardization. *Geotechnical Investigation and Testing—Identification and Classification of Soil—Part 1: Identification and Description*; Beuth: Berlin, Germany, 2018; DIN EN ISO 14688-1:2018-05.
19. Si, Z.; Herrera, C.H. Laboratory and field evaluation of base stabilization using cement kiln dust. *Transp. Res. Rec.* **2007**, *1989*, 42–49. [CrossRef]
20. Van Pelt, R.S.; Zobeck, T.M. Portable wind tunnels for field testing of soils and natural surfaces. In *Wind Tunnel Designs and Their Diverse Engineering Applications*; Ahmed, N., Ed.; InTechOpen: London, UK, 2013; ISBN 978-953-51-1047-7.
21. Neuman, C.M.; Sanderson, S. Humidity control of particle emissions in aeolian systems. *J. Geophys. Res.* **2008**, *113*, F02S14. [CrossRef]
22. Chepil, W.S. Influence of Moisture on Erodibility of Soil by Wind. *Soil Sci. Soc. Am. J.* **1956**, *20*, 288–292. [CrossRef]
23. Yuge, K.; Anan, M. Evaluation of the effect of wind velocity and soil moisture condition on soil erosion in andosol agricultural fields (model experiment). *Water* **2019**, *11*, 98. [CrossRef]
24. Bisal, F.; Hsieh, J. Influence of moisture on erodibility of soil by wind. *Soil Sci.* **1966**, *102*, 143–146. [CrossRef]
25. Chen, W.; Zhibao, D.; Zhenshan, L.; Zuotao, Y. Wind tunnel test of the influence of moisture on the erodibility of loessial sandy loam soils by wind. *J. Arid Environ.* **1996**, *34*, 391–402. [CrossRef]
26. Fécan, F.; Marticorena, B.; Bergametti, G. Parametrization of the increase of the aeolian erosion threshold wind friction velocity due to soil moisture for arid and semi-arid areas. *Ann. Geophys.* **1999**, *17*, 149–157. [CrossRef]
27. Toufigh, V.; Ghassemi, P. Control and stabilization of fugitive dust: Using eco-friendly and sustainable materials. *Int. J. Geomech.* **2020**, *20*, 4020140. [CrossRef]
28. Chen, R.; Lee, I.; Zhang, L. Biopolymer stabilization of mine tailings for dust control. *J. Geotech. Geoenviron. Eng.* **2015**, *141*, 4014100. [CrossRef]
29. Kok, J.F.; Parteli, E.J.R.; Michaels, T.I.; Karam, D.B. The physics of wind-blown sand and dust. *Rep. Prog. Phys.* **2012**, *75*, 106901. [CrossRef] [PubMed]
30. Lu, H.; Shao, Y. A new model for dust emission by saltation bombardment. *J. Geophys. Res.* **1999**, *104*, 16827–16842. [CrossRef]
31. Shao, Y.; Raupach, M.R.; Findlater, P.A. Effect of saltation bombardment on the entrainment of dust by wind. *J. Geophys. Res.* **1993**, *98*, 12719–12726. [CrossRef]
32. BeMiller, J.N. *Carbohydrate Chemistry for Food Scientists*, 3rd ed.; WP: London, UK, 2019; ISBN 978-0-12-812069-9.
33. Xu, Z. (Ed.) *Solubility of Polysaccharides*; InTechOpen: London, UK, 2017; ISBN 978-953-51-3649-1.
34. Kumitch, H.M.; Stone, A.K.; Nickerson, M.T.; Korber, D.R.; Tanaka, T. Effect of fermentation time on the physicochemical and functional properties of pea protein-enriched flour fermented by Aspergillus oryzae and Aspergillus niger. *Cereal Chem.* **2020**, *97*, 416–428. [CrossRef]

35. Zayas, J.F. *Functionality of Proteins in Food*; Springer: Berlin/Heidelberg, Germany, 1997; ISBN 978-3-642-59116-7.
36. Belnap, J.; Gillette, D.A. Vulnerability of desert biological soil crusts to wind erosion: The influences of crust development, soil texture, and disturbance. *J. Arid Environ.* **1998**, *39*, 133–142. [CrossRef]
37. Zhang, Y.M.; Wang, H.L.; Wang, X.Q.; Yang, W.K.; Zhang, D.Y. The microstructure of microbiotic crust and its influence on wind erosion for a sandy soil surface in the Gurbantunggut Desert of Northwestern China. *Geoderma* **2006**, *132*, 441–449. [CrossRef]
38. Zhang, G.; Azorin-Molina, C.; Shi, P.; Lin, D.; Guijarro, J.A.; Kong, F.; Chen, D. Impact of near-surface wind speed variability on wind erosion in the eastern agro-pastoral transitional zone of Northern China, 1982–2016. *Agric. For. Meteorol.* **2019**, *271*, 102–115. [CrossRef]
39. Shao, Y.; Lu, H. A simple expression for wind erosion threshold friction velocity. *J. Geophys. Res.* **2000**, *105*, 22437–22443. [CrossRef]
40. Ravi, S.; D'Odorico, P.; Over, T.M.; Zobeck, T.M. On the effect of air humidity on soil susceptibility to wind erosion: The case of air-dry soils. *Geophys. Res. Lett.* **2004**, *31*. [CrossRef]
41. Chang, I.; Lee, M.; Tran, A.T.P.; Lee, S.; Kwon, Y.-M.; Im, J.; Cho, G.-C. Review on biopolymer-based soil treatment (BPST) technology in geotechnical engineering practices. *Transp. Geotech.* **2020**, *24*, 100385. [CrossRef]
42. Park, J.; Kim, K.; Lee, T.; Kim, M. Tailings storage facilities (TSFs) dust control using biocompatible polymers. *Min. Metall. Explor.* **2019**, *36*, 785–795. [CrossRef]
43. Tian, K.; Bilal, M. Research progress of biodegradable materials in reducing environmental pollution. In *Abatement of Environmental Pollutants: Trends and Strategies*; Elsevier: Amsterdam, The Netherlands, 2020; pp. 313–330, ISBN 978-0-12-818095-2.
44. Göpferich, A. Mechanisms of polymer degradation and erosion. *Biomaterials* **1996**, *17*, 103–114. [CrossRef]
45. Briassoulis, D. The effects of tensile stress and the agrochemical Vapam on the ageing of low density polyethylene (LDPE) agricultural films. Part I. Mechanical behaviour. *Polym. Degrad. Stab.* **2005**, *88*, 489–503. [CrossRef]
46. Rushing, J.F.; Tingle, J.S. Evaluation of products and application procedures for mitigating dust in temperate climates. *Transp. Res. Rec.* **2007**, *1989-1*, 305–311. [CrossRef]
47. Thompson, R.J.; Visser, A.T. Selection, performance and economic evaluation of dust palliatives on surface mine haul roads. *J. S. Afr. Inst. Min. Metall.* **2007**, *107*, 435–450.
48. Chang, I.; Prasidhi, A.K.; Im, J.; Shin, H.-D.; Cho, G.-C. Soil treatment using microbial biopolymers for anti-desertification purposes. *Geoderma* **2015**, *253-254*, 39–47. [CrossRef]
49. Pylak, M.; Oszust, K.; Frąc, M. Review report on the role of bioproducts, biopreparations, biostimulants and microbial inoculants in organic production of fruit. *Rev. Environ. Sci. Biotechnol.* **2019**, *18*, 597–616. [CrossRef]
50. Chang, I.; Jeon, M.; Cho, G.-C.; Shankar, S. Application of microbial biopolymers as an alternative construction binder for earth buildings in underdeveloped countries. *Int. J. Polym. Sci.* **2015**, *2015*, 326745. [CrossRef]
51. Chang, I.; Im, J.; Cho, G.-C. Introduction of microbial biopolymers in soil treatment for future environmentally-friendly and sustainable geotechnical engineering. *Sustainability* **2016**, *8*, 251. [CrossRef]
52. Thompson, R.J.; Visser, A.T. Mine haul road fugitive dust emission and exposure characterisation. In Proceedings of the Environmental Health Risk 2003, Catania, Italy, 17–19 September 2003; Brebbia, C.A., Fayzieva, D., Eds.; WIT Press: Southampton, UK, 2003; pp. 103–112.
53. Chang, I.; Prasidhi, A.K.; Im, J.; Cho, G.-C. Soil strengthening using thermo-gelation biopolymers. *Constr. Build. Mater.* **2015**, *77*, 430–438. [CrossRef]
54. Chang, I.; Im, J.; Chung, M.-K.; Cho, G.-C. Bovine casein as a new soil strengthening binder from diary wastes. *Constr. Build. Mater.* **2018**, *160*, 1–9. [CrossRef]
55. Singh, V.; Piechota, T.; James, D. Hydrologic impacts of disturbed lands treated with dust suppressants. *J. Hydrol. Eng.* **2003**, *8*, 278–286. [CrossRef]
56. Li, P.; Karunanidhi, D.; Subramani, T.; Srinivasamoorthy, K. Sources and consequences of groundwater contamination. *Arch. Environ. Contam. Toxicol.* **2021**, *80*, 1–10. [CrossRef]
57. Jones, D. *Guidelines for the Selection, Specification and Application of Chemical Dust Control and Stabilization Treatments on Unpaved Roads*; University of California Pavement Research Center: Davis, CA, USA, 2017.
58. Kavouras, I.G.; Etyemezian, V.; Nikolich, G.; Gillies, J.; Sweeney, M.; Young, M.; Shafer, D. A new technique for characterizing the efficacy of fugitive dust suppressants. *J. Air Waste Manag.* **2012**, *59*, 603–612. [CrossRef]
59. McKenna Neuman, C.; Boulton, J.W.; Sanderson, S. Wind tunnel simulation of environmental controls on fugitive dust emissions from mine tailings. *Atmos. Environ.* **2009**, *43*, 520–529. [CrossRef]
60. Etyemezian, V.; Nikolich, G.; Ahonen, S.; Pitchford, M.; Sweeney, M.; Purcell, R.; Gillies, J.; Kuhns, H. The portable in situ wind erosion laboratory (PI-SWERL): A new method to measure PM10 windblown dust properties and potential for emissions. *Atmos. Environ.* **2007**, *41*, 3789–3796. [CrossRef]

applied sciences

MDPI

Article

Evaluation of Protein and Polysaccharide Biopolymers as Dust Suppressants on Mine Soils: Laboratory Experiments

Johannes L. Sieger *, Bernd G. Lottermoser and Justus Freer

Institute of Mineral Resources Engineering, RWTH Aachen University, 52062 Aachen, Germany; lottermoser@mre.rwth-aachen.de (B.G.L.); freer@mre.rwth-aachen.de (J.F.)
* Correspondence: sieger@mre.rwth-aachen.de

Abstract: Polysaccharide biopolymers have been shown to be alternatives to established dust suppressants. This study investigates the potential of 14 polysaccharides and proteins from diverse botanical (corn, pea, wheat, cellulose, potato, and fava bean) and animal (pig, chicken, and cow) sources as dust suppressants on two mine soils (medium-grained sand and fine-grained silica sand). Results of laboratory tests demonstrate that the type of biopolymer and its concentration have a significant effect on penetration resistance, moisture retention, and crust thickness. Depending on biopolymer type, concentration, and soil type, moisture retention range from 1.0 to 19.5 wt% (control, $C_{medium\text{-}gr.}$ = 2.5 and $C_{fine\text{-}gr.}$ = 6.9 wt%), penetration resistance from 1.7 to 37.9 N ($C_{medium\text{-}gr.}$ = 1.5 and $C_{fine\text{-}gr.}$ = 1.7 N), and crust thickness from 0.3 to 18.1 mm ($C_{medium\text{-}gr.}$ and $C_{fine\text{-}gr.}$ = 0 mm). Proteins form crusts with penetration resistances similar to polysaccharides but mainly require higher concentrations than polysaccharides. Based on the test results, xanthan gum, carboxymethyl cellulose, corn starch, fava bean protein concentrate, and plasma protein exhibit the highest potential to act as dust suppressants. This research contributes to evaluating biopolymers as environmentally friendly soil amendments that may be used to control fugitive dust emissions from large, barren surfaces.

Keywords: dust suppressant; dust control; biopolymer; penetration resistance; crust strength; moisture retention; mine soil; protein; polysaccharide

Citation: Sieger, J.L.; Lottermoser, B.G.; Freer, J. Evaluation of Protein and Polysaccharide Biopolymers as Dust Suppressants on Mine Soils: Laboratory Experiments. *Appl. Sci.* **2023**, *13*, 1010. https://doi.org/10.3390/app13021010

Academic Editor: Itzhak Katra

Received: 20 December 2022
Revised: 6 January 2023
Accepted: 7 January 2023
Published: 11 January 2023

1. Introduction

Fugitive dust emissions from industries such as mining pose a significant threat to the environment and the health of workers and surrounding communities [1–4]. These emissions can cause respiratory diseases such as pneumoconiosis, asthma, or chronic obstructive pulmonary disease (COPD) [5], increase vehicle maintenance, and reduce occupational safety by limiting visibility [6,7]. As climate change increases the frequency and severity of extreme weather events [8,9], fugitive dust emissions are expected to rise in the coming decades. This is because prolonged heat waves lead to a faster depletion of soil moisture, increasing the susceptibility of soil particles to be suspended and become fugitive dust. Industries, such as mining, quarrying, construction, and agriculture, constitute the main sources of anthropogenic fugitive industrial (non-combustion) dust [10]. By their very nature, these domains encompass large, barren surfaces with a scarce vegetative cover exposed to frequent windblow and mechanical disturbances. Hence, these industries are challenged to reduce fugitive dust emissions at their sites.

Measures such as (vegetative) barriers or encapsulations reduce fugitive dust emissions but are unsuitable for protecting large, exposed areas or mitigating emissions caused by mechanical disturbance. In mining, the spray-on application of water is the oldest yet-established means for decreasing dust emissions, but its effect rapidly diminishes upon evaporation. Dust suppressants constitute an alternative solution for controlling dust emissions. They act by either agglomerating small particles, making them less prone to be suspended in the air, or are hygroscopic, absorbing moisture from the air to increase soil moisture [11]. However, many traditional dust suppressants, such as chloride salts or petroleum-based products, can adversely affect the environment [12] or are costly, resulting in the need for environmentally friendly and cost-effective alternatives.

Recently, the potential of biopolymers as environmentally friendly alternatives to traditional soil amendments has been increasingly investigated in the fields of soil stabilisation and dust control (e.g., [13–17]). Biopolymers are produced by living organisms, such as plants, microbes, and animals, are biodegradable, and can be classified into polysaccharides, proteins, and polynucleotides (e.g., deoxyribonucleic acid (DNA) and ribonucleic acid (RNA)) [18]. Soil stabilisation refers to the process of mixing soil amendments into the soil to enhance its engineering properties, while dust control refers to the spray-on application of dust suppressants [19]. Table A1 provides a comprehensive review of previous studies on different biopolymer types, and several reviews summarise the current research in the respective fields [20–24].

Based on previous studies (cf. Table A1) and reviews, the current knowledge on biopolymers tested in soil stabilisation and dust control can be summarised as follows. (a) Most research focuses on polysaccharides, with xanthan and guar gum being the most studied, whereas only a few studies investigated proteins. Polynucleotides have not yet been investigated, likely because of their extremely high cost. (b) Many previous studies examined biopolymers from botanical sources native to tropical, arid, and partially temperate climates (e.g., guar, Persian, acacia, and locust bean gum, or soybeans), hindering the ability of regional sourcing in regions with a continental climate, such as central and eastern Europe. (c) Some tested biopolymers are only dissolvable at high temperatures (>80 °C) or in acetic or alkaline solutions. While such properties make these biopolymers more effective or resistant to degradation [25], such dissolution behaviour prevents large-scale field testing and site applications. (d) Most studies have focused on extensive testing with a single or a few biopolymers, with only a few comparative studies analysing multiple biopolymers.

Based on the current knowledge, the following research needs can be identified. (a) Research on underrepresented polysaccharides and especially proteins should be expanded to identify further biopolymers with potential as dust suppressants. (b) Research on biopolymers that can be sourced regionally in areas with continental climates should be increased to identify alternative sources that may be procurable at a lower carbon footprint. (c) Emphasis should be on biopolymers that dissolve readily in water, enabling the conduct of large field tests under real site conditions.

This study aimed to investigate the dust suppression potential of selected polysaccharides and proteins (meeting the needs above) by performing laboratory experiments on local mine soils. Penetrometer tests were performed to measure the penetration resistance of the formed crusts, as it is an established and recommended indicator for evaluating potential dust suppressants [13,17,26,27]. Moisture retention tests were conducted to evaluate the ability of biopolymer-treated soil to retain moisture, which is a further relevant indicator for analysing potential dust suppressants [17,27,28]. Crust thickness measurements were performed to assess the ability of different biopolymer types and concentrations to agglomerate particles and form crusts. The results of this research contribute to the evaluation of polysaccharide and protein biopolymers as environmentally friendly dust suppressants on large, barren surfaces.

2. Materials and Methods

2.1. Materials

2.1.1. Soils

Medium-grained sand. A 2 t bulk sample of medium-grained sand was provided by the Rheinische Baustoffwerke GmbH, which represents a sand and gravel operation located 16 km northeast of Aachen, Germany. The bulk sample is representative of overburden material that is extracted during lignite open-pit mining in the region. The material was air-dried at room temperature and homogenised by coning and quartering. Relevant soil properties (i.e., pH, specific gravity, and soil colour) and grain size distributions were determined at RWTH Aachen University. The particle-size distribution was determined according to DIN EN ISO 17892-4 [29] (Table 1 and Figure 1) and based on the unified soil classification system (USCS); the material can be classified as medium-grained, poorly graded sand (SP). The sample's geochemistry was determined at ALS Geochemistry (Loughrea, Ireland), which performed whole-rock analysis by X-ray fluorescence spectroscopy (XRF) and inductively coupled plasma mass spectrometry (ICP-MS) with four-acid digestion. Its mineralogy was established by semi-quantitative X-ray diffraction (XRD) using an X'Pert Pro (PANalytical) instrument with a data collector and an X'Pert High Score system equipped with a Co-LFF (Empyrian) tube and an automated divergence slit (Clausthal University of Technology, Germany). The sand primarily consists of quartz with plagioclase, ankerite, and rutile as accessories.

Table 1. Physical and chemical properties of mine soils used by this study.

Parameter	Unit	Medium-Grained Sand	Fine-Grained Silica Sand	Test Method
		Soil properties		
D_{60}	mm	0.75	0.24	DIN EN ISO 17892-4 [29]
D_{50}	mm	0.63	0.22	DIN EN ISO 17892-4 [29]
D_{30}	mm	0.43	0.18	DIN EN ISO 17892-4 [29]
D_{10}	mm	0.28	0.14	DIN EN ISO 17892-4 [29]
C_u	-	2.73	1.78	DIN EN ISO 17892-4 [29]
C_c	-	0.91	0.95	DIN EN ISO 17892-4 [29]
USCS classification	-	SP	SP	ASTM D-2487 [30]
Specific gravity	g/cm^3	2.59	2.63	DIN EN ISO 11508:2018-04 [31]
pH value		7.53	6.48	DIN EN 15933:2012-11 [32]
Soil colour	Munsell	9.7 YR 6.0/2.8	0.9 Y 7.1/1.3	
		Geochemistry (oxides)		
SiO_2	wt%	94.58	98.65	
Al_2O_3	wt%	2.35	0.55	
K_2O	wt%	1.15	0.04	
Fe_2O_3	wt%	0.74	0.05	
Na_2O	wt%	0.22	0.01	
CaO	wt%	0.05	0.01	
MgO	wt%	0.10	0.01	
TiO_2	wt%	0.06	0.07	
P_2O_5	wt%	0.02	<0.01	
MnO	wt%	0.01	<0.01	

Fine-grained silica sand. A 1.5 t bulk sample of fine-grained silica sand was obtained from the Quarzwerke Frechen open-pit mine, which represents a silica sand operation located 8 km west of Cologne, Germany. The sand can be classified as medium- to fine-grained, poorly graded sand (SP) and its properties, as well as its geochemistry, are listed in Table 1. Semi-quantitative XRD showed that the sand primarily consists of quartz with ankerite and clinochlore as accessories.

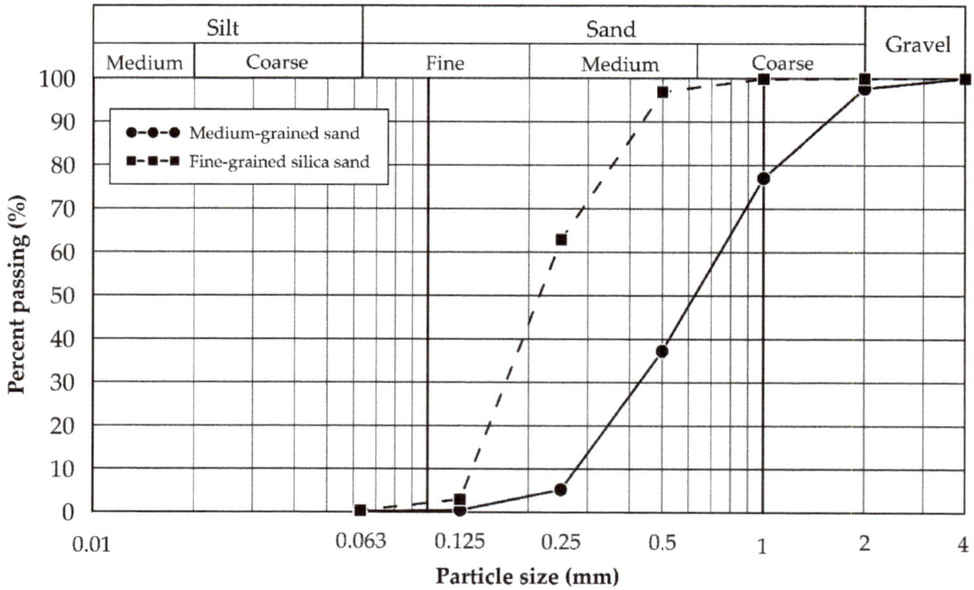

Figure 1. Particle-size distribution of medium-grained sand and fine-grained silica sand, determined according to DIN EN ISO 17892-4 [29].

2.1.2. Biopolymers

Selection methodology. Fourteen different biopolymers (seven polysaccharides and seven proteins) were selected for this study (Table 1). The polysaccharides, xanthan gum (XG), and sodium lignosulfonate (NLS) were preselected because they have already been studied in detail by previous works [13,17,33], and hence experimental results can be compared. Three qualitative criteria were used to select biopolymers relevant to this study:

1. The biopolymer should be able to be sourced regionally within central European countries where continental climate prevails.
2. The biopolymer should be commercially available at a relatively low cost compared to biopolymers investigated in previous studies, such as agar gum (avg. 18 USD/kg), alginates (avg. 12 USD/kg), carrageenan (avg. 10.5 USD/kg), chitosan (avg. 35 USD/kg), or pectin (avg. 15 USD/kg) [34–36].
3. The biopolymer should be readily soluble in water to enable large-scale field testing and potential industrial implementation. Such a requirement precludes biopolymers, whose dissolution would rely on either high temperature (e.g., agar or gellan gum) or acetic/alkaline solutions (e.g., chitosan or casein).

Selected biopolymers. Selected proteins and polysaccharides and relevant product information were obtained from the manufacturers (Table 2). Due to commercial sensitivity, the manufacturers did not allow the disclosure of their products' bulk prices. Thus, indicative bulk prices obtained from other articles are provided below. The polysaccharides used include carboxymethyl cellulose (CMC, ~1.4 USD/kg), NLS (0.2–0.5 USD/kg), XG (~2.0–3.0 USD/kg), and four different modified starches (corn, pea, potato, and wheat, typically <1.0 USD/kg in bulk) [20,37–40]. The chosen proteins comprise the plant-based fava bean protein concentrate (FBPC, ~1.4–2.5 USD/kg) and wheat protein (WP, ~1.4–2.5 USD/kg) and the animal-based proteins hen egg albumen (HEA, ~6.0–8.0 €/kg), porcine haemoglobin protein (HG, ~0.7–1.0 USD/kg), porcine plasma protein (PP, ~3.5–4.5 USD/kg), technical gelatine (TG, ~4–6 USD/kg), and whey protein concentrate (WPC, ~5.5 USD/kg) [41,42]. While technical gelatine does not strictly meet criterion (c), it was still selected because the temperature required to dissolve it (40 °C) is still modest.

Table 2. Biopolymers investigated by this study and their product data.

Biopolymer	Product Name	Specification	Manufacturer	Appearance	Moisture (wt%)
Polysaccharides					
Carboxymethyl cellulose (CMC)	DTK NV CMC	Technical grade, low viscosity	Mikro-Technik-CMC	Light-yellowish granules	8.6 [a]
Corn starch (CS)	C-Gel Instant	Pregelatinised	Cargill GmbH	White powder	5.8 [b]
Sodium lignosulfonate (NLS)	N18	Cold-water soluble	Otto Dille GmbH	Brown powder	8.6 [a]
Pea starch (PES)	Emcol EST (F11025)	Pregelatinised	Emsland Group	Off-white powder	10.4 [b]
Potato starch (POS)	KMC 18-09	Pregelatinised (acetylated)	KMC Germany	White powder	9.4 [b]
Wheat starch (WS)	Tigel	Pregelatinised	Krörner Stärke	White powder	6.3 [b]
Xanthan gum (XG)	XG TGRD	Technical grade, readily dispersible	Jungbunzlauer	White, free-flowing powder	5.1 [a]
Proteins					
Fava bean protein concentrate (FBPC)	ABPK 65%	Enzyme activated	Aloja-Starkelsen	Creamy light-yellow powder	8.8 [d]
Hen egg albumen (HEA)	Hen egg albumen	Cold-water soluble	Ovopol Sp. z. o.o.	Yellowish powder	7.0 [a]
Haemoglobin protein (HG)	HG 92P	Haemoglobin powder, porcine protein	Sonac	Dark red powder	6.6 [c]
Plasma protein (PP)	PP 70P	Plasma powder, porcine protein	Sonac	Cream-white powder	7.5 [c]
Technical gelatine (TG)	TG 330	Water soluble at ≥40 °C	Hellmann GmbH	Yellowish, free-flowing granules	13.0 [a]
Wheat protein (WP)	Glusol	Degraded, without viscoelastic properties	Kröner Stärke	Yellowish powder	6.0 [d]
Whey protein concentrate (WPC)	Instant WPC 80	From fresh cheese whey	Lactoland GmbH	White to pale yellow powder	6.2 [d]

Note. [a] = data provided by manufacturer, [b] = measured in accordance with DIN EN ISO 1666:1997 [43], [c] = measured in accordance with ISO 6496:1999 [44], [d] = measured in accordance with AOAC 930.15 [45].

2.2. Laboratory Experiments

All experiments were conducted on two substrates (medium-grained sand and fine-grained silica sand), using replicates (3x) and control (distilled water) samples and 14 different biopolymers at a fixed application rate and two different concentrations. The fixed application rate was set at 1.6 L/m^2 (20.3 mL per sample), and the biopolymer concentrations were chosen as 1.0 and 2.0 wt%. These values were chosen as they are within the range recommended by the literature and have been used in previous studies [12,13,16,17,46]. XG was applied at only 0.25 and 0.50 wt% because higher concentrations yield too viscous solutions for spraying. Each sample was subject to (a) moisture retention tests, (b) penetrometer testing, and (c) crust weight and thickness measurements.

2.2.1. Sample Preparation

Overall, 174 substrate samples were prepared and tested. Air-dried soil was placed into acryl glass cylinder moulds (127 mm diameter, 50 mm height, 126.7 cm^2 sample surface area). Samples were gently shaken to ensure slight and uniform compaction. Subsequently, sample surfaces were levelled with a ruler so that they were flush with the edge of the acryl glass cups. The resulting medium-grained sand samples weighed, on average, 1027 g (SD = 40.6 g) and the fine-grained silica sand samples 1109 g (SD = 38.6 g) (Figure 2a).

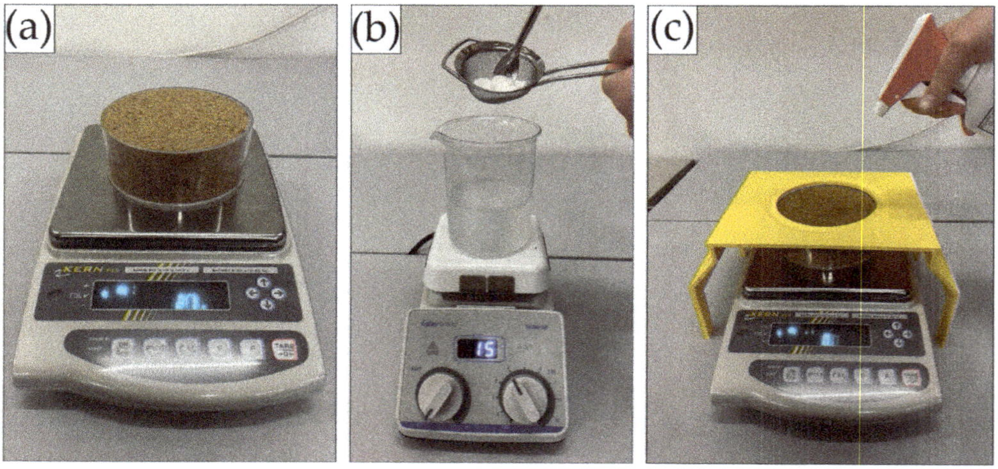

Figure 2. (**a**) Weighing of dry sample, (**b**) preparation of biopolymer solution, (**c**) gravimetric spray-on application of biopolymer solution with splash guard.

The calculation of the biopolymer mass required for the preparation of the individual solutions also accounted for the biopolymers' respective dry mass (Table 2). The biopolymer powders were dissolved at room temperature in distilled water at the specified concentrations for 10 min using a magnetic stirrer until fully dissolved (Figure 2b). To avoid clumping of the biopolymers, powders were slowly added to the distilled water through a sieve. The TG solution required preparation in 40 °C warm water.

The biopolymer solutions were sprayed onto the samples using a trigger sprayer with a nozzle suitable for viscous solutions (Ballistol hand atomizer, standard nozzle, Figure 2c). Uniform and accurate spray-on application was achieved by placing the untreated samples on a precision scale (Kern PES 4200-2M, 0.001 g resolution) and spraying the solution until the required application rate was achieved. Changes in solution density caused by the addition of biopolymer were considered negligible, so a fixed solution density of 1.0 g/mL was assumed for converting the required volumetric application rate (20.3 mL) to the gravimetric application rate (20.3 g). A 3D-printed splash guard was used to prevent the biopolymer solution from inadvertently touching the scale's weighing plate and thus distorting the scale readings (Figure 2c). After achieving the required application rate, the edges of the cylinder moulds were wiped dry. In the sample preparation process, the following weight measurements were recorded: weight of the empty sample moulds, dry sample weight, and total sample weight after biopolymer application.

2.2.2. Moisture Retention Tests

The ability of a biopolymer to enhance the soil moisture retention capacity is one determinant of its potential as a dust suppressant. Soil with increased moisture retention capacity can bind water over longer periods and slow the evaporation effect [47]. Moisture makes soil particles heavier and enhances the interparticle binding force [17], causing it to be less susceptible to being suspended in the atmosphere by erosive forces.

For the moisture retention tests performed in this study, the weights of the samples were recorded before and immediately after the biopolymer application. The treated samples were then cured in the laboratory (RWTH Aachen, spring) for four days (96 h) at ambient temperature (21 ± 1 °C) and humidity (45 ± 2.5%), their weights were recorded every 24 h (t = 0, 24, 48, 72, and 96 h post application). The moisture retention, ω (wt%), was calculated as the proportion of the initial moisture applied at t = 0 h that was retained in the sample at t = 96 h and was calculated according to Equation (1):

$$\omega = \frac{m_1 - m_{dry}}{m_0 - m_{dry}} \times 100 \quad (1)$$

where m_1 denotes the sample weight after 96 h of curing (g), m_0 is the sample weight after application of the biopolymer solution (g), and m_{dry} represents the dry sample weight prior to the application of the biopolymer solution (g).

2.2.3. Penetrometer Tests

Penetrometer tests allow measuring the maximum penetration resistance of soil crusts under controlled conditions. This analytical approach has already been applied by numerous studies investigating dust control agents [13,14,16,17,28,33,48,49]. A penetrometer is a stationary loading machine mounted with a pin to penetrate the soil crust at a set penetration angle and rate, continuously recording the penetration force and depth. Ding et al. [13] concluded that penetration resistance is a good indicator for predicting the dust control performance of a biopolymer and is even better suited than the UCS. In addition, Toufigh and Ghassemi [17] reported a strong correlation between results from the penetrometer and wind tunnel tests. Thus, penetrometer tests are a recommended and established method for evaluating potential dust suppressants.

The penetrometer tests were performed with a 6.4 mm (1/4 inch) diameter flat-ended cylindrical penetrometer pin that was mounted to a Wille UL 60/100 loading machine equipped with a 0.001 N resolution calibrated load cell (Institute of Geomechanics and Underground Technology, RWTH Aachen University, Germany). One penetration test was performed in the centre of each sample, pursuing a 4 mm penetration depth at an advance rate of 2 mm/min, a data logging interval of 5 Hz, and a fixed penetration angle of 90° (Figure 3). Since the penetrometer records penetration resistance alongside the penetration depth, the crusts' modulus of elasticity (M_e, kN m^{-1}) could be calculated by dividing the maximum penetration resistance (N) by the penetration depth reached at the moment of rupture.

Figure 3. Penetrometer testing of fine-grained silica sand sample.

2.2.4. Crust Thickness Measurements

After penetration testing, the weight and thickness of every crust were measured. Therefore, crusts were carefully removed from the sample mould using a small spatula (Figure 4a,b). Depending on crust strength and brittleness, crusts could either be recovered in one piece, broken down into multiple recoverable pieces, or were very thin, weak, and brittle and thus barely recoverable. The samples were weighed again after the removal of the crusts, which allowed the calculation of the crust weight (Figure 4c). Together with the weight recordings and known dimensions of every sample mould, each crust's average density and thickness were established.

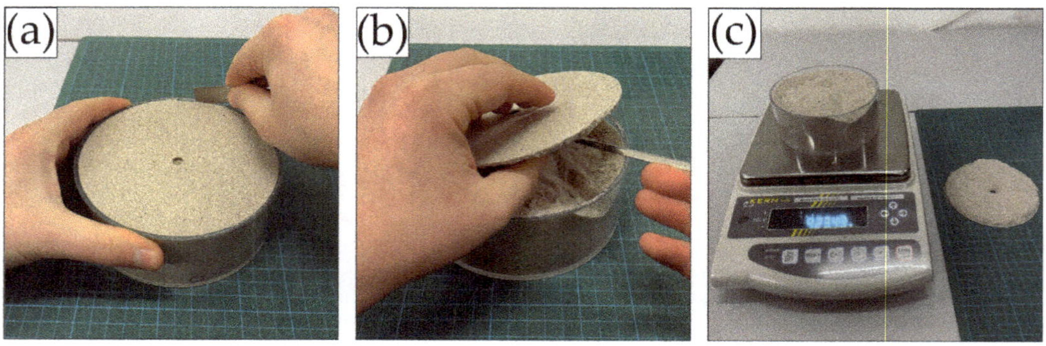

Figure 4. Process of removing the crust from the sample. (**a**) Loosening of the crust with a spatula from the mould rim, (**b**) removal of the crust with a spatula, (**c**) weighing of the sample without the crust.

2.2.5. Statistical Analysis

Two-way analysis of variance (ANOVA) with α = .05 was performed to analyse whether biopolymer type and concentration had a significant effect on the measured parameters. Two-way ANOVA is a statistical method used to analyse whether two individual independent variables (i.e., biopolymer type and concentration), as well as their interaction (biopolymer type * concentration), have a significant effect on one dependent variable (i.e., moisture retention, penetration resistance, and crust thickness) or not. If the resulting p-value is < .05, it can be concluded that the corresponding independent variable or their interaction has a significant effect on the analysed dependent variable. Separate statistical analyses were performed for the polysaccharides and proteins, medium-grained sand, and fine-grained silica sand. As recommended by the literature, any percentage data (moisture retention tests) were subject to square-root data transformation before performing the two-way ANOVA [50].

3. Results

3.1. Moisture Retention Tests

3.1.1. Medium-Grained Sand

Results of the moisture retention tests are shown in Figure 5 (exact values are appended in Table A2). Four days (96 h) after treatment, the control sample, which was treated with water, contained 6.9% (SD = 1.2) of the initially applied 20.3 g water. In the following, results of the polysaccharide and protein applications on the medium-grained sand are presented and compared.

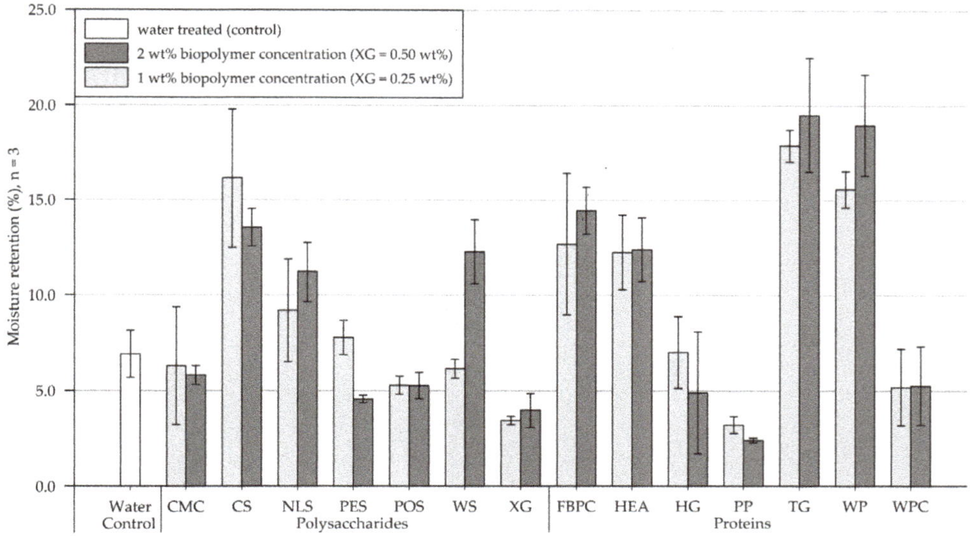

Figure 5. Mean moisture retention of medium-grained sand samples 4 days (96 h) after treatment at 1.6 L/m^2 and biopolymer concentrations of 1 and 2 wt% (XG = 0.25 and 0.50 wt%). Biopolymers are grouped into polysaccharides and proteins with water as control. Tests were performed in triplicates (n = 3), and error bars indicate the standard deviation (SD). Note. The exact values of experimental results are appended in Table A2.

Polysaccharides. Compared to the control, the 1 wt% (XG = 0.25 wt%) treatments either increased or decreased the samples' moisture retention. A slight reduction in moisture retention was observed for samples treated with CMC, POS, WS, and XG, whereas it increased for samples treated with CS, NLS, and PES. Relative to the 1 wt% treatment, the application of 2 wt% resulted in the samples' moisture retention either decreasing, increasing, or changing negligibly. It decreased slightly for CMC and CS and significantly for PES, while no change was observed for POS. By contrast, it moderately increased for NLS and XG and even doubled for WS-amended samples.

Proteins. Protein applications at 1 wt% concentration reduced the moisture retention for samples amended with PP and WPC relative to the control group, while a negligible effect was observed for HG. Doubling the concentration decreased moisture retention for HG- and PP-treated samples, increased it for FBPC-, TG-, and WP-amended samples and had a negligible effect on applications with HEA and WP.

Comparison of polysaccharide and protein treatments. At the 1 wt% concentration, the polysaccharide-treated samples had, on average, higher moisture retention than the protein-amended samples. Doubling the concentration resulted in a slight increase in the average moisture retention of polysaccharide and protein-treated substrate. A direct comparison of the polysaccharide and protein amendments that resulted in the highest moisture retention shows that the material treated with proteins achieved higher moisture retention regardless of the concentration tested.

Statistical analysis. Results of the two-way ANOVA (Table 3) show that the types of polysaccharide ($p < .001$) and protein ($p < .001$) have a significant effect on moisture retention, whereby no general trend could be identified: whether the treatments increase or decrease moisture retention. Doubling the concentration has no significant effect on moisture retention among polysaccharide- ($p = .596$) and protein-amended samples ($p = .470$), and there is no general trend, whether doubling the concentration results in the moisture retention to increase or decrease.

Table 3. Results of two-way ANOVA (α = .05) of moisture retention. Percentage values were transformed via square root transformation prior to performing two-way ANOVA.

Group	Factor	SS	df	MS	F	p
Medium-grained sand						
Polysaccharides	Type	0.01297	6	0.00216	20.578	< .001
	Concentration	0.00003	1	0.00003	0.288	.596
	Interaction	0.00219	6	0.00036	3.48	.011
	Error	0.00294	28	0.00011	0	
Proteins	Type	0.03497	6	0.00583	34.945	< .001
	Concentration	0.00009	1	0.00009	0.538	.469
	Interaction	0.00075	6	0.00013	0.752	.613
	Error	0.00467	28	0.00017		
Fine-grained silica sand						
Polysaccharides	Type	0.00241	6	0.00040	20.551	< .001
	Concentration	0.00008	1	0.00008	4.136	.052
	Interaction	0.00089	6	0.00015	7.598	< .001
	Error	0.00055	28	0.00002		
Proteins	Type	0.04336	6	0.00723	197.484	< .001
	Concentration	0.00056	1	0.00056	15.193	< .001
	Interaction	0.00103	6	0.00017	4.691	.020
	Error	0.00102	28	0.00004		

Note. SS = sum of squares, df = degrees of freedom, MS = mean square, F = F-value, *p* = *p*-value.

Conclusions. On medium-grained sand, the type of biopolymers has a significant effect on the samples' moisture retention, whereby some biopolymers significantly increased moisture retention. In contrast, others have a negligible effect or even decrease moisture retention relative to the control sample treated with water. The biopolymer's concentration does not have a significant effect on the moisture retention of the treated samples.

3.1.2. Fine-Grained Silica Sand

Results of moisture retention tests are depicted in Figure 6 (exact values are appended in Table A2). Four days (96 h) after treatment, the control sample, which was treated with water, contained 2.5% (SD = 0.7) of the initially applied 20.3 g of water. In the following, results of the polysaccharide and protein applications to the fine-grained silica sand are presented and compared.

Polysaccharides. In relation to the control sample, a 1 wt% (XG = 0.25 wt%) biopolymer treatment decreased the moisture retention of substrates treated with CS, NLS, PES, and POS, whereas an increase was observed for samples amended with CMC, WS, and XG. Increasing the concentration to 2 wt% (XG = 0.50 wt%) significantly decreased the moisture retention of XG-treated samples, and CMC had a negligible effect. By contrast, significant increases in moisture retention were observed for samples subject to applications with CS, NLS, PES, POS, and WS.

Proteins. Biopolymer applications at 1 wt% reduced the moisture retention for samples amended with HEA, PP, WP, and WPC compared to the control group. At the same time, significant increases were observed for FBPC, HG, and TG, respectively. When the concentration was doubled to 2 wt%, the moisture retention of the TG-treated silica sand decreased slightly, while the remaining proteins increased moisture retention.

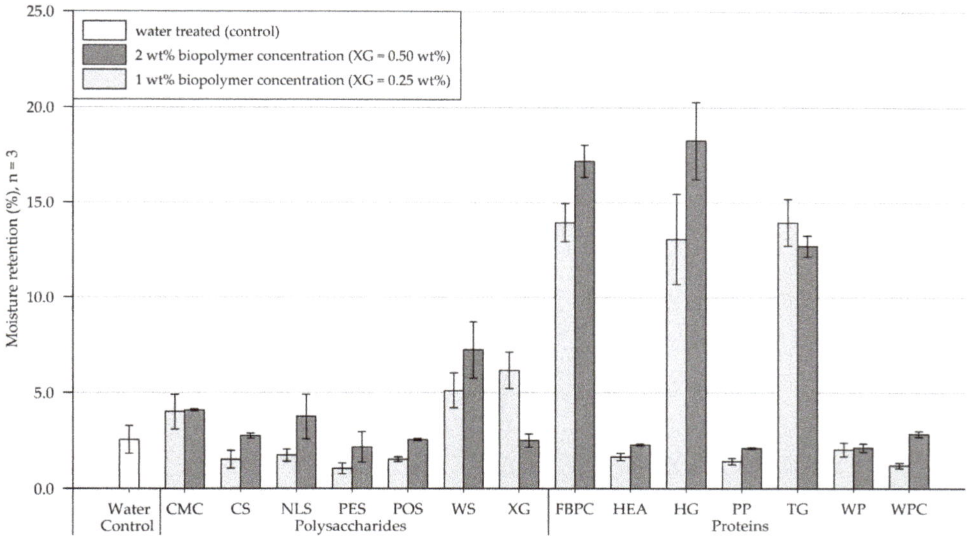

Figure 6. Mean moisture retention of fine-grained silica sand samples 4 days (96 h) after treatment at 1.6 L/m^2 and biopolymer concentrations of 1 and 2 wt% (XG = 0.25 and 0.50 wt%). Biopolymers are grouped into polysaccharides and proteins with water as control. Tests were performed in triplicates (n = 3), and error bars indicate the standard deviation (SD). Note. The exact values of experimental results are appended in Table A2.

Comparison of polysaccharide and protein treatments. At both tested biopolymer concentrations, the protein-treated samples, on average, showed significantly higher moisture retention than the polysaccharide-amended soils. A comparison of the polysaccharide- and protein-treated silica sand samples that achieved the highest moisture retention shows that the proteins performed better regardless of the tested concentration.

Statistical analysis. Results of the two-way ANOVA (Table 3) show that the types of polysaccharide ($p < .001$) and protein ($p < .001$) have a significant effect on moisture retention, with some biopolymers having a negligible effect, others decreasing moisture retention, and some resulting in the retention rate increasing significantly. Doubling the biopolymer concentration slightly or even considerably increases moisture retention for all biopolymers except XG and TG. A significant effect of concentration on moisture retention is only indicated for the tested proteins ($p < .001$) and not for the polysaccharides ($p = .052$).

Conclusions. On fine-grained silica sand, the biopolymer type has a significant effect on the samples' moisture retention, with some biopolymers increasing moisture retention, whereas others only result in minor changes or even decrease it relative to the control sample treated with water. Moisture retention of the treated silica sand is significantly influenced by the protein concentration.

3.2. Penetrometer Tests

3.2.1. Medium-Grained Sand

On medium-grained sand, all tested biopolymer applications formed crusts, and the results of the penetrometer tests are shown in Figure 7 (exact penetration resistance values and the results of the modulus of elasticity calculations are appended in Tables A3 and A4, respectively). Four days after treatment, the control group, treated with water, endured a maximum penetration resistance of 1.5 N (SD = 0.1).

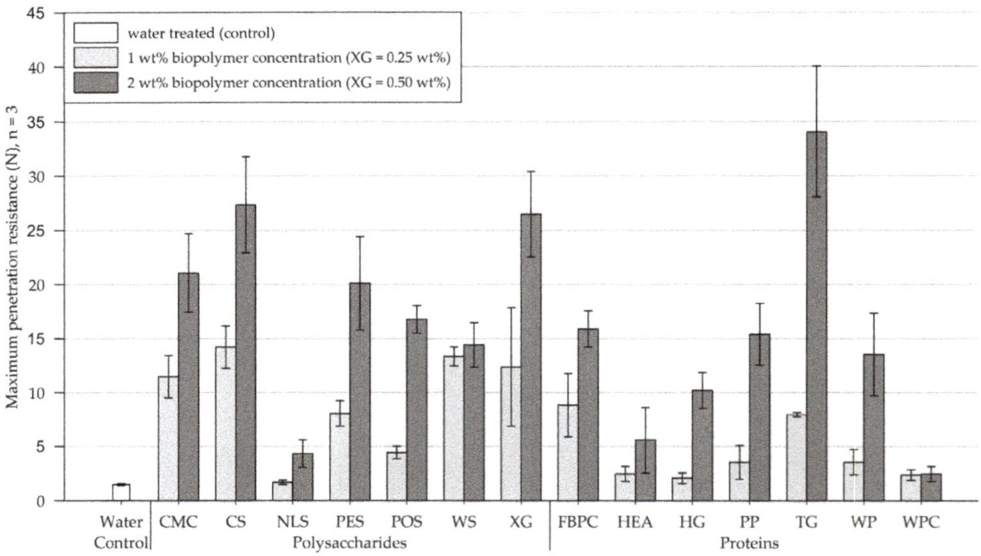

Figure 7. Mean maximum penetration resistance of crusts from medium-grained sand samples measured by penetrometer tests. Biopolymers are grouped into polysaccharides and proteins with water as control. Tests were performed in triplicates ($n = 3$), with error bars indicating the standard deviation (SD). Note. The exact values of experimental results are appended in Table A3.

Polysaccharides. Compared to the control, 1 wt% (XG = 0.25 wt%) treatments with NLS had a negligible effect on the penetration resistance. By contrast, treatments with POS resulted in a slight increase in penetration resistance, whereas the remainder of the tested polysaccharide applications showed significant increases in penetration resistance. When the concentration was increased to 2 wt% (XG = 0.50 wt%), the penetration resistance of substrate amended with WS slightly increased, while all other tested polysaccharides showed significant increases, with CS and XG achieving the highest penetration resistances among all tested biopolymers.

Proteins. The application at a concentration of 1 wt% slightly increased the penetration resistance of sand treated with HEA, HG, PP, WP, and WPC compared to the control group treated with water. By contrast, substrates amended with FBPC and TG exhibited significantly higher penetration resistances. Relative to 1 wt%, the 2 wt% protein applications increased the maximum penetration resistance of all treated samples, except for WPC. Applications with HG, PP, and TG displayed the greatest increases in penetration resistance.

Comparison of polysaccharide and protein treatments. When biopolymers were applied at concentrations of 1 wt% (XG = 0.25 wt%), polysaccharide-induced crusts were, on average, more than twice as resistant to penetration than protein-induced crusts. Doubling the concentration to 2 wt% (XG = 0.50 wt%) resulted in a disproportionate increase in the penetration resistance of the protein-induced crusts compared to the polysaccharide-induced crusts, with the polysaccharide treatments still achieving higher absolute penetration resistances.

Statistical analysis. Results of the two-way ANOVA (Table 4) show that the types of polysaccharide and protein have a significant effect on the penetration resistance ($p < .001$), with some biopolymer treatments achieving significantly higher penetration resistances than others. Among the proteins and polysaccharides, the concentration has a significant effect ($p < .001$) on the penetration resistance, whereby higher concentrations mostly enhanced the penetration resistance significantly.

Table 4. Results of two-way ANOVA (α = .05) of maximum penetration resistance measured by penetrometer.

Group	Factor	SS	df	MS	F	p
Medium-grained sand						
Polysaccharides	Type	1277.93	6	212.99	17.57	< .001
	Concentration	908.30	1	908.30	74.93	< .001
	Interaction	250.04	6	41.67	3.44	.011
	Error	339.46	28	12.12		
Proteins	Type	1386.02	6	231.00	24.74	< .001
	Concentration	952.47	1	952.47	101.99	< .001
	Interaction	623.11	6	103.85	11.12	< .001
	Error	261.49	28	9.34		
Fine-grained silica sand						
Polysaccharides	Type	582.82	6	97.14	4.91	< .001
	Concentration	108.78	1	108.78	5.50	.026
	Interaction	227.89	6	37.98	1.92	.113
	Error	554.29	28	19.80		
Proteins	Type	1753.52	6	292.25	21.93	< .001
	Concentration	798.95	1	798.95	59.95	< .001
	Interaction	298.17	6	49.70	3.73	.007
	Error	373.13	28	13.33		

Note. SS = sum of squares, df = degrees of freedom, MS = mean square, F = F-value, *p* = *p*-value.

Conclusions. On medium-grained sand, the type of biopolymer and concentration have a significant effect on the resulting penetration resistance of the cured crusts. Some biopolymers enhanced the penetration resistance only slightly or negligibly, while others had a considerable effect. For most biopolymers, doubling the concentration increased the achieved penetration resistance considerably.

3.2.2. Fine-Grained Silica Sand

Results of the penetrometer tests performed on fine-grained silica sand samples are shown in Figure 8 (exact values are appended in Table A3). Four days after treatment, the water-treated control group had a maximum penetration resistance of 1.7 N (SD = 0.2).

Polysaccharides. At biopolymer concentrations of 1 wt% (XG = 0.25 wt%), all tested polysaccharide types formed crusts with considerably enhanced penetration resistances compared to the control group treated with water. Increasing the concentration significantly enhanced the penetration resistance of all tested polysaccharides, except for WS and XG, whose penetration resistance decreased.

Proteins. Relative to the control group treated with water, all protein admixtures applied at 1 wt% concentration formed crusts with significantly increased penetration resistances. Compared to treatments at 1 wt% concentration, applications at 2 wt% resulted in the penetration resistance of all tested proteins increasing significantly, with FBPC, TG, and WP experiencing the most significant increases.

Comparison of polysaccharide and protein treatments. For amendments at 1 wt% concentration, polysaccharide-induced crusts, on average, endured slightly higher penetration resistances than protein-induced crusts. While doubling the concentration significantly increased the penetration resistance of almost all samples, the penetration resistance of protein-induced crusts displayed a disproportionate increase relative to polysaccharide crusts. At the higher biopolymer concentration, the protein-treated substrates, on average, showed higher penetration resistances than the polysaccharides.

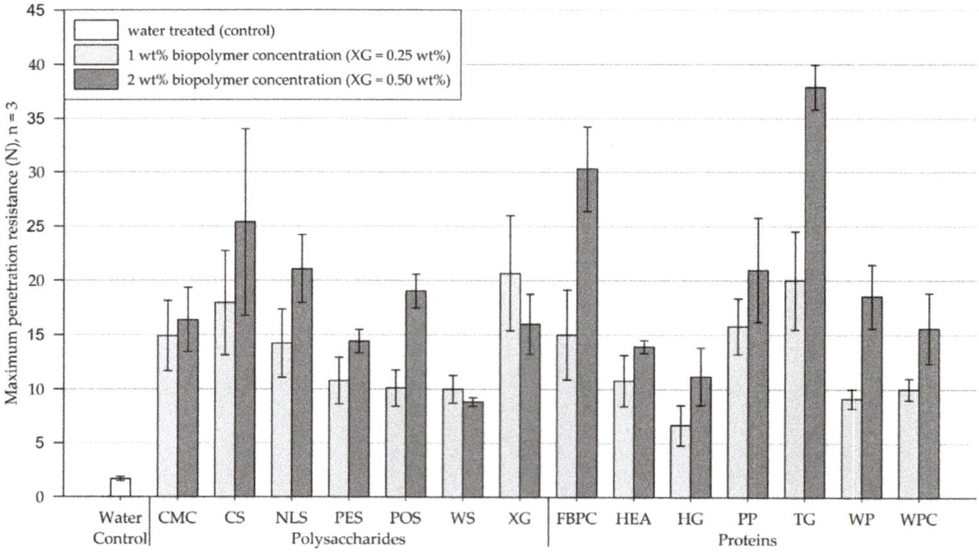

Figure 8. Mean maximum penetration resistance of crusts from fine-grained silica sand samples measured by penetrometer tests. Biopolymers are grouped into polysaccharides and proteins with water as control. Tests were performed in triplicates ($n = 3$), with error bars indicating the standard deviation (SD). Note. The exact values of experimental results are appended in Table A3.

Statistical analysis. Results of the two-way ANOVA (Table 4) show that the types of polysaccharide ($p < .001$) and protein ($p < .001$) have a significant effect on the penetration resistance, with some biopolymers forming significantly stronger crusts than others. The concentration also has a significant effect on the penetration resistance (polysaccharide: $p = .026$, proteins: $p < .001$), whereby doubling the concentration significantly increased the penetration resistance of most tested amendments.

Conclusions. On fine-grained silica sand, all biopolymers produced crusts with significantly increased penetration resistances relative to the control group, with the biopolymer type having a significant effect on the resulting penetration resistance. The effect of doubling the concentration strongly depends on the biopolymer type, whereby most biopolymer treatments exhibited significant increases.

3.3. Crust Thickness Measurements

3.3.1. Medium-Grained Sand

The results of the crust thickness measurements are shown in Figure 9 (exact values are appended in Table A5). Figures 10 and 11 show exemplary pictures of upside-down-facing crusts formed by the tested biopolymer amendments (visual classification appended in Table A6). While the control group, treated with water, formed no recoverable crust, all tested biopolymers formed crusts of varying thicknesses.

Polysaccharides. Treatments at 1 wt% (XG = 0.25 wt%) produced crusts ranging from 2 to 7.2 mm in thickness. While amendments with NLS and POS formed relatively thin to medium-thick crusts, the remaining polysaccharide applications yielded thick crusts on medium-grained sand. Compared to the 1 wt% treatment, doubling the concentration decreased the crust thickness for substrate treated with XG, CS, and WS, while crusts resulting from amendments with CMC and PES displayed a slight, and NLS and POS a significant, increase in thickness, respectively.

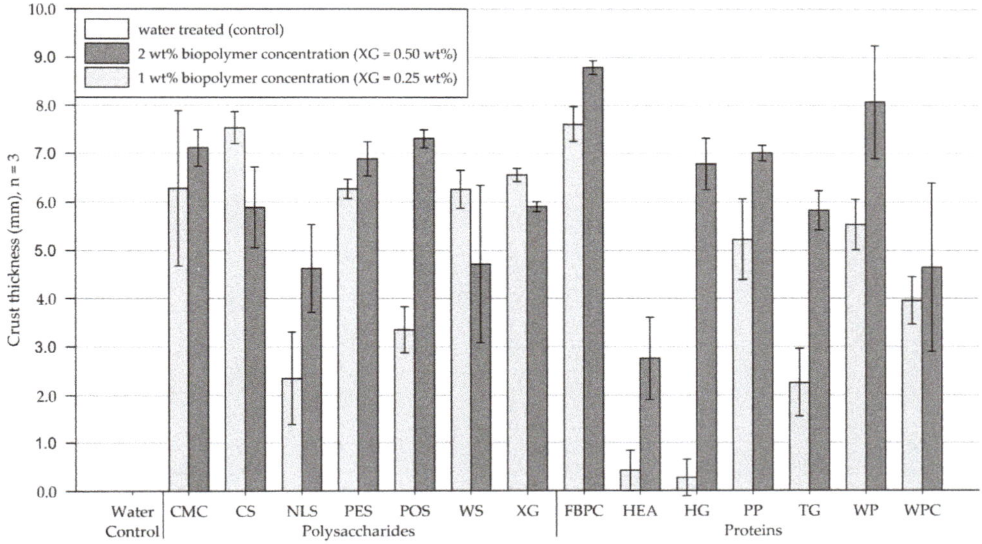

Figure 9. Mean thickness of crust formed on medium-grained sand samples. Biopolymers are grouped into polysaccharides and proteins with water as control. Tests were performed with six replicas (n = 3), with error bars indicating the standard deviation (SD). Note. The exact values of experimental results are appended in Table A5.

Proteins. Protein applications at 1 wt% concentration produced crusts of varying thickness. While treatments with HEA and HG formed fragile, brittle, and barely recoverable crusts, TG already formed slightly thicker crusts. Thick crusts formed on medium-grained sand amended with PP, WP, WPC, and FBPC, with the latter resulting in the thickest crusts of all biopolymers tested at 1 wt%. At the higher tested concentration, the crust thicknesses of all treatments increased, whereby HEA, TG, and HG displayed the most significant increases.

Comparison of polysaccharide and protein treatments. At treatments with 1 wt% (XG = 0.25 wt%) biopolymer concentration, medium-grained sand amended with polysaccharides, on average, formed slightly thicker crusts than applications with proteins. Doubling the concentration to 2 wt% (XG = 0.50 wt%) resulted in the thickness of polysaccharide-induced crusts to increase slightly, while protein-induced crusts displayed significant increases.

Conclusions. On medium-grained sand, all biopolymer treatments formed crusts, with the resulting crust thickness varying significantly among the biopolymers tested. For most polysaccharides, doubling the concentration only slightly increased the crust thickness, whereas most protein treatments exhibited significant increases in crust thickness.

Statistical analysis. Results of the two-way ANOVA (Table 5) show that the types of polysaccharide ($p < .001$) and protein ($p < .001$) have a significant effect on the crust thickness, with some biopolymers forming only very thin crusts and others thick crusts. Increasing the concentration only had a significant effect on the crust thickness of protein amendments ($p < .001$), whereby increasing the concentration caused the crust thickness of all protein-treated samples to increase.

Figure 10. Examples of polysaccharide-induced crusts recovered from the samples (cf. Table A6 for classification).

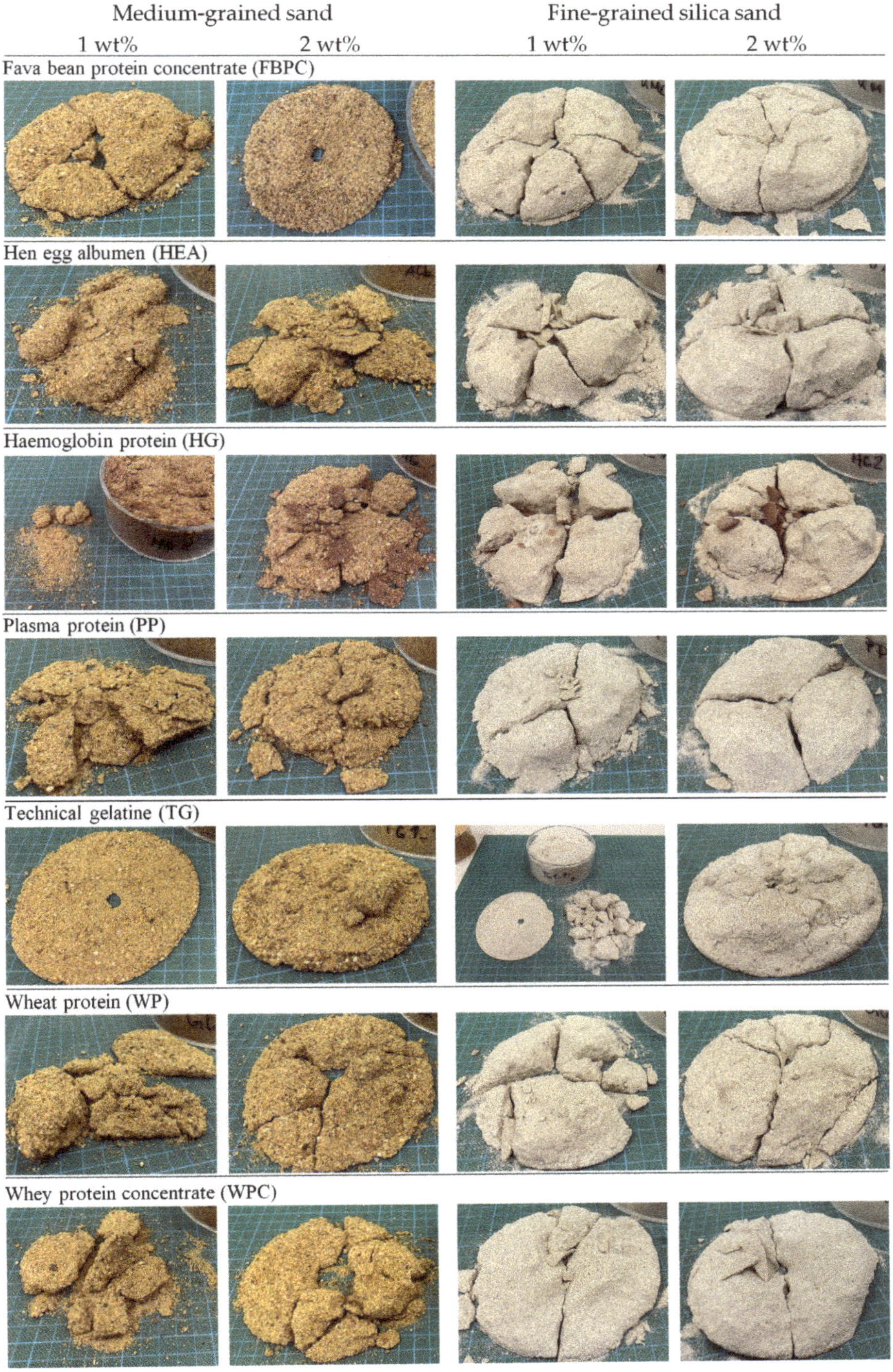

Figure 11. Examples of protein-induced crusts recovered from the samples (cf. Table A6 for classification).

Table 5. Results of two-way ANOVA (α = .05) of average calculated crust thickness.

BP Type	Factor	SS	df	MS	F	p
Medium-grained sand						
Polysaccharides	Type	48.6	6	8.1	8.86	< .001
	Concentration	3.15	1	3.15	3.44	.074
	Interaction	37.96	6	6.33	6.92	< .001
	Error	25.61	28	0.91	0	
Proteins	Type	188.1	6	31.35	38.42	< .001
	Concentration	77.67	1	77.67	95.18	< .001
	Interaction	33.83	6	5.64	6.91	< .001
	Error	22.85	28	0.82	0	
Fine-grained silica sand						
Polysaccharides	Type	636.09	6	106.01	84.27	< .001
	Concentration	1.47	1	1.47	1.17	.288
	Interaction	49.2	6	8.2	6.52	< .001
	Error	35.23	28	1.26	0	
Proteins	Type	75.25	6	12.54	8.44	< .001
	Concentration	5.08	1	5.08	3.42	.075
	Interaction	15.02	6	2.5	1.69	.161
	Error	41.6	28	1.49	0	

Note. SS = sum of squares, df = degrees of freedom, MS = mean square, F = F-value, p = p-value.

3.3.2. Fine-Grained Silica Sand

The results of the crust thickness calculation are shown in Figure 12 (exact values are appended in Table A5). Exemplary depictions of the cured crusts are displayed in Figures 10 and 11. The control group formed no recoverable crust, whereas all tested biopolymers formed crusts of varying thicknesses.

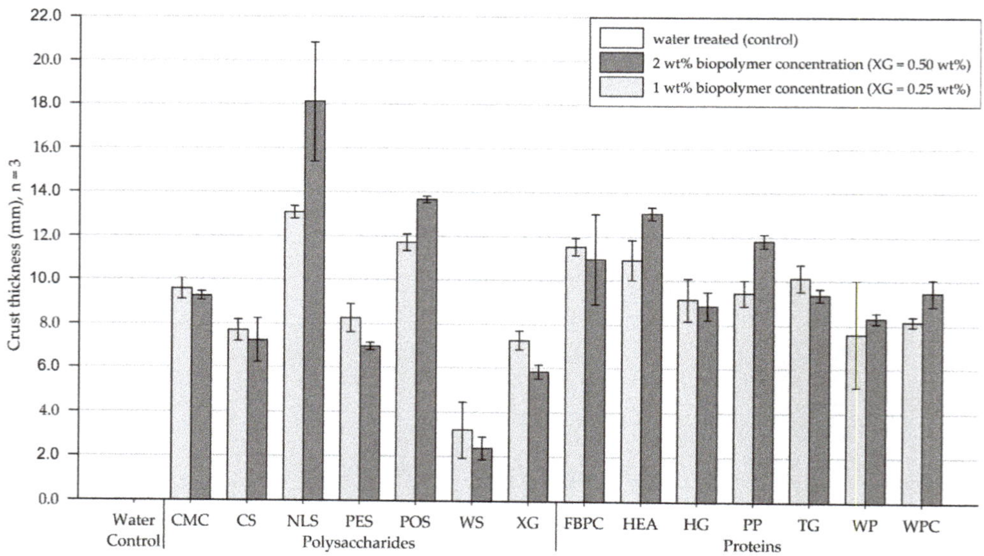

Figure 12. Mean thickness of crust formed on fine-grained silica sand samples. Biopolymers are grouped into polysaccharides and proteins with water as control. Tests were performed with six replicas (n = 3), with error bars indicating the standard deviation (SD). Note. The exact values of experimental results are appended in Table A5.

Polysaccharides. Polysaccharide applications at 1 wt% (XG = 0.25 wt%) formed crusts of varying thickness. WS formed only very thin and rather ductile crusts, while the remainder of the tested polysaccharide applications yielded significantly thicker crusts. Doubling the concentration increased the thickness of silica sand crusts amended with NLS and POS, while the other tested polysaccharide applications displayed decreased crust thicknesses. Notably, crusts formed because of treatment with WS even curled up throughout the curing period (Figure 10).

Proteins. Treatments at 1 wt% produced crusts of similar thicknesses, whereby amendments with FBPC, HEA, and TG formed the thickest crusts among the proteins tested. Doubling the concentration slightly decreased the thickness of crusts formed after treatment with FBPC, TG, and HG, whereas the other protein applications displayed slight increases in crust thickness.

Comparison of polysaccharide and protein treatments. At a biopolymer concentration of 1 wt% (XG = 0.25 wt%), protein amendments, on average, formed slightly thicker crusts than polysaccharide treatments. Doubling the concentration had a similar effect for most biopolymer applications, only slightly reducing or increasing the resulting crust thickness.

Statistical analysis. Two-way ANOVA (Table 4) shows that the biopolymer type has a significant effect on crust thickness among polysaccharides ($p < .001$) and proteins ($p < .001$), whereby the crust thickness differs significantly depending on the biopolymer type. By contrast, doubling the concentration has no significant effect (polysaccharides: $p = .288$, proteins: $p = .075$) on the resulting crust thickness, as it only slightly increased or even decreased the crust thickness.

Conclusions. All biopolymers tested formed crusts on fine-grained silica sand. While the crust thicknesses were relatively uniform among the tested proteins, the tested polysaccharide amendments resulted in a greater variability. For all biopolymers, doubling the concentration had only a small effect on the resulting crust thickness.

4. Discussion

4.1. Moisture Retention Tests

Several studies have investigated the effect of different biopolymer types and concentrations on moisture retention of biopolymer-treated soils [17,28,46,47,49,51,52]. The studies found that biopolymer treatments enhance soil moisture retention, with the biopolymer type having a significant effect on the resulting moisture retention. Most studies concluded that moisture retention increases at higher biopolymer concentrations [17,28,49], although one study did not confirm this effect [46]. Soil type also appears to significantly influence the moisture retention that can be achieved by a biopolymer amendment [17]. In the following, the key trends of this study's results are discussed in the context of the previous literature. A more detailed discussion is limited by the significant differences in soil, treatment, and experimental setup among the studies.

4.1.1. Effect of Biopolymer Type

Medium-grained sand. The water-treated control group exhibited a higher moisture retention than several biopolymer treatments, which is not consistent with the existing literature [17,46,49,53]. It is assumed that the relatively low viscosity of water allowed it to penetrate deeper into the medium-grained sand, making it less prone to evaporation. By contrast, the more viscous biopolymer solutions did not infiltrate as deeply into the sample, resulting in a higher evaporation rate.

Some biopolymer treatments significantly increased moisture retention (Figure 5), which is consistent with the findings of other studies examining the ability of biopolymers to improve soil moisture retention [17,25,28,46,49,54,55]. This increase is likely due to the high water-absorption capacity of the biopolymers, as well as the effects of the solution viscosity, bio-clogging, and crust formation. Low solution viscosity allows water molecules to infiltrate deeper into the soil, making them less susceptible to evaporation. Bio-clogging occurs when hydrated biopolymers clog the soil's pore space, reducing capillary conductivity and, thus, the evaporation rate [52,56]. Additionally, biopolymer-induced surface crusts may act as barriers between moisture and air, slowing the evaporation rate [47,57]. Biopolymers such as XG, PP, POS, and CMC did not significantly affect moisture retention (Figure 5), which does not correspond with the published literature [17,28,49]. Conversely to the biopolymer treatments that increased the moisture retention of medium-grained sand, it is likely that a low water-absorption capacity (PP and POS) or a relatively high viscosity (XG and CMC) contributed to a faster evaporation rate and thus lower moisture retention.

Fine-grained silica sand. Similar trends in moisture retention were found on fine-grained silica sand as on medium-grained sand, with some biopolymers decreasing it, some having no effect, and some increasing it significantly compared to the water-treated control group (Figure 6). In addition, as found for medium-grained sand, the control did not have the lowest moisture retention. The same reasons as identified for medium-grained sand likely caused these results.

Comparison of polysaccharides and proteins. On average, the well-performing proteins, such as FBPC, HEA, WP, and HG, achieved a higher moisture retention capacity than the well-performing polysaccharides, such as CS, NLS, and WS (Figures 5 and 6). It is likely that this trend has been caused by the lower viscosity exhibited by most protein solutions and potentially a higher water-absorption capacity.

4.1.2. Effect of Biopolymer Concentration

Medium-grained sand. Increasing the biopolymer concentration had no significant effect on moisture retention (Figure 5 and Table 3). The published literature also reported mixed findings, with most studies concluding that higher biopolymer concentrations increase moisture retention [17,28,49], while one study could not confirm this observation [46]. For biopolymer applications that increased moisture retention (Figure 5), it is assumed that doubling the concentration amplified the effects of water absorption capacity, bio-clogging [52,56], and crust formation [47,57] (as discussed in Section 4.1.1).

For some biopolymers, doubling the concentration had little effect on moisture retention or even caused it to decrease. Ding et al. [28] reported similar results when testing applications of lignosulfonates on red sand, with moisture retention only increasing noticeably up to a certain concentration. This suggests that each biopolymer has a specific threshold concentration, beyond which only minor changes or even a decrease in soil moisture retention occurs. This trend is likely related to an increase in biopolymer viscosity, which reduces infiltration depth and increases evaporation rate, resulting in the adverse effects of higher viscosity outweighing the positive effects of increased water absorption capacity, bio-clogging, and crust formation (see Section 4.1.1).

Fine-grained silica sand. On fine-grained silica sand, doubling the concentration resulted in a slight increase in moisture retention for most tested biopolymers (Figure 6 and Table 3), which agrees with previous studies [17,28,49]. It is believed that the same factors assumed for medium-grained sand caused these trends on fine-grained silica sand. Samples treated with XG and TG exhibited decreased moisture retention when the concentration was doubled, likely because of an increase in viscosity that limited the infiltration depth and, thus, increased the evaporation rate. CMC-treated samples showed no significant change, assumably because of the balancing of positive effects (e.g., increased water absorption, bio-clogging, and crust formation) and the negative effect of higher viscosity. Fine-grained silica sand samples, on average, exhibited a lower moisture-retention capacity than medium-grained sand samples (Figure 5) because of the higher hydraulic conductivity of medium-grained sand that enhanced the infiltration depth and decreased evaporation throughout the curing period.

4.2. Penetrometer Tests

Multiple studies performed penetrometer tests to investigate the effects of different biopolymers and concentrations on the penetration resistance of biopolymer-induced soil crusts [13,14,16,17,28,33,48,49]. A compilation of these studies' experimental results has been appended in Tables A7 and A8. The results suggest that the biopolymer type has a significant effect on the crusts' penetration resistance, with some biopolymers forming firmer crusts than others. Increasing the concentration generally increases the penetration resistance [33,46,49], although one study found that this relationship plateaus at a specific concentration [16]. On sandy soil, biopolymers act by coating and binding sand particles together, forming a cross-linked 3D network that increases inter-particle cohesion [21]. In the following sections, the key trends of this study's results are discussed in the context of the previous literature, as a more detailed discussion is limited by the significant differences in soil, treatment, and experimental setup among the studies.

4.2.1. Effect of Biopolymer Type

Medium-grained sand. Penetrometer test results showed that all tested biopolymers formed crusts with differing penetration resistances, with the biopolymer type significantly affecting the penetration resistance (Table 4 and Figure 7). The water-treated control group formed no crust and endured a very low penetration resistance. These trends are consistent with findings from previous studies (Tables A7 and A8), which also showed that the water-treated control endured low penetration resistances relative to the biopolymer-treated samples.

Comparison with previous studies on XG, NLS, and CMC [17,33] (cf. Tables A7 and A8) reveals that XG forms stronger crusts than NLS, which agrees with this study's results (Figure 7). By contrast, Toufigh and Ghassemi [17] reported that CMC tends to form firmer crusts than XG, which contradicts the results of this study. This discrepancy may be due to the different XG types used in the two studies. The XG used in this study formed highly viscous, not-sprayable gels at concentrations >0.5 wt%, while Toufigh and Ghassemi tested XG up to 1.5 wt% without reporting issues regarding the solution viscosity and spray-ability. Hence, it is likely that the XG used in this study had a higher gel strength and viscosity than the one tested by Toufigh and Ghassemi.

The starches (CS, POS, PES, and WS) also formed crusts, some with high penetration resistances. While the penetration resistance of starch-treated soil has not yet been investigated, other studies have already demonstrated their ability to improve soil mechanical properties. A wind tunnel study found that starch-treated soil exhibited reduced wind erosion [58], while other studies reported that starches and starch–xanthan blends enhanced the soil's geotechnical properties [59–61].

All tested proteins formed crusts, some of which had high penetration resistances (Figure 7). While previous studies have not investigated the crust strength of protein-treated soils, some studies demonstrated proteins' ability to enhance geotechnical soil properties [47,62–64]. Analogous to PP, bovine plasma protein was found to increase the compressive strength of biopolymer-bound soil composites [62]; while corresponding to WPC, cottage cheese whey was found to improve aggregate stability and reduce furrow erosion [64]. Likewise, gelatine has been used as a constituent in the formulation of dust suppressants because of its film-forming properties [47,63]. The remaining proteins tested in this study have not yet been tested regarding their ability to enhance soil properties but are commonly used in the food industry for their thickening, gelling, or water-retention abilities [65,66].

Fine-grained silica sand. The penetrometer tests revealed that all biopolymer treatments formed crusts, with the biopolymer type significantly affecting the penetration resistance (Figure 8 and Table 4). This trend corresponds with previous research (Tables A7 and A8). Compared to medium-grained sand, crusts on fine-grained silica sand, on average, had a higher penetration resistance. This can be attributed to the silica sands' smaller particle size, which results in a larger surface area, tighter packing, and lower hydraulic conductivity. As a result, the biopolymers coat a larger surface area, must bridge shorter distances between particles, and do not infiltrate as deep into the soil, and thus form a stronger crust.

4.2.2. Effect of Biopolymer Concentration

Medium-grained sand. Penetrometer test results showed that doubling the concentration significantly increased the penetration resistance of most biopolymers tested (Figure 7 and Table 4). This trend corresponds with previous research (Tables A7 and A8) [14,17,28,33]. As stated by Owji et al. [14], this is because higher biopolymer concentrations increase the thickness of the inter-particle bonds forming between the coated sand particles, which causes the crust strength to increase. Doubling the concentration of WS and WPC only slightly increased the penetration resistance (Figure 7). The minor effect of WPC suggests that it generally has limited potential to enhance soil properties. Regarding WS, it is likely that its high viscosity limited the infiltration depth, leading to the presence of more biopolymer molecules in the pores than required to form a stable 3D network. Thus, for WS doubling, the concentration likely exceeded a threshold concentration, beyond which only minor increases in crust strength occur. This corresponds with findings of Lemboye et al. [16], who observed that applications of sodium alginate and pectin only increased penetration resistance up to 2 or 3 wt%.

Fine-grained silica sand. Doubling the concentration increased the penetration resistance of most biopolymers tested (Figure 8), similar to the findings on medium-grained sand (Figure 7) and the published literature (Tables A7 and A8). WS and XG exhibit a different trend, as doubling their concentration resulted in decreased penetration resistances. This can be attributed to the high viscosities of their respective solutions, which limited infiltration depth and crust thickness, causing the crust to endure less load. For WS, the low infiltration and high local biopolymer concentration even caused the crust to curl up because of tension forces that occurred as the swollen biopolymers dehydrated and shrank (Figure 10).

Comparison of polysaccharides and proteins. On average, protein-treated samples exhibited larger incremental increases in penetration resistance than polysaccharide-treated samples upon doubling the concentration (Figures 7 and 8). Thereby, on medium-grained sand, some protein treatments only formed stable crusts at 2 wt%, while most tested polysaccharides formed stable crusts at 1 wt% on both soils (XG = 0.25 wt%). This suggests that for the tested polysaccharides, doubling the concentration likely surpassed a threshold beyond which only small increases in crust strength occur. As a result, proteins must be applied at higher concentrations to achieve comparable penetration resistances as polysaccharide-induced crusts on medium-grained sand.

4.3. Crust Thickness Measurements

Several studies have shown that the type and concentration of biopolymers significantly affect the thickness of biopolymer-induced soil crusts, and a compilation of their results has been appended for reference (Table A9) [13,14,16,28,46,67]. Thereby, viscosity and the ability to enhance inter-particle cohesion influence the resulting crust thickness and depend on the biopolymer type and concentration. In the following, the main trends of this study's results are discussed in the context of previous studies. A more detailed discussion is limited by the significantly different soil types, treatments, and experimental setups used in previous studies.

4.3.1. Effect of Biopolymer Type

Medium-grained sand. The crust thickness measurements showed that all tested biopolymers produced crusts, with the biopolymer type significantly affecting the resulting crust thickness (Figure 9 and Table 5). These findings correspond with the published literature (Table A9). In contrast to the other biopolymers, NLS, HEA, HG, and TG partially formed very thin and brittle crusts (Figures 10 and 11). Regarding NLS, this can be explained as previous studies found that lignosulfonates must be applied at concentrations >2 wt% to enhance soil mechanical properties effectively [19,33,68]. The ability of HEA and HG to enhance soil mechanical properties has not yet been studied, but their treatments at

1 wt% and 2 wt% (only HEA) did not increase inter-particle cohesion sufficiently to form a competent crust.

Fine-grained silica sand. The test results showed that the biopolymer type significantly affects the crust thickness (Figure 12 and Table 5), which agrees with the existing literature (Table A9). Compared to the other tested biopolymers, NLS, POS, HEA, and PP formed relatively thick crusts, likely because of a favourable combination of viscosity and bonding strength, allowing for a high infiltration depth and sufficient inter-particle cohesion to form a thick crust. By contrast, WS and XG only formed thin crusts because of their high viscosity. Hence, on fine-grained silica sand, biopolymer viscosity is a limiting factor regarding the achievable crust thickness. Treatments on fine-grained silica sand generally formed thicker crusts than on medium-grained sand, as the finer particle size and tighter packing of the silica sand favour particle agglomeration and crust formation. In this context, NLS, HG, and HEA exhibited significant differences, forming only thin and brittle crusts on medium-grained sand but achieving considerably higher crust thicknesses on fine-grained silica sand.

4.3.2. Effect of Biopolymer Concentration

Medium-grained sand. The results of crust thickness measurements revealed that doubling the biopolymer concentration significantly affected crust thickness (Figure 9 and Table 5). For most polysaccharides, crust thickness only increased slightly or even decreased upon doubling the concentration, which corresponds with the published literature (Table A9). This is likely because the increased viscosity limited the infiltration depth, resulting in a thinner crust. By contrast, NLS and POS treatment benefitted significantly from doubling the concentration, which increased the inter-particle cohesion sufficiently to form a thick crust.

In contrast to the polysaccharide treatments, doubling the concentration significantly increased the thickness of protein-induced crusts. This is because doubling the concentration increased inter-particle cohesion without limiting the infiltration depth. While most polysaccharides formed relatively thick crusts at 1 wt%, most proteins required concentrations of 2 wt% to achieve crusts of similar thickness. Thus, proteins must be applied at higher concentrations to achieve similar crust thicknesses as polysaccharide treatments.

Fine-grained silica sand. Results showed that doubling the concentration had no significant effect on the resulting crust thickness, leading only to slight increases or decreases (Figure 12 and Table 5). It appears that concentrations of 1 wt% are already sufficient to form a thick, stable crust, so doubling the concentration mainly resulted in adverse effects caused by the increase in viscosity.

4.4. Evaluation of Tested Biopolymers as Dust Suppressants

This study found that the tested biopolymers formed crusts with significantly increased penetration resistances relative to the water-treated control and, in part, enhanced moisture retention. As penetration resistance [13,17,26] and moisture retention [28] are commonly used indicators for assessing potential dust suppressants, it is concluded that most tested polysaccharides and proteins show potential as dust suppressants. XG was incorporated as a benchmark in the study, as it is the most widely studied biopolymer in soil stabilisation and dust control [22,69]. The results on both tested soil types showed that some polysaccharides (CMC, CS, and WS) and proteins (FBPC, PP, and WP) achieved similar crust penetration resistances as XG (Figures 7 and 8). However, it must be considered that XG was only tested at a quarter of the concentration of the other biopolymers.

Commercially available dust control products, such as chloride salts, petroleum-based products, and synthetic polymers, are effective dust suppressants but remain costly and can have adverse environmental effects [12,70]. Thus, there is a need for affordable, environmentally friendly alternatives that are easy to apply. While approaches such as microbially induced carbonate precipitation show potential to mitigate wind erosion [71–73], their cultivation and application are challenging [23], rendering them not easy-to-apply off-the-shelf solutions. Alternatively, by-products and wastes from food processing have shown potential as dust suppressants [74,75], but their inconsistent composition and dry-matter content may limit their application potential. By contrast, biopolymers show potential to meet the needs mentioned above.

This study and previous research [13–16] have shown that biopolymers have potential as dust suppressants. Moreover, they are biodegradable [76], well-studied, and frequently applied in the food industry and medical applications [40,66,77]. The tested biopolymers have relatively low cost, are easily dissolvable in water, and can be applied using conventional water-spraying equipment. In addition, the tested biopolymers originate from botanical (corn, pea, wheat, cellulose, potato, and fava bean) and animal (pig, chicken, and cow) sources that are widely cultivated and bred. Consequently, biopolymers show potential as environmentally benign, readily available, low-cost, and easy-to-use alternatives to traditional dust suppressants. Further laboratory and field studies are needed to investigate their potential at different dosages and under realistic operating conditions to raise industry awareness.

5. Conclusions

This study examined the potential of 14 polysaccharides and proteins as dust suppressants by testing spray-on treatments at two different biopolymer concentrations on a medium-grained sand and a fine-grained silica sand. Moisture retention tests, penetrometer tests, and crust thickness measurements were performed, and the following conclusions were drawn:

1. Penetrometer test results on biopolymer-treated medium-grained sand ranged from 1.7 to 34.0 N (control = 1.5 N) and on fine-grained silica sand from 6.7 to 37.9 N (control = 1.7 N), respectively. The results showed that all tested biopolymers formed crusts with significantly differing penetration resistances depending on the biopolymer type ($p < .05$). Increasing the biopolymer concentration significantly increased the penetration resistance on medium-grained sand ($p < .001$). In contrast, on fine-grained silica sand, it only increased the penetration resistance of protein treatments significantly ($p < .001$). Proteins achieved similar penetration resistances as polysaccharides but required higher concentrations.
2. Moisture-retention test results on medium-grained sand ranged from 3.4 to 19.5 wt% (control = 6.9 wt%) and on fine-grained silica sand from 1.0 to 18.2 wt% (control = 2.5 wt%). On both tested soil types, the biopolymer type had a significant effect ($p < .001$) on the samples' moisture retention, resulting in it decreasing or increasing relative to the water-treated control. Increasing the concentration increased moisture retention of protein-treated fine-grained silica sand samples significantly ($p < .001$).
3. The thicknesses of crusts formed on biopolymer-amended medium-grained sand samples ranged from 0.3 to 8.8 mm (control = 0 mm) and on fine-grained silica sand from 3.2 to 18.1 mm (control = 0 mm). The results showed that the different biopolymers formed crusts of varying thicknesses, with the biopolymer type significantly affecting the crust thickness ($p < .001$). On medium-grained sand, doubling the concentration only had a significant effect for protein amendments ($p < .001$) and resulted in the crust thickness increasing. On fine-grained sand, increasing the concentration slightly reduced the crust thickness of most treatments because of a lower infiltration depth.

The results of this study demonstrate that the tested polysaccharides and proteins have the potential to be applied as dust suppressants in industries such as mining. Thereby, biopolymers can contribute to reducing the industry's environmental and human health impacts.

Author Contributions: Conceptualisation, J.L.S. and J.F.; methodology, J.L.S. and J.F.; formal analysis, J.L.S.; investigation, J.L.S.; data curation, J.L.S.; writing—original draft preparation, J.L.S.; writing—review and editing, J.L.S., B.G.L. and J.F.; visualisation, J.L.S.; project administration, J.L.S. All authors have read and agreed to the published version of the manuscript.

Funding: This research received no external funding.

Data Availability Statement: Data supporting the findings of this study will be made available from the corresponding author upon reasonable request.

Acknowledgments: We thank Rheinische Baustoffwerke GmbH and Quarzwerke Frechen AG for providing the mine soils used in this study. We thank Devrim Gürsel, Martin Knippertz, and Stephanie Lohmeier for assisting in characterising the soil samples and Johannes J. Emontsbotz for designing and printing the 3D-printed splash guard. Last, the authors would like to thank the anonymous reviewers for their helpful comments and suggestions, which improved the manuscript.

Conflicts of Interest: The authors declare no conflict of interest.

Appendix A

Table A1. Compilation of previous studies investigating biopolymer applications in the disciplines of soil stabilisation and dust control.

Biopolymer	Source	Solubility	Type of Application	References
Polysaccharides				
Arabic gum (Acacia gum)	Exudate from acacia trees	Cold-water soluble	Spray-on (dust control) Mix-in (soil stabilisation)	[16] [16,78]
Agar gum	Red algae (Gelidium and Gracilaria)	Hot-water soluble (>86 °C)	Mix-in (soil stabilisation)	[25,79,80]
β-glucan	Extracted from cells of yeast, fungi, certain bacteria and cereals	Cold-water soluble	Mix-in (soil stabilisation)	[25,81–86]
Carrageenan	Red algae (Chondracanthus)	Cold-water soluble	Spray-on (dust control) Mix-in (soil stabilisation)	[87,88] [58,88]
Chitosan	Chitin shells of crustaceans	Soluble in acetic solutions	Mix-in (soil stabilisation) Spray-on (dust control)	[55,89–93] [87,94,95]
Carboxymethyl cellulose	Cellulose derivative	Cold-water soluble	Mix-in (soil stabilisation) Spray-on (dust control)	[96–99] [14,17]
Dextran	Microbial		Mix-in (soil stabilisation) Spray-on (dust control)	[100] [101]
Gellan gum	Bacteria	Poor solubility at low temperature (fully dissolvable > 80 °C)	Mix-in (soil stabilisation)	[37,96,102–106]
Guar gum	Guar beans	Cold-water soluble	Mix-in (soil stabilisation) Spray-on (dust control)	e.g., [13,33,82,83,96,107–111] [14,15,46]
Lignosulfonate	By-product of wood pulp production	Cold-water soluble	Spray-on (dust control)	[19,26,33,112]
Locust bean gum	Carob tree seeds		Mix-in (soil stabilisation)	[113]
Pectin	Citrus fruit lamella and cell walls	Cold-water soluble	Spray-on (dust control)	[16]
Persian gum	Exudate from almond tree trunk and branches	Cold-water soluble (30% soluble, 70% insoluble)	Mix-in (soil stabilisation)	[114]
Sodium alginate	Brown algae	Cold-water soluble	Spray-on (dust control) Mix-in (soil stabilisation)	[15,16,67,115] [16,91,116–120]
Starch				
Corn	Corn	Cold-water soluble	Spray-on (dust control) Mix-in (soil stabilisation)	[121] [58,80]
Potato	Potato	Cold-water soluble	Spray-on (dust control)	[47]
Xanthan gum	Bacteria	Cold-water soluble	Spray-on (dust control) Mix-in (soil stabilisation)	[17,122] e.g., [82,107,123–128]
Proteins				
Casein	Milk	Soluble in alkaline solutions	Mix-in (soil stabilisation)	[129–134]
Bovine blood plasma	Bovine blood			[62]
Gelatin	Collagen from animal bones and tissues	Warm-water soluble (>40 °C)	Spray-on (dust control)	[47,63]
Soybean	Soybean	Cold-water soluble	Spray-on (dust control)	[135,136]

Table A2. Moisture retention of medium-grained sand and fine-grained silica sand samples treated at biopolymer concentrations of 1 and 2 wt% (XG = 0.25 and 0.50 wt%). Tests were performed in triplicates ($n = 3$), with M = mean and SD = standard deviation. On medium-grained sand the water-treated control group achieved a mean moisture retention of 6.9 wt% (SD = 1.2) and on fine-grained silica sand 2.5 wt% (SD = 0.7).

	Moisture Retention (wt%)							
	Medium-Grained Sand				**Fine-Grained Silica Sand**			
	1 wt%		**2 wt%**		**1 wt%**		**2 wt%**	
	M	**SD**	**M**	**SD**	**M**	**SD**	**M**	**SD**
Polysaccharides								
Carboxymethyl cellulose	6.3	3.1	5.8	0.5	4.0	0.9	4.1	0.1
Corn starch	16.1	3.6	13.6	1.0	1.5	0.5	2.7	0.1
Sodium lignosulfonate	9.2	2.7	11.2	1.6	1.7	0.3	3.7	1.2
Pea starch	7.8	0.9	4.6	0.2	1.0	0.3	2.1	0.8
Potato starch	5.3	0.5	5.3	0.7	1.5	0.1	2.5	0.1
Wheat starch	6.2	0.5	12.3	1.7	5.1	0.9	7.2	1.5
Xanthan gum	3.4	0.2	4.0	0.9	6.2	0.9	2.5	0.3
Average	7.8	3.8	8.1	3.8	3.0	1.9	3.6	1.6
Proteins								
Fava bean protein concentrate	12.7	3.7	14.5	1.2	14.0	1.0	17.2	0.8
Hen egg albumen	12.3	2.0	12.4	1.7	1.6	0.2	2.3	0.0
Haemoglobin protein	7.0	1.9	4.9	3.2	13.1	2.4	18.2	2.0
Plasma protein	3.2	0.4	2.4	0.1	1.4	0.2	2.1	0.0
Technical gelatin	17.9	0.9	19.5	3.0	14.0	1.2	12.7	0.6
Wheat protein	15.6	1.0	19.0	2.7	2.0	0.4	2.1	0.2
Whey protein concentrate	5.2	2.0	5.3	2.1	1.2	0.1	2.8	0.1
Average	10.5	5.1	11.1	6.5	6.7	6.0	8.2	7.0

Table A3. Penetration resistance of medium-grained and fine-grained silica sand samples treated at biopolymer concentrations of 1 and 2 wt% (XG = 0.25 and 0.50 wt%). Tests were performed in triplicates ($n = 3$), with M = mean and SD = standard deviation. The water-treated control group achieved a penetration resistance of 1.5 N (SD = 0.1) on medium-grained sand and 1.7 N (SD = 0.2) on fine-grained silica sand.

	Penetration Resistance (N)							
	Medium-Grained Sand				**Fine-Grained Silica Sand**			
	1 wt%		**2 wt%**		**1 wt%**		**2 wt%**	
	M	**SD**	**M**	**SD**	**M**	**SD**	**M**	**SD**
Polysaccharides								
Carboxymethyl cellulose	11.5	2.0	21.1	3.6	14.9	3.2	16.4	2.9
Corn starch	14.2	2.0	27.4	4.4	18.0	4.8	25.4	8.6
Sodium lignosulfonate	1.7	0.2	4.3	1.3	14.2	3.1	21.1	3.1
Pea starch	8.0	1.2	20.1	4.3	10.8	2.2	14.4	1.1
Potato starch	4.4	0.6	16.8	1.3	10.1	1.7	19.0	1.6
Wheat starch	13.3	0.9	14.4	2.1	10.0	1.3	8.8	0.4
Xanthan gum	12.4	5.5	26.5	3.9	20.7	5.3	16.0	2.8
Average	9.4	4.5	18.7	7.3	14.1	3.8	17.3	4.9
Proteins								
Fava bean protein concentrate	8.8	2.9	15.9	1.7	15.0	4.1	30.3	3.9
Hen egg albumen	2.5	0.7	5.6	3.0	10.8	2.4	13.9	0.6
Haemoglobin protein	2.1	0.5	10.2	1.7	6.7	1.9	11.2	2.7
Plasma protein	3.5	1.6	15.4	2.8	15.8	2.6	21.0	4.8
Technical gelatin	8.0	0.2	34.0	6.0	20.0	4.5	37.9	2.1
Wheat protein	3.5	1.2	13.5	3.8	9.1	0.9	18.5	2.9
Whey protein concentrate	2.4	0.5	2.5	0.7	10.0	1.0	15.6	3.2
Average	4.4	2.6	13.9	9.5	12.5	4.3	21.2	8.9

Table A4. Mean modulus of elasticity (M_e) of crusts formed on medium-grained and fine-grained silica sand samples treated at biopolymer concentrations of 1 and 2 wt% (XG = 0.25 and 0.50 wt%). Tests were performed in triplicates (n = 3), with M = mean and SD = standard deviation. The water-treated control group formed no crust, so no M_e could be derived.

	Modulus of Elasticity (kN m^{-1})							
	Medium-Grained Sand				Fine-Grained Silica Sand			
	1 wt%		2 wt%		1 wt%		2 wt%	
	M	SD	M	SD	M	SD	M	SD
Polysaccharides								
Carboxymethyl cellulose	17.2	4.0	21.7	5.7	20.7	4.3	20.7	6.3
Corn starch	27.7	14.1	27.2	5.9	69.1	50.5	38.7	26.2
Sodium lignosulfonate	0.6	0.1	3.7	2.1	22.2	9.1	28.7	6.4
Pea starch	6.8	1.4	20.1	3.2	24.2	8.3	43.4	18.8
Potato starch	4.8	0.3	20.2	6.2	25.7	5.9	65.0	13.9
Wheat starch	15.9	1.5	13.2	3.0	8.6	2.4	7.4	0.7
Xanthan gum	23.9	9.3	36.0	11.4	41.2	12.1	20.7	3.5
Average	13.8	9.4	20.3	9.4	30.2	18.2	32.1	17.5
Proteins								
Fava bean protein concentrate	13.0	8.6	21.1	5.0	36.3	7.8	87.2	44.8
Hen egg albumen	1.3	0.3	7.4	3.3	19.7	9.3	26.3	3.7
Haemoglobin protein	1.2	0.2	13.1	5.6	14.1	7.1	27.3	12.8
Plasma protein	4.3	3.1	18.3	9.3	22.7	6.1	30.9	3.0
Technical gelatin	9.4	1.4	31.2	5.3	32.8	9.3	33.7	6.2
Wheat protein	3.7	2.8	13.8	3.8	14.1	3.7	47.0	16.6
Whey protein concentrate	1.8	1.6	1.5	0.5	19.9	2.6	22.4	6.2
Average	5.0	4.2	15.2	8.9	22.8	8.0	39.3	20.9

Medium-grained sand. For 1 wt% treatments, the M_e of polysaccharides-induced crusts was on average 2.8 times greater than the M_e of protein-induced crusts (Table A4). Amendments with CMC, CS, and XG resulted in the highest M_e among the tested polysaccharides and FBPC and TG among the tested proteins, respectively. Doubling the concentration resulted in the M_e of most biopolymer-induced crusts increasing, whereby protein-induced crusts exhibited a disproportionate increase relative to the polysaccharide-induced crusts. Fine-grained silica sand. On samples treated at 1 wt% biopolymer concentration, polysaccharide-induced crusts on average exhibited a larger M_e than protein-induced crusts, with CS, XG, FBPC, and TG achieving the highest M_e within their respective groups. Treatment at 2 wt% increased the M_e of most biopolymer treatments tested, with protein-induced crusts increasing disproportionately in thickness relative to the polysaccharide-induced crusts.

Table A5. Mean crust thickness of medium-grained sand and fine-grained silica sand samples treated at biopolymer concentrations of 1 and 2 wt% (XG = 0.25 and 0.50 wt%). Tests were performed in triplicates (*n* = 3), with M = mean and SD = standard deviation. Water-treated control groups formed no crusts.

	Crust Thickness (mm)							
	Medium-Grained Sand				Fine-Grained Silica Sand			
	1 wt%		2 wt%		1 wt%		2 wt%	
	M	SD	M	SD	M	SD	M	SD
Polysaccharides								
Carboxymethyl cellulose	6.3	1.6	7.1	0.4	9.6	0.5	9.3	0.2
Corn starch	7.5	0.3	5.9	0.8	7.7	0.5	7.3	1.0
Sodium lignosulfonate	2.4	1.0	4.6	0.9	13.1	0.3	18.1	2.7
Pea starch	6.3	0.2	6.9	0.4	8.3	0.6	7.0	0.2
Potato starch	3.4	0.5	7.3	0.2	11.7	0.4	13.7	0.2
Wheat starch	6.3	0.4	4.7	1.6	3.2	1.3	2.4	0.5
Xanthan gum	6.6	0.1	5.9	0.1	7.3	0.4	5.8	0.3
Average	5.5	1.8	6.1	1.0	8.7	3.0	9.1	4.9
Proteins								
Fava bean protein concentrate	7.6	0.4	8.8	0.1	11.6	0.4	11.0	2.1
Hen egg albumen	0.4	0.4	2.8	0.9	11.0	0.9	13.1	0.3
Haemoglobin protein	0.3	0.4	6.8	0.5	9.2	1.0	8.9	0.6
Plasma protein	5.2	0.8	7.0	0.2	9.4	0.6	11.8	0.3
Technical gelatin	2.3	0.7	5.8	0.4	10.1	0.6	9.4	0.3
Wheat protein	5.5	0.5	8.1	1.2	7.6	2.4	8.3	0.2
Whey protein concentrate	4.0	0.5	4.7	1.7	8.2	0.2	9.5	0.6
Average	3.6	2.6	6.3	1.9	9.6	1.3	10.3	1.6

Table A6. Qualitative classification of biopolymers according to their crust's visual appearance (cf. Figures 10 and 11).

Classification	Biopolymers	Description
Medium-grained sand		
Solid crusts	Polysaccharides: CS, CMC, WS, XG Proteins: FBPC, TG	1 and 2 wt% (XG = 0.25 and 0.50 wt%) concentrations. Crusts were recoverable in a single piece or up to four fully recoverable pieces.
Mediocre crusts	Polysaccharides: PES, POS Proteins: PP, WP, WPC	1 wt% concentration. Crusts broke into multiple large pieces, whereby some pieces were only partially recoverable and crumbled into countless pieces. 2 wt% concentration. Crusts were almost fully recoverable in several pieces.
Weak crusts	Polysaccharides: NLS Proteins: HEA, HG	1 wt% concentration. Crusts were very weak and brittle, crumbling into countless unrecoverable pieces. 2 wt% concentration. Crusts had increased stability but were still extremely fragile.
Fine-grained silica sand		
Solid crusts	Polysaccharides: CS, CMC, PES, XG	1 and 2 wt% (XG = 0.25 and 0.50 wt%) concentrations. Crusts were recoverable in a single piece or up to four fully recoverable pieces.
Mediocre crusts	Polysaccharides: NLS, POS Proteins: FBPC, HEA, HG, PP, TG, WP, WPC	1 and 2 wt% concentrations. Crusts were thick and almost fully recoverable in a few pieces at both tested concentrations. (TG exhibited a unique characteristic, as the uppermost part of its crust peeled off from the lower part while recovering the crust.)
Ductile crusts	Polysaccharides: WP	1 and 2 wt% concentrations. Crusts were very thin and ductile and even curled up during the curing period.

Table A7. Compilation of results from previous studies performing penetrometer testing on different soils to assess the penetration resistance of spray-on biopolymer applications.

Biopolymer	Soil	D_{50} (mm)	C_u	Penetrometer Shape	Penetrometer d (mm)	AR (L/m²)	Maximum Penetration Resistance (N) at Different Tested Concentrations (wt%) 0	0.3	0.5	0.6	0.7	0.8	1.0	1.5	1.6	2.0	3.0	5.0	Reference
Acacia gum	Poorly graded sand (SP)	0.15 [a]	2.1 [a]	flat	6	1.3	1.1 [a]		10.0 [a]				15.0 [a]			21.0 [a]	30.0 [a]	50.0 [a]	[16]
		0.15 [a]	2.1 [a]	flat	6	3.5	6.4 [a]		18.0 [a]				20.0 [a]			50.0 [a]	70.0 [a]	145.0 [a]	[16]
Sodium alginate		0.15 [a]	2.1 [a]	flat	6	1.3	1.1 [a]		9.0 [a]				17.5 [a]			18.0 [a]	15.0 [a]	N/A	[16]
		0.15 [a]	2.1 [a]	flat	6	3.5	6.4 [a]		7.5 [a]				21.0 [a]			25.0 [a]	N/A	N/A	[16]
Pectin		0.15 [a]	2.1 [a]	flat	6	1.3	1.1 [a]		15.0 [a]				20.0 [a]			28.0 [a]	22.0 [a]	15.0 [a]	[16]
		0.15 [a]	2.1 [a]	flat	6	3.5	6.4 [a]		20.0 [a]				39.0 [a]			33.0 [a]	30.0 [a]	N/A	[16]
Carboxymethyl cellulose	Poorly graded sand with silt (SP-SM)	0.16 [a]	2.2 [a]	flat	6	1	0.0 [b]	1.0 [b]	2.2 [b]		2.3 [b]								[14]
		0.16 [a]	2.2 [a]	flat	6	2	0.0 [b]	2.8 [b]	3.6 [b]		5.7 [b]								[14]
Guar gum		0.16 [a]	2.2 [a]	flat	6	1	0.0 [b]	0.7 [b]	1.4 [b]		2.4 [b]								[14]
		0.16 [a]	2.2 [a]	flat	6	2	0.0 [b]	2.3 [b]	2.8 [b]		3.5 [b]								[14]
Guar gum	Mine tailings	0.15 [a]	33.9 [a]	flat	6	1.9	212.8			263.0			358.8		428.0				[49]
Xanthan gum		0.15 [a]	33.9 [a]	flat	6	1.9	212.8			250.8			304.5		340.6				[49]
Xanthan gum	Mine tailings	0.13 [a]	56.3 [a]	flat	6	1.9	213.8	278.4	312.5			331.8							[137]
Xanthan gum	Hooralazim lagoon sand	0.22 [a]	7.5 [a]	flat	6	1.9	5.0 [a]		22.0 [a]				28.0 [a]	32.0 [a]					[17]
	Urmia lake sand	0.13 [a]	2.1 [a]	flat	6	1.9	5.0 [a]		6.0 [a]				6.0 [a]	11.0 [a]					[17]
	Mine tailings	0.28 [a]	9.4 [a]	flat	6	1.9	12.0 [a]		14.0 [a]				27.0 [a]	30.0 [a]					[17]
Carboxymethyl cellulose	Hooralazim lagoon sand	0.22 [a]	7.5 [a]	flat	6	1.9	5.0 [a]		15.0 [a]				38.0 [a]	58.0 [a]					[17]
	Urmia lake sand	0.13 [a]	2.1 [a]	flat	6	1.9	5.0 [a]		14.0 [a]				16.0 [a]	28.0 [a]					[17]
	Mine tailings	0.28 [a]	9.4 [a]	flat	6	1.9	12.0 [a]		18.0 [a]				27.0 [a]	30.0 [a]					[17]
Guar gum	Hooralazim lagoon sand	0.22 [a]	7.5 [a]	flat	6	1.9	5.0 [a]		8.0 [a]				28.0 [a]	39.0 [a]					[17]
	Urmia lake sand	0.13 [a]	2.1 [a]	flat	6	1.9	5.0 [a]		10.0 [a]				14.0 [a]	20.0 [a]					[17]
	Mine tailings	0.28 [a]	9.4 [a]	flat	6	1.9	12.0 [a]		16.0 [a]				25.0 [a]	31.0					[17]
Sodium alginate + $CaCl_2$	Poorly graded sand (SP)	0.24 [a]	1.7	flat	6	2.2	N/A		2.8 [c]				5.4 [c]			5.4 [c]			[67]

Note. AR = application rate, N/A = not available, SP = poorly graded sand, SP-SM = poorly graded sand with silt, [a] = values extracted from diagram in source, [b] = original values given in kPA and converted into N by multiplying with pin area, [c] = results from pocket penetrometer tests. Original values given in kg/cm² and converted into N using penetrometer pin area and g = 9.81 m/s².

Table A8. Compilation of results from previous studies performing penetrometer testing on different types of red sand (bauxite residue) to assess the penetration resistance of spray-on biopolymer applications.

Biopolymer	Soil	D_{50} (mm)	C_u	Penetrometer		AR	Maximum Penetration Resistance (N) at Different Tested Concentrations (wt%)													Reference
				shape	d (mm)	(L/m^2)	0	0.4	0.8	1.0	1.2	1.6	2.0	3.0	4.0	5.0	6.0	8.0	10.0	
Polyacrylamide	Red loam sand (SP)	N/A	<5	cone	2	2	3.0 [a]	5.5 [a]	7.5 [a]		9.5 [a]	15.0 [a]								[46]
Guar gum		N/A	<5	cone	2	2	3.0 [a]	5.5 [a]	7.0 [a]		7.5 [a]	8.5 [a]								[46]
Xanthan gum		N/A	<5	cone	2	2	3.0 [a]	5.5 [a]	6.5 [a]		7.0 [a]	7.5 [a]								[46]
Polyacrylamide	Red loam sand (SP)	N/A	<5	cone	2	2	N/A	6.0 [a]	8.0 [a]		9.8 [a]									[48]
Guar gum		N/A	<5	cone	2	2	N/A	5.5 [a]	7.0 [a]		8.0 [a]									[48]
Xanthan gum		N/A	<5	cone	2	2	N/A	5.3 [a]	6.5 [a]		7.0 [a]									[48]
Sodium lignosulfonate	Red loam sand (SP)	N/A	<5	cone	2	2	3.0 [a]			4.8 [a]			5.0 [a]	5.3 [a]	6.5 [a]	7.5 [a]	7.8 [a]	8.0 [a]	9.5 [a]	[28]
Calcium lignosulfonate		N/A	<5	cone	2	2	3.0 [a]			3.5 [a]			4.0 [a]	4.5 [a]	5.0 [a]	5.3 [a]	5.8 [a]	6.3 [a]	7.3 [a]	[28]
Polyacrylamide	Red sand (d < 0.15)	N/A	N/A	cone	2	2	3.2	6.2	7.8		9.6	12.3	15.3							[13]
	Red sand (0.15 < d < 0.3)	N/A	N/A	cone	2	2	3.2	7.4	9.6		11.4	14.2	16.5							[13]
	Red sand (0.3 < d < 0.45)	N/A	N/A	cone	2	2	3.3	8.1	11.1		13.2	16.0	18.0							[13]
Guar gum	Red sand (d < 0.15)	N/A	N/A	cone	2	2	3.2	5.7	7.0		8.3	9.3	10.2							[13]
	Red sand (0.15 < d < 0.3)	N/A	N/A	cone	2	2	3.2	6.7	9.0		10.2	11.7	14.2							[13]
	Red sand (0.3 < d < 0.45)	N/A	N/A	cone	2	2	3.3	7.1	9.9		11.0	12.8	14.4							[13]
Xanthan gum	Red sand (d < 0.15)	N/A	N/A	cone	2	2	3.2	5.4	6.4		7.2	7.7	8.1							[13]
	Red sand (0.15 < d < 0.3)	N/A	N/A	cone	2	2	3.2	5.7	6.9		8.6	10.8	13.5							[13]
	Red sand (0.3 < d < 0.45)	N/A	N/A	cone	2	2	3.3	6.5	8.0		10.0	12.3	13.4							[13]
Sodium lignosulfonate	Red sand (d < 0.15)	N/A	N/A	cone	2	2	3.0 [a]						5.0 [a]		7.0 [a]		7.5 [a]	8.0 [a]	9.0 [a]	[33]
	Red sand (0.15 < d < 0.3)	N/A	N/A	cone	2	2	3.0 [a]						6.5 [a]		8.0 [a]		9.0 [a]	10.5 [a]	13.0 [a]	[33]
	Red sand (d > 0.3)	N/A	N/A	cone	2	2	3.0 [a]						6.0 [a]		8.5 [a]		10.0 [a]	12.5 [a]	14.0 [a]	[33]
Calcium lignosulfonate	Red sand (d < 0.15)	N/A	N/A	cone	2	2	3.0 [a]						3.5 [a]		4.0 [a]		5.5 [a]	6.5 [a]	7.5 [a]	[33]
	Red sand (0.15 < d < 0.3)	N/A	N/A	cone	2	2	3.0 [a]						4.0 [a]		5.5 [a]		7.0 [a]	8.5 [a]	11.5 [a]	[33]
	Red sand (d > 0.3)	N/A	N/A	cone	2	2	3.0 [a]						5.0 [a]		6.0 [a]		8.0 [a]	10.0 [a]	12.5 [a]	[33]

Note. AR = application rate, N/A = not available, SP = poorly graded sand, SP-SM = poorly graded sand with silt, [a] = values extracted from diagram in source. Original values given in kg/cm^2 and converted into N using penetrometer pin area and $g = 9.81$ m/s^2.

Table A9. Compilation of results from previous studies analysing the crust thickness of biopolymer-treated soils.

Biopolymer	Soil Type	AR (L/m^2)	Crust Thickness (mm) at Different Tested Concentrations (%)																		Reference
			0	0.3	0.4	0.5	0.7	0.8	1.0	1.2	1.6	2.0	3.0	4.0	5.0	6.0	7.0	8.0	9.0	10.0	
Acacia gum	Poorly graded sand (SP)	1.3				7.4–11.9			5.1–11.4			5.2–11.5	3.8–10.9		2.8–10.9						[16]
Sodium alginate		1.3				8.5–15.3			7.6–15.2			2.2–10.3	3.0–6.7								[16]
Pectin		1.3				3.8–8.4			3.6–7.3			4.6–6.1	2.4–3.0		0.9–1.2						[16]
Acacia gum		3.5				15.1–29.8			12.6–29.3			11.5–22.1	11.5–22.0		9.7–21.6						[16]
Sodium alginate		3.5				9.0–12.1			2.6–4.7			1.6–3.2	1.2–2.1								[16]
Pectin		3.5				8.5–15.3			7.6–15.2			2.2–10.3	3.0–6.7		0.7–1.9						[16]
Sodium alginate + $CaCl_2$	Poorly graded sand (SP)	2.2				5.2–7.8			4.9–8.0			5.9–7.7									[67]
Carboxymethyl cellulose	Poorly graded sand with silt (SP-SM)	1.0		4.0		3.8	3.4														[14]
Guar gum		2.0		3.9		4.5	4.2														[14]
Carboxymethylcellulose		1.0		7.9		6.3	6.0														[14]
Guar gum		2.0		6.6		7.8	10.3														[14]
Sodium lignosulfonate	Red sand (SP)	2.0							11.0 [a]			10.5 [a]	11.0 [a]	8.5 [a]	7.5 [a]	7.5 [a]	7.0 [a]	6.5 [a]	6.5 [a]	6.0 [a]	[28]
Calcium lignosulfonate		2.0							11.0 [a]			10.5 [a]	10.5 [a]	10.0 [a]	10.0 [a]	9.5 [a]	9.0 [a]	8.5 [a]	8.0 [a]	7.5 [a]	[28]
Polyacrylamide	Red sand (SP)	2.0	30.0 [a]		26.0 [a]			15.0 [a]		14.5 [a]	12.0 [a]										[46]
Guar gum		2.0	30.0 [a]		19.0 [a]			23.0 [a]		27.0 [a]	22.5 [a]										[46]
Xanthan gum		2.0	30.0 [a]		13.0 [a]			12.0 [a]		10.0 [a]	8.0 [a]										[46]
Polyacrylamide	Red sand (d < 0.15)	2.0	30.0 [a]		27.5 [a]			22.5 [a]		19.0 [a]	15.0 [a]	15.0 [a]									[33]
Guar gum		2.0	30.0 [a]		26.0 [a]			20.0 [a]		14.0 [a]	14.5 [a]	14.0 [a]									[33]
Xanthan gum		2.0	30.0 [a]		13.5 [a]			11.5 [a]		12.0 [a]	12.5 [a]	10.5 [a]									[33]

Note. AR = application rate, SP = poorly graded sand, SP-SM = poorly graded sand with silt, [a] = values extracted from diagram in source.

References

1. Entwistle, J.A.; Hursthouse, A.S.; Marinho Reis, P.A.; Stewart, A.G. Metalliferous mine dust: Human health impacts and the potential determinants of disease in mining communities. *Curr. Pollut. Rep.* **2019**, *5*, 67–83. [CrossRef]
2. Zota, A.R.; Riederer, A.M.; Ettinger, A.S.; Schaider, L.A.; Shine, J.P.; Amarasiriwardena, C.J.; Wright, R.O.; Spengler, J.D. Associations between metals in residential environmental media and exposure biomarkers over time in infants living near a mining-impacted site. *J. Expo. Sci. Environ. Epidemiol.* **2016**, *26*, 510–519. [CrossRef] [PubMed]
3. Tian, S.; Liang, T.; Li, K. Fine road dust contamination in a mining area presents a likely air pollution hotspot and threat to human health. *Environ. Int.* **2019**, *128*, 201–209. [CrossRef] [PubMed]
4. Zanetta-Colombo, N.C.; Fleming, Z.L.; Gayo, E.M.; Manzano, C.A.; Panagi, M.; Valdés, J.; Siegmund, A. Impact of mining on the metal content of dust in indigenous villages of northern Chile. *Environ. Int.* **2022**, *169*, 107490. [CrossRef]
5. Khan, R.K.; Strand, M.A. Road dust and its effect on human health: A literature review. *Epidemiol. Health* **2018**, *40*, e2018013. [CrossRef]
6. du Plessis, J.J.; Janse van Rensburg, L. Effectiveness of applying dust suppression palliatives on haul roads. *J. Mine Vent. Soc. S. Afr.* **2015**, *69*, 15–19.
7. Thompson, R.J.; Visser, A.T. Selection, performance and economic evaluation of dust palliatives on surface mine haul roads. *J. S. Afr. Inst. Min. Metall.* **2007**, *107*, 435–450.
8. Clarke, B.; Otto, F.; Stuart-Smith, R.; Harrington, L. Extreme weather impacts of climate change: An attribution perspective. *Environ. Res. Clim.* **2022**, *1*, 12001. [CrossRef]
9. IPCC. *Climate Change 2022: Impacts, Adaption and Vulnerability: Contribution of Working Group II to the Sixth Assessment Report of the Intergovernmental Panel on Climate Change*; IPCC: Geneva, Switzerland, 2022. [CrossRef]
10. Philip, S.; Martin, R.V.; Snider, G.; Weagle, C.L.; van Donkelaar, A.; Brauer, M.; Henze, D.K.; Klimont, Z.; Venkataraman, C.; Guttikunda, S.K.; et al. Anthropogenic fugitive, combustion and industrial dust is a significant, underrepresented fine particulate matter source in global atmospheric models. *Environ. Res. Lett.* **2017**, *12*, 44018. [CrossRef]
11. Parvej, S.; Naik, D.L.; Sajid, H.U.; Kiran, R.; Huang, Y.; Thanki, N. Fugitive dust suppression in unpaved roads: State of the art research review. *Sustainability* **2021**, *13*, 2399. [CrossRef]
12. Piechota, T.C.; van Ee, J.; Stave, K.; James, D. *Potential Environmental Impacts of Dust Suppressants: "Avoiding Another Times Beach"*; U.S. Environmental Protection Agency: Las Vegas, NV, USA, 2002.
13. Ding, X.; Luo, Z.; Xu, G.; Chang, P. Characterization of red sand dust pollution control performance via static and dynamic laboratorial experiments when applying polymer stabilizers. *Environ. Sci. Pollut. Res. Int.* **2021**. [CrossRef] [PubMed]
14. Owji, R.; Habibagahi, G.; Nikooee, E.; Afzali, S.F. Wind erosion control using carboxymethyl cellulose: From sand bombardment performance to microfabric analysis. *Aeolian Res.* **2021**, *50*, 100696. [CrossRef]
15. Wade, E.; Zowada, R.; Foudazi, R. Alginate and guar gum spray application for improving soil aggregation and soil crust integrity. *Carbohydr. Polym.* **2021**, *2*, 100114. [CrossRef]
16. Lemboye, K.; Almajed, A.; Alnuaim, A.; Arab, M.; Alshibli, K. Improving sand wind erosion resistance using renewable agriculturally derived biopolymers. *Aeolian Res.* **2021**, *49*, 100663. [CrossRef]
17. Toufigh, V.; Ghassemi, P. Control and stabilization of fugitive dust: Using eco-friendly and sustainable materials. *Int. J. Geomech.* **2020**, *20*, 4020140. [CrossRef]
18. Runnels, C.M.; Lanier, K.A.; Williams, J.K.; Bowman, J.C.; Petrov, A.S.; Hud, N.V.; Williams, L.D. Folding, assembly, and persistence: The essential nature and origins of biopolymers. *J. Mol. Evol.* **2018**, *86*, 598–610. [CrossRef]
19. Katra, I. Comparison of diverse dust control products in wind-induced dust emission from unpaved roads. *Appl. Sci.* **2019**, *9*, 5204. [CrossRef]
20. Jang, J. A review of the application of biopolymers on geotechnical engineering and the strengthening mechanisms between typical biopolymers and soils. *Adv. Mater. Sci. Eng.* **2020**, *2020*, 1465709. [CrossRef]
21. Fatehi, H.; Ong, D.E.L.; Yu, J.; Chang, I. Biopolymers as green binders for soil improvement in geotechnical applications: A review. *Geosci. J.* **2021**, *11*, 291. [CrossRef]
22. Mendonça, A.; Morais, P.V.; Pires, A.C.; Chung, A.P.; Oliveira, P.V. A Review on the importance of microbial biopolymers such as xanthan gum to improve soil properties. *Appl. Sci.* **2021**, *11*, 170. [CrossRef]
23. Choi, S.-G.; Chang, I.; Lee, M.; Lee, J.-H.; Han, J.-T.; Kwon, T.-H. Review on geotechnical engineering properties of sands treated by microbially induced calcium carbonate precipitation (MICP) and biopolymers. *Constr. Build. Mater.* **2020**, *246*, 118415. [CrossRef]
24. Chang, I.; Im, J.; Cho, G.-C. Introduction of microbial biopolymers in soil treatment for future environmentally-friendly and sustainable geotechnical engineering. *Sustainability* **2016**, *8*, 251. [CrossRef]
25. Chang, I.; Prasidhi, A.K.; Im, J.; Shin, H.-D.; Cho, G.-C. Soil treatment using microbial biopolymers for anti-desertification purposes. *Geoderma* **2015**, *253–254*, 39–47. [CrossRef]
26. Xu, G.; Ding, X.; Kuruppu, M.; Zhou, W.; Biswas, W. Research and application of non-traditional chemical stabilizers on bauxite residue (red sand) dust control, a review. *Sci. Total Environ.* **2018**, *616–617*, 1552–1565. [CrossRef] [PubMed]

27. Liu, J.; Wang, T.; Jin, L.; Li, G.; Wang, S.; Wei, Y.; Ou, S.; Wang, Y.; Xu, J.; Lin, M.; et al. Suppression Characteristics and Mechanism of Molasses Solution on Coal Dust: A Low-Cost and Environment-Friendly Suppression Method in Coal Mines. *Int. J. Environ. Res. Public Health* **2022**, *19*, 16472. [CrossRef] [PubMed]
28. Ding, X.; Xu, G.; Kizil, M.; Zhou, W.; Guo, X. Lignosulfonate treating bauxite residue dust pollution: Enhancement of mechanical properties and wind erosion behavior. *Water Air Soil Pollut.* **2018**, *229*, 1084. [CrossRef]
29. German Institute for Standardization. *DIN EN ISO 17892-4*; Geotechnical Investigation and Testing-Laboratory Testing of Soil-Part 4: Determination of Particle Size Distribution. German Institute for Standardization: Berlin, Germany, 2017.
30. *AST D2487-17*; Standard Practice for Classification of Soils for Engineering Purposes (Unified Soil Classification System). ASTM International: West Conshohocken, PA, USA, 2018.
31. German Institute for Standardization. *DIN EN ISO 11508-2017*; Soil Quality-Determination of Particle Density. German Institute for Standardization: Berlin, Germany, 2017.
32. *DIN EN 15933:2012-11*; Sludge, Treated Biowaste and Soil—Determination of pH. German Institute for Standardization: Berlin, Germany, 2012.
33. Ding, X.; Xu, G.; Zhang, Y.; Luo, Z.; Deng, J. Reduction of airborne bauxite residue dust pollution by enhancing the structural stability via the application of non-traditional stabilizers. *Water Air Soil Pollut.* **2021**, *232*, 100. [CrossRef]
34. Bixler, H.J.; Porse, H. A decade of change in the seaweed hydrocolloids industry. *J. Appl. Phycol.* **2011**, *23*, 321–335. [CrossRef]
35. Zuorro, A.; Moreno-Sader, K.A.; González-Delgado, Á.D. Economic Evaluation and Techno-Economic Sensitivity Analysis of a Mass Integrated Shrimp Biorefinery in North Colombia. *Polymers* **2020**, *12*, 2397. [CrossRef]
36. Criminna, R.; Fidalgo, A.; Delisi, R.; Ilharco, L.M.; Pagliaro, M. Pectin production and global market. *Agro Food Ind. Hi-Tech* **2016**, *27*, 17–20.
37. Chang, I.; Im, J.; Cho, G.-C. Geotechnical engineering behaviors of gellan gum biopolymer treated sand. *Can. Geotech. J.* **2016**, *53*, 1658–1670. [CrossRef]
38. Gosselink, R.J.A. Lignin as a Renewable Aromatic Resource for the Chemical Industry. Ph.D. Thesis, Wageningen University, Wageningen, The Netherlands, 2011.
39. Hailu, A. Production and Optimisation of Carboxymethylcellulose from Waste Cartons. Master Thesis, Addis Ababa University, Addis Ababa, Ethiopia, 2016.
40. Phillips, G.O.; Edwards, C.A.; Garcia, A.L.; Williams, P.A.; Dickinson, E.; Armisén, R.; Taggart, P.; Mitchell, J.R.; Haug, I.J.; Draget, K.I.; et al. *Handbook of Hydrocolloids*; CRC: Boca Raton, FL, USA; Woodhead: Cambridge, UK, 2009; ISBN 978-1-84569-414-2.
41. Mulder, W.; van der Peet-Schwering, C.; Hua, N.-P.; van Ree, R. Proteins for Food, Feed and Biobased Applications: Biorefining of Protein Containing Biomass. 2016. Available online: https://www.ieabioenergy.com/wp-content/uploads/2016/10/Proteins-for-Food_Feed_and_Biobased_Applications_-IEA-Bioenergy-Task42_September-2016.pdf (accessed on 6 January 2023).
42. Heusala, H.; Sinkko, T.; Sözer, N.; Hytönen, E.; Mogensen, L.; Knudsen, M.T. Carbon footprint and land use of oat and faba bean protein concentrates using a life cycle assessment approach. *J. Clean. Prod.* **2020**, *242*, 118376. [CrossRef]
43. German Institute for Standardization. *EN ISO 1666-1997*; Starch-Determination of Moisture Content-Oven-Drying Method. German Institute for Standardization: Berlin, Germany, 1998.
44. International Organization for Standardization. *ISO 6496:1999*; Animal Feeding Stuffs-Determination of Moisture and Other Volatile Matter Content. International Organization for Standardization: Genève, Switzerland, 1999.
45. Association of Official Analytical Chemists. *Official Methods of Analysis-Agricultural Chemicals; Contaminants Drugs*; Association of Official Analytical chemists: Arlington, TX, USA, 1990.
46. Ding, X.; Xu, G.; Zhou, W.; Kuruppu, M. Effect of synthetic and natural polymers on reducing bauxite residue dust pollution. *Environ. Technol.* **2018**, *41*, 556–565. [CrossRef] [PubMed]
47. Hu, Y.; Shi, L.; Shan, Z.; Dai, R.; Chen, H. Efficient removal of atmospheric dust by a suppressant made of potato starch, polyacrylic acid and gelatin. *Environ. Chem. Lett.* **2020**, *18*, 1701–1711. [CrossRef]
48. Ding, X.; Xu, G.; Liu, W.V.; Yang, L.; Albijanic, B. Effect of polymer stabilizers' viscosity on red sand structure strength and dust pollution resistance. *J. Powder Technol.* **2019**, *352*, 117–125. [CrossRef]
49. Chen, R.; Lee, I.; Zhang, L. Biopolymer stabilization of mine tailings for dust control. *J. Geotech. Geoenviron. Eng.* **2014**, *141*, 4014100. [CrossRef]
50. Maidapwad, S.L.; Sananse, S.L. On analysis of two-way ANOVA using data transformation techniques. *Int. J. Sci. Res.* **2014**, *3*, 480–483.
51. Tran, T.P.A.; Cho, G.-C.; Ilhan, C. Water retention characteristics of biopolymer hydrogel containing sandy soils. *Hue Univ. J. Sci. Earth Sci. Environ.* **2020**, *129*, 5–17. [CrossRef]
52. Tran, T.P.; Cho, G.C.; Lee, S.J.; Chang, I. Effect of xanthan gum biopolymer on the water retention characteristics of unsaturated sand. In Proceedings of the UNSAT2018 the 7th International Conference on Unsaturated Soils, Hong Kong, China, 3–6 August 2018.
53. Huang, J.; Ho, C.-H.; Gao, Y.; Wu, Z.; Zhang, Y. Evaluation of polymer based dust suppressant mixed with clayey soil in unpaved road: Lab experiment. In *Transportation and Geotechniques: Materials, Sustainability and Climate: Proceedings of the 5th GeoChina International Conference 2018-Civil Infrastructures Confronting Severe Weathers and Climate Changes: From Failure to Sustainability, Held on July 23 to 25, 2018 in HangZhou, China*, 1st ed.; Barman, M., Zaman, M., Chang, J.-R., Eds.; Springer International Publishing: Cham, Switzerland, 2019; pp. 1–9. ISBN 978-3-319-95768-5.

54. Parameswaran, K.; Ekholm, J.; Zhang, L. Evaluation of mine tailings dust control. In Proceedings of the Geoenvironmental Engineering, Selected Papers from the 2014 Geoshanghai International Congress, Shanghai, China, 26–28 May 2014; Reddy, K.R., Ed.; Curran: New York, NY, USA, 2014; pp. 80–89, ISBN 978-0-784-41343-2.
55. Hataf, N.; Ghadir, P.; Ranjbar, N. Investigation of soil stabilization using chitosan biopolymer. *J. Clean. Prod.* **2018**, *170*, 1493–1500. [CrossRef]
56. Tran, T.P.A.; Im, J.; Cho, G.-C.; Chang, I. Soil-water characteristics of xanthan gum biopolymer containing soils. In Proceedings of the UNSAT2018 the 7th International Conference on Unsaturated Soils, Hong Kong, China, 3–6 August 2018.
57. Liu, J.; Shi, B.; Lu, Y.; Jiang, H.; Huang, H.; Wang, G.; Kamai, T. Effectiveness of a new organic polymer sand-fixing agent on sand fixation. *Environ. Earth Sci.* **2012**, *65*, 589–595. [CrossRef]
58. Ayeldeen, M.; Negm, A.; El Sawwaf, M.; Gädda, T. Laboratory study of using biopolymer to reduce wind erosion. *Int. J. Geo-Eng.* **2017**, *12*, 228–240. [CrossRef]
59. Seo, S.; Lee, M.; Im, J.; Kwon, Y.-M.; Chung, M.-K.; Cho, G.-C.; Chang, I. Site application of biopolymer-based soil treatment (BPST) for slope surface protection: In-situ wet-spraying method and strengthening effect verification. *Constr. Build. Mater.* **2021**, *307*, 124983. [CrossRef]
60. Kwon, Y.-M.; Ham, S.-M.; Kwon, T.-H.; Cho, G.-C.; Chang, I. Surface-erosion behaviour of biopolymer-treated soils assessed by EFA. *Géotechnique Lett.* **2020**, *10*, 106–112. [CrossRef]
61. Vishweshwaran, M.; Padmashree, S.; Kalambari, R.; Sathya Bhaarathi, C.R.; Ramani Sujatha, E. Pre-gelatinized starch-a sustainable soil stabilizer. *J. Adv. Res. Dyn. Control Syst.* **2018**, *10*, 543–546.
62. Rosa, I.; Roedel, H.I.; Allende, M.D.; Lepech, M.J.; Loftus, D. On Designing Biopolymer-Bound Soil Composites (BSC) for Peak Compressive Strength. *J. Renew. Mater.* **2020**, *8*, 845–861. [CrossRef]
63. Dang, X.; Shan, Z.; Chen, H. Usability of oxidized corn starch-gelatin blends for suppression and prevention of dust. *J. Appl. Polym. Sci.* **2017**, *134*, 1–9. [CrossRef]
64. Brown, M.J.; Robbins, C.W.; Freeborn, L.L. Combining cottage cheese whey and straw reduces erosion while increasing infiltration in furrow irrigation. *J. Soil Water Conserv.* **1998**, *53*, 152–156.
65. Sharan, S.; Zanghelini, G.; Zotzel, J.; Bonerz, D.; Aschoff, J.; Saint-Eve, A.; Maillard, M.-N. Fava bean (Vicia faba L.) for food applications: From seed to ingredient processing and its effect on functional properties, antinutritional factors, flavor, and color. *Compr. Rev. Food Sci. Food Saf.* **2021**, *20*, 401–428. [CrossRef] [PubMed]
66. Phillips, G.O.; Williams, P.A.; O'Kennedy, B.T.; Boland, M.; Tarté, R.; Mayer, O.; Haug, I.J.; Draget, K.I.; Tahergorabi, R.; Hosseini, S.V.; et al. *Handbook of Food Proteins*; Woodhead Publishing: Oxford, UK, 2011; ISBN 978-1-84569-758-7.
67. Almajed, A.; Lemboye, K.; Arab, M.G.; Alnuaim, A. Mitigating wind erosion of sand using biopolymer-assisted EICP technique. *Soils Found.* **2020**, *60*, 356–371. [CrossRef]
68. Blanck, G.; Cuisinier, O.; Masrouri, F. Soil treatment with organic non-traditional additives for the improvement of earthworks. *Acta Geotech.* **2014**, *9*, 1111–1122. [CrossRef]
69. Moghal, A.A.B.; Vydehi, K.V. State-of-the-art review on efficacy of xanthan gum and guar gum inclusion on the engineering behavior of soils. *Innov. Infrastruct. Solut.* **2021**, *6*, 108. [CrossRef]
70. Kunz, B.K.; Little, E.E.; Barandino, V.L. Aquatic toxicity of chemical road dust suppressants to freshwater organisms. *Arch. Environ. Contam. Toxicol.* **2021**, *82*, 294–305. [CrossRef]
71. Katebi, H.; Fahmi, A.; Ouria, A.; Babaeian Amini, A.; Kafil, H.S. Microbial surface treatment of sand with sporosarcina pasteurii to improve the wind erosion resistance in Urmia Lake. *Appl. Environ. Soil Sci.* **2021**, *2021*, 1–11. [CrossRef]
72. Liu, J.; Li, G.; Li, X. Geotechnical engineering properties of soils solidified by microbially induced $CaCO_3$ precipitation (MICP). *Adv. Civ. Eng.* **2021**, *2021*, 6683930. [CrossRef]
73. Meng, H.; Gao, Y.; He, J.; Qi, Y.; Hang, L. Microbially induced carbonate precipitation for wind erosion control of desert soil: Field-scale tests. *Geoderma* **2021**, *383*, 114723. [CrossRef]
74. Freer, J.; Lübeck, M.; Sieger, J.L.; Lottermoser, B.G.; Braun, M. Effectiveness of food processing by-products as dust suppressants for exposed mine soils: Results from laboratory experiments and field trials. *Appl. Sci.* **2022**, *12*, 11551. [CrossRef]
75. Freer, J.; Bucher, P.G.; Braun, M.; Lottermoser, B.G. Food processing by-products and wastes as potential dust suppressants at mine sites: Results from unconfined compressive strength testing. *J. Air Waste Manag. Assoc.* **2022**, *72*, 1012–1026. [CrossRef] [PubMed]
76. Niaounakis, M. Definitions of terms and types of biopolymers. In *Biopolymers: Applications and Trends*; Niaounakis, M., Ed.; Elsevier: Oxford, UK, 2015; pp. 1–90. ISBN 9780323353991.
77. Thomas, S.; Durand, D.; Chassenieux, C.; Jyotishkumar, P. *Handbook of Biopolymer-Based Materials*; Wiley-VCH Verlag GmbH & Co. KGaA: Weinheim, Germany, 2013; ISBN 9783527652457.
78. Rimbarngaye, A.; Mwero, J.N.; Ronoh, E.K. Effect of gum arabic as partial replacement of cement on the durability properties of compressed laterite blocks. *Open J. Civ. Eng.* **2021**, *11*, 398–410. [CrossRef]
79. Smitha, S.; Sachan, A. Use of agar biopolymer to improve the shear strength behavior of sabarmati sand. *Int. J. Geo. Eng.* **2016**, *10*, 387–400. [CrossRef]

80. Khatami, H.R.; O'Kelly, B.C. Improving mechanical properties of sand using biopolymers. *J. Geotech. Geoenviron. Eng.* **2013**, *139*, 1402–1406. [CrossRef]
81. Vishweshwaran, M.; Sujatha, E.R. β-glucan as a sustainable alternative to stabilize pavement subgrade. *Polymers* **2022**, *14*, 2850. [CrossRef]
82. Soldo, A.; Aguilar, V.; Miletić, M. Macroscopic stress-strain response and strain-localization behavior of biopolymer-treated soil. *Polymers* **2022**, *14*, 997. [CrossRef] [PubMed]
83. Anandha Kumar, S.; Sujatha, E.R. An appraisal of the hydro-mechanical behaviour of polysaccharides, xanthan gum, guar gum and β-glucan amended soil. *Carbohydr. Polym.* **2021**, *265*, 118083. [CrossRef] [PubMed]
84. Chang, I.; Im, J.; Cho, G.-C. An environmentally-friendly geotechnical approach for soil erosion reduction using microbial biopolymers. In Proceedings of the Geo-Chicago 2016, Sustainability and Resiliency in Geotechnical Engineering, Chicago, IL, USA, 14–18 August 2016; Zekkos, D., Farid, A., De, A., Reddy, K.R., Yesiller, N., Eds.; American Society of Civil Engineers: Reston, VA, USA, 2016; pp. 17–24, ISBN 9780784480120.
85. Chang, I.; Cho, G.-C. Geotechnical behavior of a beta-1,3/1,6-glucan biopolymer-treated residual soil. *Geomech. Eng.* **2014**, *7*, 633–647. [CrossRef]
86. Chang, I.; Cho, G.-C. Strengthening of Korean residual soil with β-1, 3/1, 6-glucan biopolymer. *Constr. Build. Mater.* **2012**, *30*, 30–35. [CrossRef]
87. Donayre, A.; Sanchez, L.F.; Kim, S.; Aguilar, R.; Nakamatsu, J. Eco-friendly improvement of water erosion resistance of unstable soils with biodegradable polymers. *IOP Conf. Ser. Mater. Sci. Eng.* **2018**, *416*, 12044. [CrossRef]
88. Nakamatsu, J.; Kim, S.; Ayarza, J.; Ramírez, E.; Elgegren, M.; Aguilar, R. Eco-friendly modification of earthen construction with carrageenan: Water durability and mechanical assessment. *Constr. Build. Mater.* **2017**, *139*, 193–202. [CrossRef]
89. Bocheńska, M.; Bujko, M.; Dyka, I.; Srokosz, P.; Ossowski, R. Effect of chitosan solution on low-cohesive soil's shear modulus G determined through resonant column and torsional shearing tests. *Appl. Sci.* **2022**, *12*, 5332. [CrossRef]
90. Shariatmadari, N.; Reza, M.; Tasuji, A.; Ghadir, P.; Javadi, A.A. Experimental study on the effect of chitosan biopolymer on sandy soil stabilization. *E3S Web Conf.* **2020**, *195*, 6007. [CrossRef]
91. Soldo, A.; Miletić, M.; Auad, M.L. Biopolymers as a sustainable solution for the enhancement of soil mechanical properties. *Sci. Rep.* **2020**, *10*, 267. [CrossRef]
92. Aguilar, R.; Nakamatsu, J.; Ramírez, E.; Elgegren, M.; Ayarza, J.; Kim, S.; Pando, M.A.; Ortega-San-Martin, L. The potential use of chitosan as a biopolymer additive for enhanced mechanical properties and water resistance of earthen construction. *Constr. Build. Mater.* **2016**, *114*, 625–637. [CrossRef]
93. Taytak, B.; Pulat, H.F.; Yukselen-Aksoy, Y. Improvement of engineering properties of soils by biopolymer additives. In Proceedings of the 3. ICNDSMGE-ZM 2012 3rd International Conference on New Developments, Nicosia, Cyprus, 28–30 June 2012.
94. Alsand, A. Novel Biopolymer Treatment for Wind Induced Soil Erosion. Ph.D. Thesis, Arizona State University, Tempe, AZ, USA, 2011.
95. Kavazanjian, E.; Iglesias, E.; Karatas, I. Biopolymer soil stabilization for wind erosion control. In Proceedings of the 17th International Conference on Soil Mechanics and Geotechnical Engineering; Hamza, M., Shahien, M., El-Mossallamy, Y., Eds.; IOS Press: Amsterdam, The Netherlands, 2009. ISBN 978-1-60750-031-5.
96. Armistead, S.J.; Rawlings, A.E.; Smith, C.C.; Staniland, S.S. Biopolymer stabilization/solidification of soils: A rapid, micro-macro, cross-disciplinary approach. *Environ. Sci. Technol.* **2020**, *54*, 13963–13972. [CrossRef] [PubMed]
97. Ojuri, O.O.; Ramdas, V.; Aderibigbe, E.A.; Williams, C.G.; Ramchuran, S.; Al-Nageim, H. Improving strength and hydraulic characteristics of regional clayey soils using biopolymers. *Case Stud. Constr. Mater.* **2022**, *17*, e01319. [CrossRef]
98. Yang, Q.; Pei, X.; Huang, R. Impact of polymer mixtures on the stabilization and erosion control of silty sand slope. *J. Mt. Sci.* **2019**, *16*, 470–485. [CrossRef]
99. Inyang, H.I.; Bae, S.; Pando, M.A. Contaminant dust suppression materials: A cost-effectiveness estimation methodology. *Measurement* **2016**, *93*, 563–571. [CrossRef]
100. Ham, S.-M.; Chang, I.; Noh, D.-H.; Kwon, T.-H.; Muhunthan, B. Improvement of surface erosion resistance of sand by microbial biopolymer formation. *J. Geotech. Geoenviron. Eng.* **2018**, *144*, 6018004. [CrossRef]
101. Ramachandran, A.L.; Mukherjee, A.; Dhami, N.K. Nanoscale to macroscale characterization of in-situ bacterial biopolymers for applications in soil stabilization. *Front. Mater.* **2022**, *8*, 546. [CrossRef]
102. Bitar, L. Optimum Mixing Design of Xanthan and Gellan Treated Soils for Slope Stabilization for Weathered Shales and Glacial Tills in Nebraska. Master's Thesis, University of Nebraska, Lincoln, NE, USA, 2020.
103. Chang, I.; Cho, G.-C. Shear strength behavior and parameters of microbial gellan gum-treated soils: From sand to clay. *Acta Geotech.* **2019**, *14*, 361–375. [CrossRef]
104. Chang, I.; Im, J.; Lee, S.-W.; Cho, G.-C. Strength durability of gellan gum biopolymer-treated Korean sand with cyclic wetting and drying. *Constr. Build. Mater.* **2017**, *143*, 210–221. [CrossRef]
105. Im, J.; Tran, A.T.; Chang, I.; Cho, G.-C. Dynamic properties of gel-type biopolymer-treated sands evaluated by Resonant Column (RC) Tests. *Geomech. Eng.* **2017**, *12*, 815–830. [CrossRef]
106. Ferruzzi, G.; Pan, N.; Casey, W.H. Mechanical properties of gellan and polyacrylamide gels with implications for soil stabilization. *Soil Sci.* **2000**, *165*, 778–792. [CrossRef]

107. Bonal, N.S.; Prasad, A.; Verma, A.K. Effect of microbial biopolymers on mechanical properties of bauxite residue. *KSCE J. Civ. Eng.* **2021**, *25*, 2437–2450. [CrossRef]
108. Bozyigit, I.; Javadi, A.; Altun, S. Strength properties of xanthan gum and guar gum treated kaolin at different water contents. *J. Rock Mech. Geotech. Eng.* **2021**, *13*, 1160–1172. [CrossRef]
109. Muguda, S.; Hughes, P.N.; Augarde, C.E.; Perlot, C.; Walter Bruno, A.; Gallipoli, D. Cross-linking of biopolymers for stabilizing earthen construction materials. *Build. Res. Inf.* **2021**, *50*, 502–514. [CrossRef]
110. Bonal, N.S.; Prasad, A.; Verma, A.K. Use of biopolymers to enhance the geotechnical properties of coal mine overburden waste. *Géotech. Lett.* **2020**, *10*, 179–185. [CrossRef]
111. Muguda, S.; Lucas, G.; Hughes, P.N.; Augarde, C.E.; Perlot, C.; Bruno, A.W.; Gallipoli, D. Durability and hygroscopic behaviour of biopolymer stabilised earthen construction materials. *Constr. Build. Mater.* **2020**, *259*, 119725. [CrossRef]
112. Zhao, X.; Zhao, X.; Han, F.; Song, Z.; Wang, D.; Fan, J.; Jia, Z.; Jiang, G. A research on dust suppression mechanism and application technology in mining and loading process of burnt rock open pit coal mines. *J. Air Waste Manag. Assoc.* **2021**, *71*, 1568–1584. [CrossRef] [PubMed]
113. Armistead, S.J.; Smith, C.C.; Staniland, S.S. Sustainable biopolymer soil stabilization in saline rich, arid conditions: A 'micro to macro' approach. *Sci. Rep.* **2022**, *12*, 2880. [CrossRef]
114. Ghasemzadeh, H.; Modiri, F. Application of novel Persian gum hydrocolloid in soil stabilization. *Carbohydr. Polym.* **2020**, *246*, 116639. [CrossRef]
115. Yan, J.; Nie, W.; Xiu, Z.; Yuan, M.; Zhou, W.; Bao, Q.; Peng, H.; Niu, W.; Yu, F. Development and characterization of a dust suppression spray agent based on an adhesive NaAlg−gln−poly/polysaccharide polymer. *Sci. Total Environ.* **2021**, *785*, 147192. [CrossRef]
116. Soltani, A.; Raeesi, R.; Taheri, A.; Deng, A.; Mirzababaei, M. Improved shear strength performance of compacted rubberized clays treated with sodium alginate biopolymer. *Polymers* **2021**, *13*, 764. [CrossRef] [PubMed]
117. Arab, M.G.; Mousa, R.A.; Gabr, A.R.; Azam, A.M.; El-Badawy, S.M.; Hassan, A.F. Resilient behavior of sodium alginate–treated cohesive soils for pavement applications. *J. Mater. Civ. Eng.* **2019**, *31*, 4018361. [CrossRef]
118. Fatehi, H.; Bahmani, M.; Noorzad, A. Strengthening of dune sand with sodium alginate biopolymer. In Proceedings of the Geo-Congress 2019 8th International Conference on Case Histories in Geotechnical Engineering, Philadelphia, PA, USA, 24–27 March 2019; Meehan, C.L., Kumar, S., Pando, M.A., Coe, J.T., Eds.; American Society of Civil Engineers: Reston, VA, USA, 2019; pp. 157–166, ISBN 9780784482117.
119. Zhao, Y.; Zhuang, J.; Wang, Y.; Jia, Y.; Niu, P.; Jia, K. Improvement of loess characteristics using sodium alginate. *Bull. Eng. Geol. Environ.* **2019**, *79*, 1879–1891. [CrossRef]
120. Bouazza, A.; Gates, W.P.; Ranjith, P.G. Hydraulic conductivity of biopolymer-treated silty sand. *Géotechnique* **2009**, *59*, 71–72. [CrossRef]
121. Zhou, Y.; Wu, L.; Tong, L.; Liu, Y.; Yang, X.; Zhang, B.; Liao, Z.; Chen, A.; Hu, Y. Preparation of high-temperature resistant and environment friendly dust suppressant. *J. Phys. Conf. Ser.* **2021**, *2076*, 12054. [CrossRef]
122. Chen, R.; Ding, X.; Lai, H.; Zhang, L. Improving dust resistance of mine tailings using green biopolymer. *Environ. Geotech.* **2019**, *8*, 382–391. [CrossRef]
123. Chen, C.; Peng, Z.; Gu, J.; Peng, Y.; Huang, X.; Wu, L. Exploring environmentally friendly biopolymer material effect on soil tensile and compressive behavior. *Int. J. Environ. Res. Public Health* **2020**, *17*, 9032. [CrossRef]
124. Ni, J.; Hao, G.-L.; Chen, J.-Q.; Ma, L.; Geng, X.-Y. The optimisation analysis of sand-clay mixtures stabilised with xanthan gum biopolymers. *Sustainability* **2021**, *13*, 3732. [CrossRef]
125. Ramachandran, A.L.; Dubey, A.A.; Dhami, N.K.; Mukherjee, A. Multiscale study of soil stabilization using bacterial biopolymers. *J. Geotech. Geoenviron. Eng.* **2021**, *147*, 4021074. [CrossRef]
126. Sujatha, E.R.; Atchaya, S.; Sivasaran, A.; Keerdthe, R.S. Enhancing the geotechnical properties of soil using xanthan gum—An eco-friendly alternative to traditional stabilizers. *Bull. Eng. Geol. Environ.* **2021**, *80*, 1157–1167. [CrossRef]
127. Joga, J.R.; Varaprasad, B.J.S. Effect of xanthan gum biopolymer on dispersive properties of soils. *World J. Eng.* **2020**, *17*, 563–571. [CrossRef]
128. Singh, S.P.; Das, R. Geo-engineering properties of expansive soil treated with xanthan gum biopolymer. *Geomech. Eng.* **2020**, *15*, 107–122. [CrossRef]
129. Theyab, A.F.; Muhauwiss, F.M.; Alabdraba, W.M. Enhancing gypseous soil behavior using casein from milk wastes. *J. Mech. Behav. Mater.* **2022**, *31*, 306–313. [CrossRef]
130. Park, S.-S.; Woo, S.-W.; Jeong, S.-W.; Lee, D.-E. Durability and strength characteristics of casein-cemented sand with slag. *Materials* **2020**, *13*, 3182. [CrossRef] [PubMed]
131. Gopika, A.S.; Mohandas, T.V. Soil strengthening using caseinate: A protein based biopolymer. *Int. J. Res. Eng. Sci. Manag.* **2019**, *2*, 538–540.
132. Chang, I.; Im, J.; Chung, M.-K.; Cho, G.-C. Bovine casein as a new soil strengthening binder from diary wastes. *Constr. Build. Mater.* **2018**, *160*, 1–9. [CrossRef]
133. Fatehi, H.; Abtahi, S.M.; Hashemolhosseini, H.; Hejazi, S.M. A novel study on using protein based biopolymers in soil strengthening. *Constr. Build. Mater.* **2018**, *167*, 813–821. [CrossRef]

134. Im, J.; Cho, G.-C.; Chang, I. A new soil treatment method using casein from bovine milk. In Proceedings of the Geo-Chicago 2016, Sustainability and Resiliency in Geotechnical Engineering, Chicago, IL, USA, 14–18 August 2016; Zekkos, D., Farid, A., De, A., Reddy, K.R., Yesiller, N., Eds.; American Society of Civil Engineers: Reston, VA, USA, 2016; pp. 1–6, ISBN 9780784480120.
135. Jin, H.; Nie, W.; Zhang, Y.; Wang, H.; Zhang, H.; Bao, Q.; Yan, J. Development of environmental friendly dust suppressant based on the modification of soybean protein isolate. *Processes* **2019**, *7*, 165. [CrossRef]
136. Cruse, R.M.; Berghoefer, B.E.; Mize, C.W.; Ghaffarzadeh, M. Water drop impact angle and soybean protein amendment effects on soil detachment. *Soil Sci. Soc. Am. J.* **2000**, *64*, 1474–1478. [CrossRef]
137. Chen, R.; Ding, X.; Ramey, D.; Lee, I.; Zhang, L. Experimental and numerical investigation into surface strength of mine tailings after biopolymer stabilization. *Acta Geotech.* **2016**, *11*, 1075–1085. [CrossRef]

applied sciences

MDPI

Article

Evaluation of Groundwater Salinization Risk Following Application of Anti-Dust Emission Solutions on Unpaved Roads in Arid and Semiarid Regions

Meni Ben-Hur [1,2,*], Reut Cohen [1], Michael Danon [3], Uri Nachshon [1] and Itzhak Katra [2]

[1] Institute of Soil, Water & Environmental Sciences, ARO, Volcani Center, Rishon Lezion 7505101, Israel; reutlevy100@gmail.com (R.C.); urina@volcani.agri.gov.il (U.N.)
[2] Department of Geography and Environmental Development, Ben Gurion University of the Negev, Beer Sheva 8410501, Israel; katra@bgu.ac.il
[3] Mines and Quarries Division, Natural Resources Administration, Ministry of Energy, Jerusalem 9136002, Israel; michaeld@energy.gov.il
* Correspondence: meni@volcani.agri.gov.il

Abstract: Unpaved roads could be a significant source of dust emission. A common and effective practice to suppress this emission is the application of brine solution on these roads. However, this application could increase the risk of water source salinization in arid and semiarid regions, such as Israel. The general objective of the present study was to investigate the potential effects of treated wastewater (TWW), fresh water (FW), and brine applications as anti-dust emission solutions on water source salinization in these regions. A rainfall simulator experiment and a mass balance model were used for this goal. The TWW loaded the highest amounts of Cl, Na, and Ca+Mg on the unpaved roads, while the brine loaded higher amounts of Cl and Ca+Mg than the FW, and ~0 Na. In the rainfall experiment, runoff was not formed, and ~100% of the loaded amounts were leached downwards by rain, indicating a negligible salinization risk to surface water. We estimated that the average increases in the Cl concentrations in the modeled aquifer, following TWW, brine, and FW applications, were low: 1.2–1.6, 0.58–0.8, and 0.32–0.4 mg L^{-1}, respectively. Thus, the solution selection for preventing dust emission should be based on the total cost of the solution application.

Keywords: rainfall simulator; dust emission; salinization; groundwater; wind erosion; brine solution; treated wastewater

Citation: Ben-Hur, M.; Cohen, R.; Danon, M.; Nachshon, U.; Katra, I. Evaluation of Groundwater Salinization Risk Following Application of Anti-Dust Emission Solutions on Unpaved Roads in Arid and Semiarid Regions. *Appl. Sci.* **2021**, *11*, 1771. https://doi.org/10.3390/app11041771

Academic Editor: Mauro Marini
Received: 23 January 2021
Accepted: 13 February 2021
Published: 17 February 2021

1. Introduction

Dust emission to the atmosphere, caused by wind erosion, could have significant effects on the environment and human health [1–3]. This depends on the characteristics of the winds and the eroded surface area [4,5]. Anthropogenic activities, such as mining, quarrying, excavating, and the use of unpaved roads, increase the terrain's sensitivity to dust emission. For example, heavy vehicles traveling on unpaved roads could grind the roads' surfaces to fine particles that become available to wind erosion [6–8].

Environmental regulators, such as the U.S. Environmental Protection Agency (EPA), recommend and enforce the use of dust control techniques. A common practice to prevent dust emission from unpaved roads is the application of anti-dust emission solutions. The two main types of such solutions are as follows: (i) Natural or synthetic polymers, e.g., lignin, resin, bitumen, and polyvinyl acetate (PVA) [9]. These materials contain long organic chains with functional groups that could adsorb and adhere to dust particles, forming bigger and heavier aggregates [10]. (ii) Brines with high concentrations of hydrous calcium–magnesium chloride solution, which decreases the dust emission by a hygroscopic mechanism [11]. For both cases, the increased weights of these aggregates limit their lifting by the erosive forces of the wind, leading to a reduction in dust emission.

Katra [12] tested the effectiveness of the lignin, resin, bitumen, PVA, and brine application in preventing dust emission by wind from unpaved roads. These tests were conducted under controlled laboratory and field conditions, using wind tunnels with varied wind velocities and time durations. It was found in this experiment that among the tested anti-dust emission solutions, brine from the Dead Sea was the most effective solution in preventing the dust emission; the brine application decreased the dust emission by >90% of the dust emissions under the control treatment [12].

Arid and semiarid regions are characterized by a long dry season and a short wet season with limited precipitation. Thus, natural water resources, such as groundwater and non-salty surface water (fresh water, FW) in these regions, are scarce [13]. Moreover, the climate conditions and the high population growth in these regions lead to an escalation of irrigation with low quality water, such as treated wastewater (TWW) and saline water [13], which increases the salinization of the water sources in the region [13].

Most of the haul roads in mines and quarries are unpaved roads covered with a layer (usually 20–30 cm thick) of calcareous granules (granular material) that sensitive to dust emission. A common practice to prevent the dust emission from these unpaved roads in arid and semiarid regions, such as Israel, is the application of hyper-saline solution (brine) on the unpaved roads. In recent years, concerns have been raised regarding the application of the brine solution on unpaved roads for controlling dust emission. It was claimed that the application of the brine solution with very high salinity will further increase the salinization of the water resources. Indeed, some quarries use FW instead of brine solution in their anti-dust practices. However, since FW is scarce in arid and semiarid regions, TWW is also used for the purpose of anti-dust emission on unpaved roads.

The TWW is characterized by moderate salt concentration, particularly Cl and Na ions; high concentrations of toxic, inorganic macro- and micro-elements, such as NO_3, boron and heavy metals; and high contents of suspended and soluble organic matters, including organic pollutants and pathogens [13]. The brine solution, however, is characterized by very high salt concentration, particularly of Cl, Ca, and Mg ions, and a high concentration of toxic heavy metals, such as Ni, Co, Cd, and Pb. Therefore, during the rainfall season, high concentrations of salts and toxic elements and compounds could be leached from the unpaved roads treated with TWW or brine solutions and salt, and contaminate the water sources.

The present study is focused on the salinization of water sources following the application of anti-dust emission solutions on unpaved roads. The trend of using FW or TWW as anti-dust emission solutions instead of brine solution in arid and semiarid regions is based on the assumption that a decrease in the salinity of the anti-dust emission solution will necessarily decrease the salinization of the water sources in the region. However, the salinization of the water sources, under the application of anti-dust emission solutions, is dependent on varied factors other than the salinity of the anti-dust emission solutions. Therefore, the general objective of the present study was to investigate the effects of the application of TWW, FW, and brine solutions on unpaved roads as anti-dust emission solutions on the salinity of the water resources in arid and semiarid regions, such as Israel. The specific objectives were to determine or estimate the effects of the application of TWW, FW and brine solutions on the following: (i) The accumulated amounts of the solutes on the unpaved roads. (ii) The leaching rates of the ions, Cl, Na, and Ca+Mg, from the treated granular material by consecutive rainstorms. (iii) The distribution movement of the rainwater and solutes during the rainstorm via lateral movement on the terrine surface as surface runoff, or vertical movement with the infiltrating water; (iv) the long-term average of the salinization rates of the water sources.

2. Materials and Methods

The study comprises two main parts: (i) a rainfall simulator experiment to study the leaching rate of the ions, Cl, Na, and Ca+Mg, from the treated granular material during consecutive rainstorms, and (ii) an estimation of the water sources' salinization following the application of the anti-dust emission solutions, brine, TWW, and FW, on the unpaved roads.

2.1. Rainfall Simulator Study

Chemical analysis of the studied materials: Granules of calcareous material (granular material), containing mainly particles with 5–10 mm sizes, were obtained from Modiim Quarry Ltd, Israel. These calcareous granules are used as a cover layer on the unpaved roads. The brine solution and secondary TWW were obtained from the Dead Sea Works and the Taoz treated wastewater plant near the city of Beit-Shemes in Israel, respectively. These solutions were used in the rainfall simulator experiment. A sample of the brine solution was obtained a couple of days before the beginning of the experiment, and the TWW samples were taken <48 h before their uses in the experiment. Filtrated samples of <0.45 μm of the brine and TWW solutions were analyzed using the following standard methods: (i) Electrical conductivity (EC) and pH values by standard EC and pH meters, respectively. (ii) Cl concentration by Digital Chloridometer, 442500, Labconco, Kansas City, MO. (iii) Concentrations of Na by flame photometer model 420 Clinical Flame Photometer, Sherwood Scientific Ltd, Cambridge, UK. (iv) Concentration of various elements by Dual-View High-Resolution ICP-OES Plasma Quant 9000 Elite, Analytik Jena, Germany. The chemical properties of the brine and TWW solutions that were used in the rainfall simulator study are presented in Table 1.

Table 1. Electrical conductivity (EC), pH values, and the concentrations of various ions in the studied treatment solutions: brine from the Dead Sea and treated wastewater (TWW) from the Taoz wastewater treatment plant.

Treatment Solutions	pH	EC	Cl	Ca	Mg	Na	K	HCO_3
		dS m^{-1}	mg/L					
Brine	6.0	130	250×10^3	23.2×10^3	62.2×10^3	2.5×10^3	2.8×10^3	13.7×10^3
TWW	7.8	1.6	233.4	34.4	11.2	72.7	41.3	1018.7

The treatments: The granular material was packed and leveled in metal trays with a perforated bottom and dimensions of 0.3×0.5 m^2 area and 0.02 m depth for each tray. The trays with the granular material were treated by the following treatments, with 4 replicates (4 trays) for each treatment:

1. Control: four trays with untreated granular material were subjected to 3 consecutive rainstorms by a rainfall simulator, 1 storm a day, as described below. After each rainstorm, the trays were left for 3 days in a net house for drying to air-dry.
2. TWW treatment: four trays with the granular material were sprayed 10 times with 4 mm of TWW in each time. After each spray, the trays were left for 3 days in a net house for drying to air-dry. Three days after the last TWW application, the trays were subjected to 3 consecutive rainstorms, with no further TWW addition between the rainstorms, as described above for the control treatment.
3. Brine treatment: four trays with the granular material were sprayed 1 time with 4 mm of brine solution, and then were left for 3 days at a net house for drying to air-dry. After the drying period, the trays were subjected to 6 consecutive rainstorms, with 3 days of drying between the rainstorms, as described above for the control treatment, and with no further brine addition between the consecutive rainstorms.

Simulated rainstorms runs: A laboratory rotary disc-type rainfall simulator [14] was used to determine the leakage of the ions, Cl, Na, and Ca+Mg (the studied ions), from the pretreated granular materials during consecutive rainstorms. Before each rainstorm, 4 trays with the pretreated granular materials were placed in a carousel below the rainfall rotary disc. Each tray was placed at a 9% gradient on an 8 cm layer of coarse gravel in a box. The trays were exposed to a rainstorm of 50 mm of deionized water, with the following mechanical parameters: rain intensity = 48 mm h^{-1}, raindrop mean diameter = 1.9 mm, drop velocity = 6.02 m s^{-1}, and kinetic energy of 18.1 J mm^{-1} m^{-2}. During the rainstorm, the water volumes percolating through the granular material in each tray (outflow leachate) were recorded at different times, and the infiltration rates were calculated with respect to the cumulative rainfall. In addition, the outflow leachates from each tray were collected in 5 separated fractions (every 10 mm of rainfall) along each rainstorm, and their volumes were measured. Subsamples were taken from each leached fraction, filtered through a <0.45 μm filter, and the EC values and concentrations of Cl, Na, Ca, and Mg were measured as described above. Surface runoff was not formed during the rainstorms, and therefore was not measured.

2.2. Estimation of the Water Sources Salinization

The effects of the anti-dust emission solutions (brine, TWW, and FW) on water sources' salinization were estimated based on the following assumptions and measurements:

1. A mass balance model, under long-term and steady state conditions, was used to estimate the effects of the application of the treatments on the groundwater salinization.
2. Because many quarries in Israel are located above the western mountain aquifer (Yarqon-Taninim aquifer), the mass balance model was based on the characteristics of this aquifer and on the activities and the environmental conditions at its basin. The dominant rocks in the Yarqon-Taninim aquifer are cracked limestone with heterogeneous paths for water and solutes movement. The recharge of the aquifer occurs through annual average rainfall ranging from 200 to 600 mm. This rainfall falls in the short winter (4 months), while the rest of the months are completely dry.
3. The leaching rates of the studied ions from the treated granular materials during consecutive rainstorms were determined by the rainfall simulator experiment.
4. The chemical compositions of the TWW, brine, and FW solutions, which were used in the rainfall simulator experiment and in the model, are presented in Tables 1 and 2.

Table 2. Typical electrical conductivity (EC), pH values, and concentrations of various ions in the treated wastewater (TWW) and fresh water (FW) that were used in the region of the Yarkon-Taninim aquifer [15].

Treatment Solutions	pH	EC	Cl	Ca	Mg	Na	K	HCO_3
		dS m^{-1}	mg/L					
TWW	7.6	1.3	259	127.5	42.5	81.1	15.1	$ NA
FW	6.7	0.67	78.1	59.8	19.8	48	2.2	333.8

$ Not available.

3. Results and Discussion

The evaluation of the salinization risks of the water sources following the application of the anti-dust emission solutions (brine, TWW, and FW) was based on two main processes: (i) The annual loaded amounts of the solutes and their accumulations on the unpaved roads following the applications of the treatment solutions in the dry season (summer). (ii) Leaching rates by rainfall of the studied ions (Cl, Na, and Ca+Mg) from the granular material, which was pretreated with the treatment solutions.

The leaching rates of the Cl, Na, and Ca+Mg ions from the pretreated granular materials were determined using a rainfall simulator. Runoff was not formed in the different treated granular materials during the consecutive rainstorms. The packaging of the granular materials in the surface of the unpaved roads and in the rainfall simulator trays was similar in general. Thus, the infiltration rates of the granular materials are >48 mm h^{-1} (the rainstorms' intensity), and most of the movement of the solutes with the rainwater is vertical. Therefore, the salinization risk to the surface water sources near the treated unpaved roads by the treatment solutions is likely negligible, unless the infiltration rate of the surface area of the unpaved roads below the granular material layer is lower than the rainfall intensity. Otherwise, on the sublayer, the lateral movement of water with soluble ions could occur. However, this movement is very slow, and usually infiltrates through the cracks into the rocks, so would not be a predominant flow route.

The ion concentrations in the outflow (leachate) from the trays with the treated granular material as functions of the cumulative rainfall are presented in Figures 1–3 for Cl, Na, and Ca+Mg, respectively. In these figures, the first and third (last) consecutive rainstorms for the control and TWW treatments, and the first and sixth (last) for the brine treatment are presented. Because the concentrations of each ion in the leachate in the second and the last consecutive rainstorms were fairly similar, only the first and the last consecutive rainstorm are presented in the figures. In addition, in the control treatment, the concentrations of the analyzed ions in the leachate reached steady state values quite fast, and no later than the end of the third consecutive rainstorm.

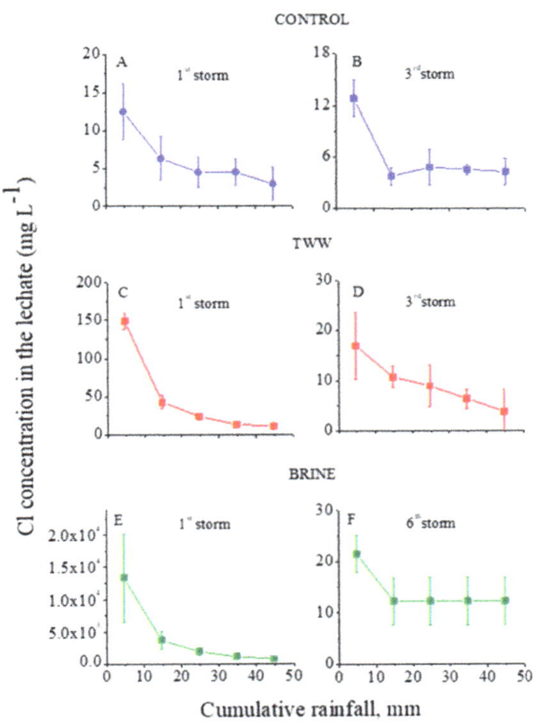

Figure 1. Cl^- concentrations in the outflow leachate from the trays with treated granular materials as functions of the cumulative rainfall of the 1st (**A**) and 3rd (**B**) consecutive rainstorms for the control, of the 1st (**C**) and 3rd (**D**) consecutive rainstorms for the treated wastewater (TWW) treatment, and of the 1st (**E**) and 6th (**F**) consecutive rainstorms for the brine treatment. Vertical lines near the symbols are standard deviations (the Y-axes' scales are different).

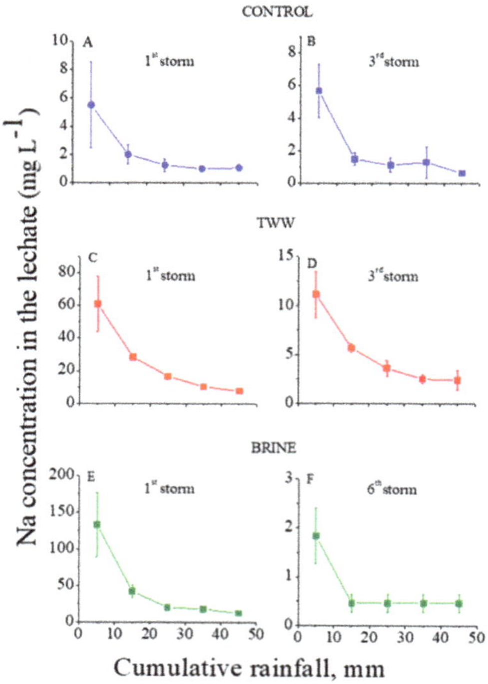

Figure 2. Na^{+} concentrations in the outflow leachate from the trays with treated granular materials as functions of the cumulative rainfall of the 1st (**A**) and 3rd (**B**) consecutive rainstorms for the control, of the 1st (**C**) and 3rd (**D**) consecutive rainstorms for the treated wastewater (TWW) treatments, and of the 1st (**E**) and 6th (**F**) consecutive rainstorms for the brine treatment. Vertical lines near the symbols are standard deviations (the Y-axes' scales are different).

In the control treatment, the concentrations of Cl and Na in the leachate at the first and third consecutive rainstorms were relatively low: <13 and <6 mg L^{-1}, respectively (Figures 1A and 2A). These low concentrations suggest that the amounts of highly soluble minerals with Na and Cl in the granular material were low. In contrast, the dissolution of the lime in the calcareous granular material in the control treatment kept the concentrations of the Ca+Mg in the leachate relatively high and quite constant, with an average value of ~7.5 mg L^{-1} along the consecutive rainstorms (Figure 3A,B). For the TWW and brine treatments, in the beginning of the first rainstorms (Figures 1–3) the concentrations of the ions (Cl, Na, and Ca+Mg) in the leachate were higher than in the control treatment. These high concentrations were a result of the previous applications of the TWW and brine solutions on the granular materials. However, these high concentrations of Cl, Na, and Ca+Mg decreased sharply with the cumulative rainfall, and near the end of the last rainstorm, the concentrations of these ions were similar to their corresponding concentrations in the control treatment (Figures 1–3).

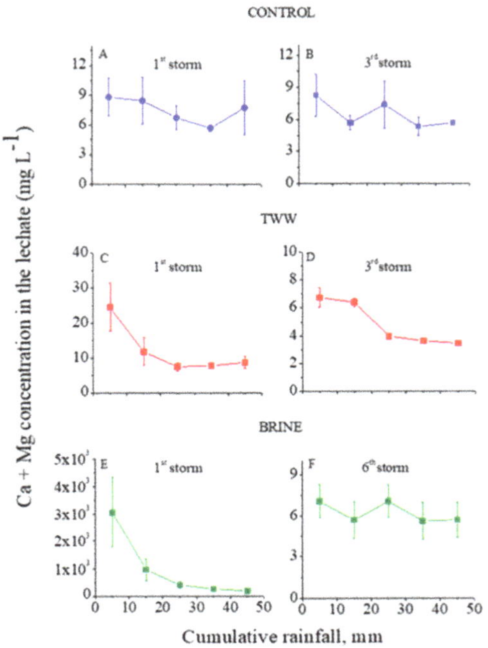

Figure 3. Ca+Mg concentrations in the outflow leachate from the trays with treated granular materials as functions of the cumulative rainfall of the 1st (**A**) and 3rd (**B**) consecutive rainstorms for the control, of the 1st (**C**) and 3rd (**D**) consecutive rainstorms for the treated wastewater (TWW) treatment, and of the 1st (**E**) and 6th (**F**) consecutive rainstorms for the brine treatment. Vertical lines near the symbols are standard deviations (the Y-axes' scales are different).

According to the rainfall simulator results (Figures 1–3), for each studied ion and treatment solution the total leaching amounts ($£_j$) of ion j (Cl, Na, or Ca+Mg) from the treated granular material during the consecutive rainstorms, as a percentage of the total added amount of ion j by the treatment solution, were calculated by Equation (1):

$$£_j = \left(\frac{\sum_{k=1}^{k=m}\left(\sum_{i=1}^{i=n}(Vd_i \cdot Cd_i) - \sum_{i=1}^{i=n}(Vc_i \cdot Cc_i)\right)}{L \cdot S_j}\right) \cdot 100 \quad (1)$$

where k is the consecutive serial number of the consecutive rainstorm; Vd_i and Cd_i are the leachate volumes (L m^{-2}) and concentrations (gr L^{-1}) of ion j, respectively, for brine or TWW applications; and the Vc_i and Cc_i are the leachate volumes (L m^{-2}) and concentrations (gr L^{-1}) of ion j, respectively, for the control treatment, when I is the consecutive serial number of the rainfall fractions in each consecutive rainstorm. S_j is the amount of ion j (mg m^{-2}) added to the granular material with each application event of the brine or TWW, when L is the total number of the application events. For determining the $£_j$ values for brine treatment by Equation (1), the values of Vc_i and Cc_i of the fourth, fifth and sixth consecutive rainstorms in the control treatment were, most likely, equal, corresponding with the values of the third rainstorm in the control treatment (Figures 1A–3A). This is because the values of Vc_i and Cc_i at the third rainstorm in the control treatment reached steady state values (Figures 1A–3A).

The $£_j$ values for Cl, Na, and Ca+Mg ions indicated that >99.5% of the amounts of these ions added onto the granular material by the WWT or brine solutions were leached down during the rainstorms. These results strengthen the previous observation that the calcareous granular material has a negligible capacity to adsorb the studied ions, and to prevent their vertical movement.

The total loaded amounts of the solutes on the unpaved roads by the treatment solutions could be leached down by the rainstorm and increase the salinity of the groundwater. Therefore, the load of each studied ion by each solution treatment should be determined. The total annual load amount (P_j) (kg m^{-2}) of ion j, where j could be Cl, Na, or Ca+Mg, on 1 m^2 of the surface area of the unpaved road by the application of brine, TWW, or FW, was determined by Equation (2):

$$P_j = \beta \cdot \sum_{i=1}^{i=N} (V_{ji} \cdot C_{ji}) \quad (2)$$

where C_{ji} is the concentration (mg L^{-1}) of the applied ion j (Cl, Na, or Ca+Mg), and V_{ji} is the volume (L m^{-2}) of the brine, TWW, or FW solution containing the ion j, which is applied in the application event i on a specific application day. β is the number of application days in a year. According to the Israeli regulation, for the three treatment solutions (brine, TWW, and FW), the V_{ji} equals 4 L m^{-2}; for the FW and TWW treatment solutions, the N equals 10 and β equals 192; and for the brine solution the N equals 1 and β equals 1.

The P_j values of Cl, Na, and Ca+Mg on the unpaved roads by the TWW, brine, and FW applications are presented in Figure 4. It can be seen from this figure that (i) the TWW application loaded the highest annual amounts of Cl, Na, and Ca+Mg on the unpaved roads, and (ii) the brine application loaded higher annual amounts of Cl and Ca+Mg than the FW application, and the lowest annual amount of Na (~0). In addition, different mechanisms were responsible for the dust emission control in the various studied solutions (brine, TWW, and FW). The mechanism of the brine solution controlling dust emission is based on the high hygroscopic properties of the Ca and Mg elements. The concentrations of the Ca and Mg ions in the brine solution are very high, 23.2 and 62.2 g L^{-1}, respectively (Table 1). The application of the brine solution causes the accumulation of high contents of Ca and Mg on the unpaved road, which absorb large amounts of H_2O molecules from the air. These H_2O molecules are adsorbed on the dust particles by adhesion forces, forming big and heavy aggregates which are beyond the lifting capability of the wind. Consequently, the dust emission under the brine application decreases.

In contrast, the application of the TWW and FW solutions with relatively low concentrations of Ca and Mg, <127.5 and <59.8 mg L^{-1}, respectively (Table 2), leads to the lower accumulation of these ions on the unpaved roads. Furthermore, the frequent applications of the TWW and FW on the unpaved roads, 10 times in each day during the dry season, leach down the Ca and Mg ions from the unpaved roads' surfaces, which in turn decreases their accumulations on the unpaved roads. In this case, the hygroscopic mechanism is negligible. Therefore, the mechanism of the WWT and FW application in controlling dust emission is based on the "active wetting" mechanism, i.e., repetitive and frequent wetting of the unpaved roads throughout the day to keep the roads wet during the quarries' activities.

The effects of the vertical leaching of the solutes, which accumulated on the unpaved roads following the application of the treatment solutions, on the groundwater salinization could be indicated by the Cl^- concentration increase in the groundwater. The Cl^- is a conservative ion, very mobile in soil/rock systems, and is the most dominant anion in groundwater. Therefore, the concentration of Cl^- is used as a parameter of water salinity; the higher the Cl^- concentration, the higher the water salinity is [16].

Many quarries in Israel are located above the Yarkon-Taninim aquifer, where the dominant rocks in this aquifer are cracked limestone. The paths for water and solutes movement in this aquifer are heterogeneous, complicated, and difficult to describe and predict. Nevertheless, the comparative effects of the treatment solutions (brine, TWW, and FW) on groundwater salinity could be fairly estimated on the basis of the following assumptions and simplifications: (1) Most of the added solutes on the unpaved roads during the dry season reach the groundwater. (2) Under long-term and steady state conditions, the average amounts of the Cl^- that were added on the unpaved roads or pumped with the water from the aquifer are equal. (3) The annual average volume of the water that is pumped from the aquifer and that which recharges the aquifer are equal.

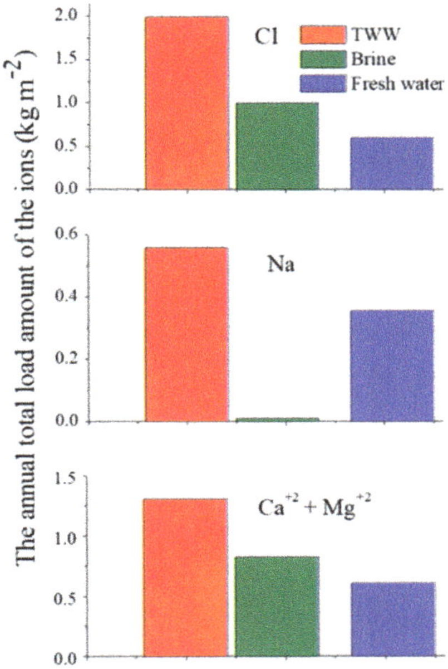

Figure 4. The annual loaded amounts (P_j) of Cl (upper part), Na (middle part), or Ca+Mg (bottom part), on the unpaved roads following the treated wastewater (TWW), brine and fresh water (FW) applications (the Y-axes' scales are different).

According to the above conditions, a mass balance model (Equation (3)) was used to predict the effects of brine, TWW, or FW applications on the long-term annual average salinity increase (ΔC_d) in the pumped water from the Yarkon-Taninim aquifer.

$$\Delta C_{cl} = \mathrm{A}\cdot\frac{P_{cl}}{WR} \quad (3)$$

where A is the long-term average area (m^2) of all the unpaved roads located above the Yarkon-Taninim aquifer and treated with the treatment solutions (brine, TWW, or FW). The P_{cl} is the long-term average amount (mg m^{-2}/yr) of the Cl^- that was added on the A area, and is calculated by Equation (2). WR is the long-term average of the water volume (L/yr) that was pumped from the Yarkon-Taninim aquifer.

The ΔC_{cl} values of the pumped water from the Yarkon-Taninim aquifer, which were calculated by Equation (3), are presented in Figure 5 for the various treatment solutions (brine, TWW, FW), and two different A values. For these calculations, a WR value of 400×10^9 L per a year was used (a long-term annual average of the pumped water from this aquifer), as were A values in the years 2016 and 2017, which were used as two examples. The long-term annual average values of the ΔC_{cl} in the pumped water from the Yarkan-Taninim aquifer following the TWW, brine, and FW applications were 1.6, 0.8, and 0.4 mg/L, respectively, in the first example (2016), and 1.2, 0.58, and 0.32 mg/L, respectively, in the second example (2017) (Figure 5). These ΔC_{cl} values in the first and second examples under the application of TWW were approximately two and four times higher than under brine and FW applications, respectively. In spite of these large differences in the ΔC_{cl} values between the various treatments (Figure 5), the long-term annual average additions of Cl^- concentration in the groundwater following the three treatment solutions were relatively low and ranged between 0.32 to 1.6 mg L^{-1} in the two examples (Figure 5).

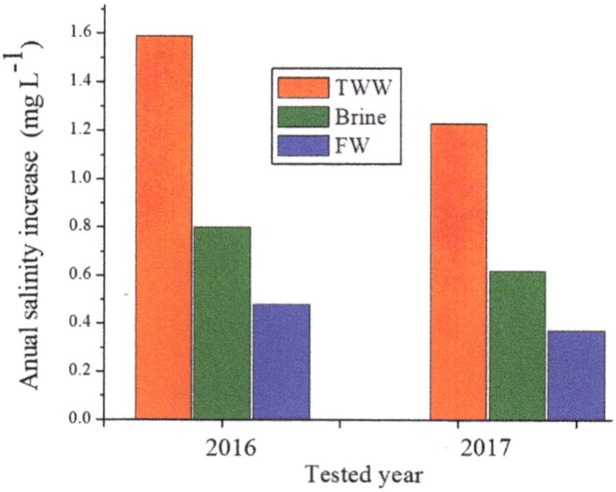

Figure 5. The long-term average of the salinity increase (ΔC_{cl}) of the pumped water from the Yarkon-Taninim aquifer following the treated wastewater (TWW), brine, or fresh water (FW) applications on the total area of the unpaved roads located above the aquifer in the years 2016 and 2017 as examples.

4. Conclusions

The estimation of the long-term average increases in Cl^- concentrations in the pumped water from the groundwater, following the TWW, brine, or FW applications, expressed as percentages of the Cl concentration in the groundwater without the application of anti-dust emission, ranged between 0.5 and 2.2%. In such Cl concentration increases, using brine solutions with high concentrations of Ca and Mg ions could provide an environmentally sound solution for application on unpaved roads for dust control, and no regulatory restrictions are required on this use. Therefore, the selection of the solution to prevent dust emission should be basically based on the cost, including labor, transport and the solutions themselves.

Author Contributions: M.B.-H, original draft preparation; M.B.-H. and I.K., conceptualization, writing—review and editing; M.D. and U.N., conceptualization, review and editing; R.C., conceptualization, methodology, analyses. All authors have read and agreed to the published version of the manuscript.

Funding: The study was supported by a grant from the Ministry of Energy, Israel (No. 215-17-014).

Data Availability Statement: Not applicable.

Conflicts of Interest: The authors declare no conflict of interest.

References

1. Nenes, A.; Murray, B.; Bougiatioti, A. Mineral Dust and its Microphysical Interactions with Clouds. In *Mineral Dust: A Key Player in the Earth System*; Knippertz, P., Stuut, J.B., Eds.; Springer: New York, NY, USA, 2014; pp. 287–325. [CrossRef]
2. Kok, J.F.; Ridley, D.A.; Zhou, Q.; Miller, R.L.; Zhao, C.; Heald, D.A.R.C.L.; Ward, D.S.; Albani, S.; Haustein, K. Smaller desert dust cooling effect estimated from analysis of dust size and abundance. *Nat. Geosci.* **2017**, *10*, 274–278. [CrossRef] [PubMed]
3. Jickells, T.D.; Anderson, K.; Andersen, K.K.; Baker, A.R.; Bergametti, G.; Brooks, N.; Cao, J.J.; Boyd, P.W.; Duce, R.A.; Hunter, K.A.; et al. Global Iron Connections Between Desert Dust, Ocean Biogeochemistry, and Climate. *Science* **2005**, *308*, 67–71. [CrossRef] [PubMed]
4. Kok, J.F.; Parteli, E.J.; Michaels, T.I.; Karam, D.B. The physics of wind-blown sand and dust. *Rep. Prog. Phys.* **2012**, *75*, 106901. [CrossRef] [PubMed]
5. Shao, Y.; Raupach, M.R.; Findlater, P.A. Effect of saltation bombardment on the entrainment of dust by wind. *J. Geophys. Res. Space Phys.* **1993**, *98*, 12719–12726. [CrossRef]
6. Gillies, J.; Etyemezian, V.; Kuhns, H.; Nikolic, D.; Gillette, D. Effect of vehicle characteristics on unpaved road dust emissions. *Atmos. Environ.* **2005**, *39*, 2341–2347. [CrossRef]

7. Goossens, D.; Buck, B. Dust emission by off-road driving: Experiments on 17 arid soil types, Nevada, USA. *Geomorphology* **2009**, *107*, 118–138. [CrossRef]
8. Yulevitch, G.; Danon, M.; Krasovitov, B.; Fominykh, A.; Swet, N.; Tsesarsky, M.; Katra, I. Evaluation of wind-induced dust-PM emission from unpaved roads varying in silt content by experimental results. *Atmos. Pollut. Res.* **2020**, *11*, 261–268. [CrossRef]
9. Castellanos, A. The relationship between attractive interparticle forces and bulk behavior in dry and uncharged fine powders. *Adv. Phys.* **2005**, *54*, 263–376. [CrossRef]
10. Ben-Hur, M. Using synthetic polymers as soil conditioners to control runoff and soil loss in arid and semiarid regions: A review. *Aust. J. Soil Res.* **2006**, *44*, 191–204. [CrossRef]
11. Bustos, M.; Cordo, O.; Girardi, P.; Pereyra, M. Evaluation of the Use of Magnesium Chloride for Surface Stabilization and Dust Control on Unpaved Roads. *Transp. Res. Rec.* **2015**, *2473*, 13–22. [CrossRef]
12. Katra, I. Comparison of various dust control products in wind-induced dust emission from unpaved roads. *Appl. Sci.* **2019**, *9*, 5204. [CrossRef]
13. Ben-Hur, M. Sewage water treatments and reuse in Israel. In *Water in the Middle East and in North Africa: Resources, Protection, and Management*; Zereini, F., Jaeschke, W., Eds.; Springer: Berlin/Heidelberg, Germany, 2004; pp. 167–180.
14. Morin, J.; Goldberg, D.; Seginer, A.I. A Rainfall Simulator with a Rotating Disk. *Trans. ASAE* **1967**, *10*, 0074–0077. [CrossRef]
15. Tarchitki, J.; Inbar, Y.; Shmoali, L. *Environment Impact of Application of Plant Nutrient and Organic Wastes on Arable Land in Israel*; Final Report; The Ministry of Environment Production: Israel, 2016; submitted. (In Hebrew)
16. Edelstein, M.; Plaut, Z.; Ben-Hur, M. Water salinity and sodicity effects on soil structure and hydraulic properties—A review. *Adv. Hort. Sci.* **2010**, *24*, 154–160.

applied sciences

MDPI

Article

Dust Emission Thresholds in Loess Soil Under Different Saltation Fluxes

Aviv Rubinstein [1], Meni Ben-Hur [2] and Itzhak Katra [1,*]

[1] Department of Geography and Environmental Development, Ben Gurion University, Beersheba 8410501, Israel; avivru@post.bgu.ac.il

[2] Institute of Soil, Water & Environmental Sciences, Agricultural Research Organization—The Volcani Center, Rishon LeZion 7505101, Israel; meni@volcani.agri.gov.il

* Correspondence: katra@bgu.ac.il

Received: 7 August 2020; Accepted: 26 August 2020; Published: 27 August 2020

Abstract: Soil-derived dust particles produced by aeolian (wind) processes have significant impacts on humans and the Earth's systems. The soil particle size distribution is a major soil characteristic in dust emission models. Yet empirical information on the dependence of dust emission thresholds on soil particle size distribution is still lacking. The main goal of this study was to explore the dust emission threshold from semi-arid loess soil samples by a targeted wind-tunnel experiment. The results clearly show that the dust emission threshold is associated with the saltation threshold with no distinct direct aerodynamic lifting of the loose dust particle. The dust flux depends on the amount of the clay-silt fraction in the soil, the shear velocity, and the saltation flux under certain shear velocity. The study aimed to advance our understating of the dust emission processes, and to provide empirical information for parametrization in dust emission models and for management strategy of soils in preventing dust emission.

Keywords: wind tunnel; particulate matter; soil erosion; particle size distribution

1. Introduction

The majority of aerosol mass in the atmosphere is related to soil-derived dust particles produced by aeolian (wind) processes [1]. Annual global dust emissions from soils into the atmosphere are estimated to be as high as 3000 million tons, including particulate matter (PM) that is less than 10 micrometers in diameter (PM10). The emission of PM10 from soils is directly associated with loss of soil nutrients (clays and organic matter) [2], air pollution, and substantial health risks [3,4]. Other environmental impacts of atmospheric dust refer to the Earth's global energy balance and climate sensitivity through radiative effects and climate feedbacks [1]. Considering these various important impacts of soil dust on the Earth system, it is critical to estimate accurately the dust emission. Although models that estimate the dust emission from soils have been improved [5,6], still there is large disagreement between the empirical results and their estimated values by the models [1]. These discrepancies are partly a result of our gap in understanding the threshold wind friction velocity needed to mobilize soil particles.

Dust emission depends on the soil properties and on the wind shear stress exerted on the soil surface. The shear stress is characterized by the friction velocity (u_*), which express the velocity gradient in boundary layer flow. The threshold friction velocity (u_{*t}) is then the minimum value of the shear velocity (u_*) for which the soil experiences dust emission [7,8]. This threshold friction velocity has been the subject of many experimental studies, starting with the pioneering work of Bagnold [9]. His work revealed that, for a bed of monodisperse (single particle size) beds, the threshold friction velocity has a minimum around 100 μm. Larger particles are more difficult to mobilize because of the increased gravitational force, and smaller particles are more difficult to mobilize because of the increased cohesive forces [10]. The cohesive forces of the clay and fine silt particles limit the dust emission form soil by direct aerodynamic lifting of loose particle [8]. Consequently, dust is normally emitted by the impacts of more easily mobilized sand-sized particles. These "saltators" undergo ballistic trajectories that impact the soil surface. The resulted saltation process enables the entrainment of cohesive dust particles by the disintegration of the soil aggregates [7,8,11]. However, the threshold of motion of monodisperse beds of sand particles is unrealistic because dust-emitting soils contain a wide range of particle sizes (<1 μm to ~1000 μm).

Many soils throughout the world are subjected to increase human activities, and dust emission, in particular, in semi-arid climates [12]. The soil disturbance resulted in disintegration of the soil aggregates, which produce small aggregates and loose sand particles that are available for saltation [11]. It is generally assumed that soils with a higher amount of large aggregates have stronger resistance against erosive forces [13]. Yet, the effect of the soil particle size distribution on the saltation and dust emission threshold is poorly understood, and hence poorly parameterized in dust emission models. Consequently, predictions of future environmental changes and the resulting changes in soil properties and wind erosion are limited by this missing information of how dust emission threshold are affected by the particle size distribution.

The main goal of this study was to examine the dependence of dust emission threshold on soils with different distribution of dust and sand-sized particles. To this end, we analyzed soil samples that are characteristics of a semi-arid loess soil and conducted a targeted experiment on dust emission with a boundary-layer wind tunnel to provide empirical information on the link between soil particle size distribution and saltation and dust emission thresholds.

2. Materials and Methods

2.1. Soil Sample Setup

Two different materials, loess soil and dune sand, were used to form the experimental soil samples with specific dust (<63 μm) and sand contents (see Section 2.2). Those materials were collected from two sites in the northwestern Negev, Israel, where sand transport and dust emission have been extensively studied. The loess material (Figure 1a) was sampled from the topsoil (0–~2 cm) of a natural area with no human disturbance [11]. The soil contained a mix of clay, silt, and sand particles, which is an ideal case for dust emission by saltation. The sand material was collected from an active sand dune in the northwestern Negev. The Negev dunefield is located in the eastern part of the Sinai– Negev erg [14]. The sand has a typical size (mode at ~250 μm), and it contains less than 2% of clay- and silt-sized particles [15]. The bulk materials of the loess and sand samples were analyzed to determine the particle size distribution (PSD) by the ANALYSETTE 22 MicroTec Plus (Fritsch, Idar-Oberstein, Germany) laser diffraction, which measures particles in the size range of 0.08–2000 μm. The replicates (100 mg) of each sample were dispersed in Na-hexametaphosphate solution (0.5%) by sonication (38 kHz). PSD data was calculated using the Fraunhofer diffraction model with a size resolution of 1 μm using MasControl software (Version 1.8, Idar-Oberstein, Germany) (Figure 1b).

The loess material was first crushed for aggregate disintegration, and then was sieved by an electronic sieving apparatus with horizontal and vertical motions (RETSCH AS 300 Control, Haan city, Germany) to receive the dust material (<63 μm). The sieved material was analyzed by the laser diffractometer to ensure that the particle size is less than 63 μm. The sand dune was added to the loess-sieved material (<63 μm) in a known weighted ratio to form samples with specific weight percentage of silt-clay and sand fractions. The samples were kept at air-dry state (<1.5% gravimetric water content) to eliminate effect of soil-water mass [16]. Four reference samples of 15 kg each with various sand/dust percentage were prepared: "sand 100%" (0% dust); "sand 90%" (10% dust); "sand 75%" (25% dust); "sand 50%" (50% dust). The samples were analyzed by the laser diffractometer to test the sand and dust fractions (Figure 1b).

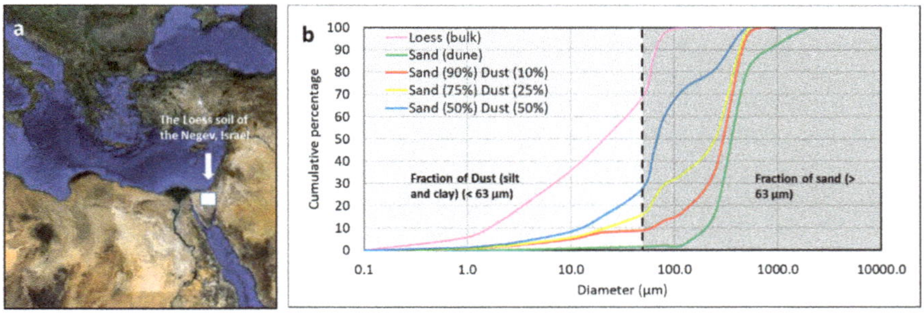

Figure 1. (**a**) Location of the loess soil in the Negev (Israel), Eastern Mediterranean. (**b**) Particle size distribution of the soil samples used for the experiments: sand dune and various mixes of sand and dust fractions that were used for the wind tunnel experiment. The dust fraction, which used for the preparation the mixed samples, was extracted from a loess (bulk) soil.

2.2. Wind Tunnel Experiment

The experiment on dust emission threshold was conducted with a boundary layer wind tunnel [16–18]. The wind tunnels enable aeolian simulations under standardized quasi-natural wind conditions and provide quantitative information on aeolian particle transport in the field and dust emission rates from soils [19–22]. The wind tunnel has a cross sectional area 0.5 × 0.5 m with open-floored working sections of up to 10 m length. The tunnel fan was operated at nine fan frequencies (8, 11, 14, 17, 23, 26, 29, 32, 35, and 38 Hz) to represent a wide range of wind velocities.

The wind velocity profile was measured for each fan frequency at different heights above the tunnel bed: 0.02, 0.035, 0.05, 0.075, 0.10, 0.15, 0.20, 0.25, 0.30, 0.35, 0.40, and 0.45 m. The wind measurement was conducted with a micro-vane anemometer with vane diameter of 14 mm that measure wind velocities at the range of ~0–30 m s^{-1} with reading resolution of 0.1 m s^{-1} (KIMO vt 200, Ontario, Canada). The data are logged at time interval of 5 s for each test. The data were analyzed to determine the height of the boundary layer, the average wind velocity in the boundary layer, and the wind shear velocity (u_*). The wind shear velocity is expressed by the Prandtl–von Karman equation [8]:

$$\frac{u}{u_*} = \frac{1}{K}\ln\left(\frac{z}{z_0}\right) \tag{1}$$

where u is the wind velocity (m s^{-1}) at height z (m), z_0 is the aerodynamic roughness length of the surface (m), u_* is the shear velocity (m s^{-1}), and K is von Karman's constant (≈0.4).

Each soil sample was tested under all the fan frequencies/wind velocities. Overall 108 tests were conducted in this study (4 soil samples, 9 wind velocities, 3 replicas). Each run was last 30 s to record the trend of sand transport and dust emission. A series of traps were placed in the tunnel along the wind direction. The transported particles were collected at height of 0–0.10 m above the tunnel bed. After the experiment, the sediments in the traps were analyzed for the relative contents of sand and dust (<63 µm) fractions. In addition, the mass of the transported particles calculated as saltation flux (Q, kg m^{-1} s^{-1}) that cross a width of 1 m [15].

Dust concentrations (µg m^{-3}) PM10 were recorded by a light-scattering device, DustTrak DRX 8534 (TSI, Shoreview, MN, USA), in the range of 0.001–150 µg m^{-3} (±0.1% of reading) at sec-1 intervals, was placed at 15 cm above the tunnel bed. The recorded PM10 concentrations were converted into mass flux (Fpm) emitted from the soil surface (kg m^{-2} s^{-1}) based on the wind tunnel dimensions and the area of the sand bed [15]:

$$F_{PM} = (C_{PM} - C_{bg}) \times V_{air} \times A_{cs}/A_p \tag{2}$$

where C_{PM} is the recorded PM concentrations (kg m^{-3}) from the soil, C_{bg} is the average PM background concentration (kg m^{-3}), V_{air} is the mean horizontal wind velocity in m s^{-1} (average over height), A_{cs} is the cross-section (height × width) of the wind tunnel (m^2), and A_p is the area (m^2) of the experimental plot (length × width).

The recorded PM10 flux (kg m^{-2} s^{-1}) and sand fluxes (kg m^{-1} s^{-1}) were used to calculate the sandblasting efficiency a (m^{-1}), which is the vertical dust flux produced by a unit horizontal sand saltation flux, and is an important property to inform the dust emission by saltation process [23]:

$$a = F_{PM}/Q \quad (3)$$

where F_{PM} is the average PM10 vertical mass flux (kg m^{-2} s^{-1}), and Q (kg m^{-1} s^{-1}) is the averaged horizontal sand flux integrated over the hole experiment time for all sand grain sizes.

3. Results

3.1. Wind Profile

The profiles of the wind measured in the wind tunnel under all fan frequencies are characterized by a logarithmic increase in wind velocity of each fan frequency as an increase in height (Figure 2a). The wind shear velocity (u_*) for each wind profile was calculated by the von Karman's equation (Section 2.2). The shear velocities are at the range of 0.17 to 0.56 m s^{-1} (Figure 3). This include shear velocities that are below the saltation threshold (<0.28 m s^{-1}) in laboratory and field experiments [15,23]. Then, the wind shear velocity was predicted by the fan frequency (R^2 = 0.98) (Figure 2b).

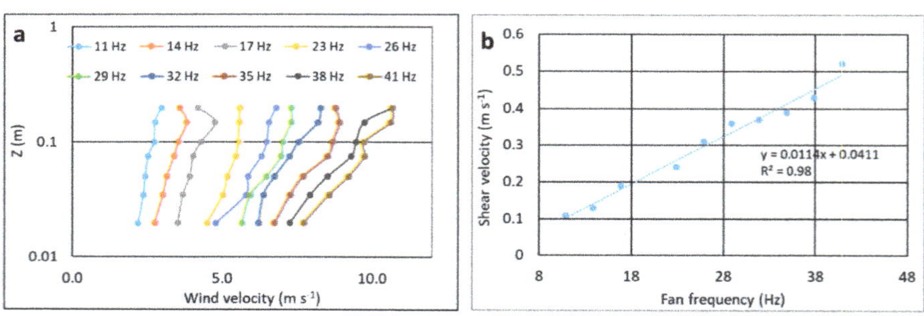

Figure 2. (**a**) Wind velocity (m s^{-1}) measured at different heights above the tunnel bed (z) under different fan frequencies (Hz) (**b**) Correlation between the fan frequencies applied in the wind tunnel and the calculated shear velocity (u_*).

3.2. PM10 Concentration

The PM10 concentrations resulted from dust emission are presented in Figure 3. With the sand dune sample, no dust concentration, which is above the background value (30 µg m^{-3}), was recorded in all wind velocities. The background value was determined as the maximum atmospheric PM10 concentration in the wind tunnel before the experiment. With all the soil samples, there is a trend of increase in the dust concentrations as the shear velocity increases. The dust emission in the "sand 90%" sample resulted in a range of PM10 concentrations from 30 µg m^{-3} (background value) to a maximum value of 12,856 µg m^{-3} under the shear velocity of 0.52 m s^{-1}. At low wind shear velocities (<0.24 m s^{-1}), no distinct PM10 concentration was recorded. In the sample "sand 75%", only a very low PM10 concentration (44 µg m^{-3}) was recorded at 0.19 m s^{-1}. As in the 90% sand sample, the significant increase in concentration was obtained in the transition from 0.31 to 0.39 m s^{-1}, in which the mean PM concentration was raised from 944 to 6429 µg m^{-3}, which is higher than at 90% sample. The highest mean concentration (14,309 µg m^{-3}) was recorded at 0.52 m s^{-1}, almost two times higher compared with the "sand 90%" sample under the same shear velocity. In the sample "sand 50%", no PM10 concentration was recorded at the low wind shear velocities as in the other samples. The values in this sample run from 30 μg m^{-3} (background value) to a maximum value of 43,266 μg m^{-3}. At shear velocities of 0.31 to 0.39 m s^{-1} the mean concentration increased from 1070 to 5404 μg m^{-3} which is at similar range to the sample of "sand 75%." The highest mean concentration (29,073 μg m^{-3}) was recorded at 0.52 m s^{-1}, which is a significantly higher value than the one obtained in the "sand 75%."

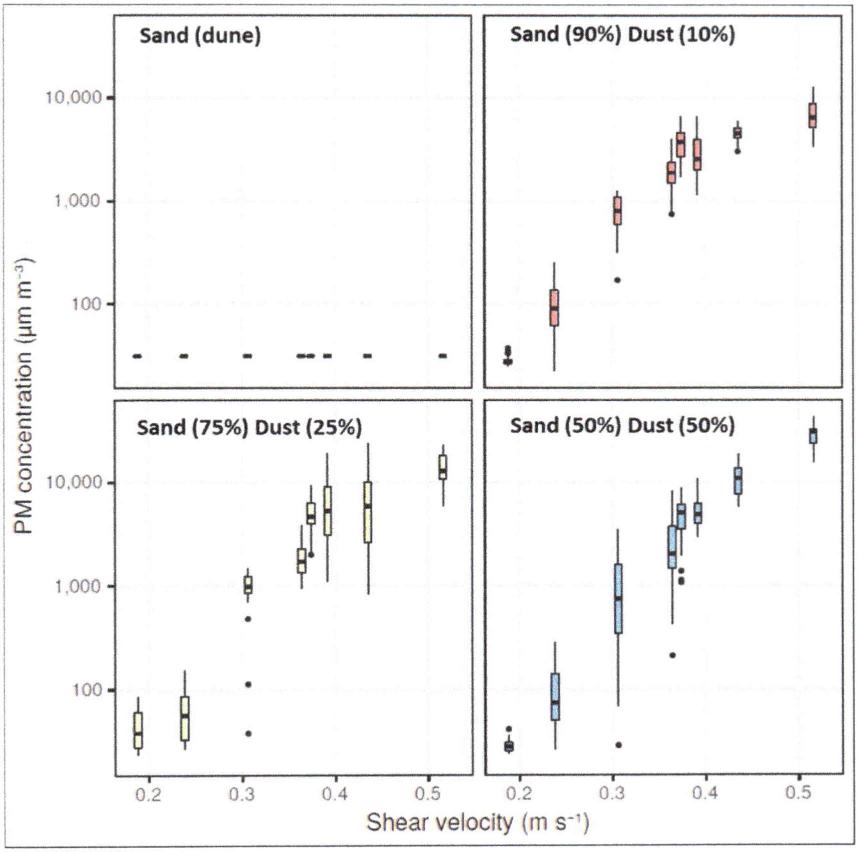

Figure 3. Particulate matter (PM) concentrations measured in the wind tunnel as a result of dust emission from the reference samples under various shear velocities. The reference soils are presented by the weight ratio of clay and silt and sand 10%, 25%, and 50%. No dust concentration was recorded for the sand (dune) sample. The average concentration is presented by the solid line inside the minimum-maximum box, and standard deviations are in bar lines. Note that the black dots represent extreme values.

3.3. Saltation Mass

The saltation mass measured during the experiment is presented in Figure 4. In most cases of the experiment (total of 120 runs), a vast amount of the material was collected only in the lower traps close to the tunnel bed (<0.05 m). The saltation threshold at all sand samples is at $u_* = 0.31$ m s^{-1} as no distinct saltation mass was recorded in the lower shear velocities (0.11–0.24 m s^{-1}). The saltation mass in the "sand dune" sample resulted in a range of 0.63 g (at the saltation threshold) to a maximum value of 8.14 g at shear velocity of 0.52 m s^{-1} during the a 30-s run of the experimental procedure. In the "sand 90%" sample, the saltation mass run from 0.48 g at 0.31 m s^{-1} to a maximum mass of 6.07 g at shear velocity of 0.52 m s^{-1}. These values are significantly lower than the "sand dune" sample at the same shear velocities. Unlike the 100% sand sample, the dust fraction received in the saltation mass was 4–36% (Table 1). The saltation mass of the "sand 75%" sample was much lower than the "sand 90%" sample, and ranged from 0.17 to 4.79 g. From wind shear velocity of 0.31 to 0.43 m s^{-1}, the mass increased only 10%, which is at the same level (7%) as in the sample of "sand 50%". The maximum value of the dust fraction in the "sand 75%" sample was measured at a wind velocity of 0.39 m s^{-1}. The maximum amount of the dust fraction in the saltation (26%) was measured from wind velocity of 0.39 to 0.52 m s^{-1} (Table 1).

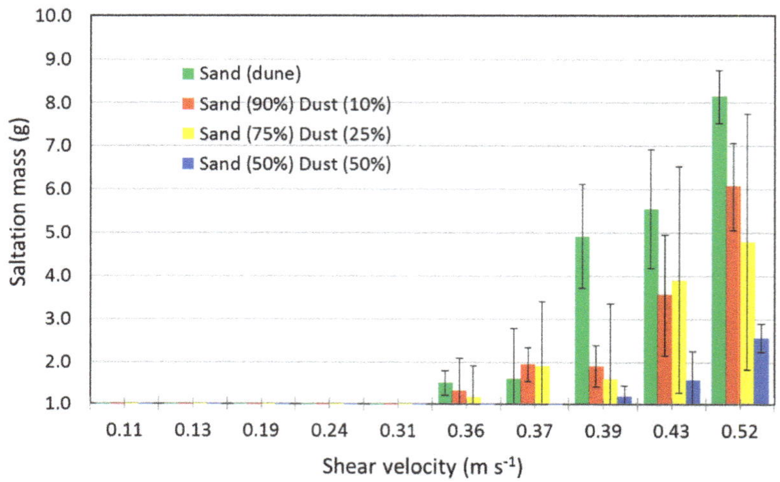

Figure 4. Saltation mass from the wind tunnel measurements of the reference samples under various shear velocities. Note that no distinct saltation mass was recorded in shear velocity below 0.31 m s^{-1}.

Table 1. The dust fraction received in the saltation mass calculated as the class weight of the dust particles (<63 μm) in the total weight of the saltation: [1 − (dust (g)/sand (g) + dust (g)]. Note that no saltation was recorded in shear velocity below 0.31 m s^{-1}.

Shear Velocity (m s^{-1})	Sand (Dune)	Sand (90%) Dust (10%)	Sand (75%) Dust (25%)	Sand (50%) Dust (50%)
0.11	N/A	N/A	N/A	N/A
0.13	N/A	N/A	N/A	N/A
0.19	N/A	N/A	N/A	N/A
0.24	N/A	N/A	N/A	N/A
0.31	0	0.04	0.12	0.18
0.36	0	0.36	0.19	0.22
0.37	0	0.08	0.19	0.17
0.39	0	0.11	0.34	0.26
0.43	0.01	0.11	0.15	0.25
0.52	0	0.11	0.18	0.26

The calculated saltation flux refers to the mass of all the particles; sand, silt, and clay, that pass a specific distance over the surface (Section 2.2, Equation (2)). With all soil samples, the saltation flux was started at shear velocity 0.31 m s^{-1} as shown in the record of the saltation mass (Figure 4).

The calculated PM10 flux refers to transported particles from the surface (Section 2.2, Equation (3)). The results show a linear positive correlation between the saltation flux and the PM10 flux (Figure 5) in all soil samples. As the amount of dust in the sample decreases, both the saltation flux and the PM10 emission decrease. The highest correlation was recorded in the "sand 90%" ($R^2 = 0.99$), while the (relatively) lowest correlation was recorded in the sample of "sand 50%" ($R^2 = 0.91$). Since no dust was recorded in the sand dune sample, a correlation with saltation is not presented.

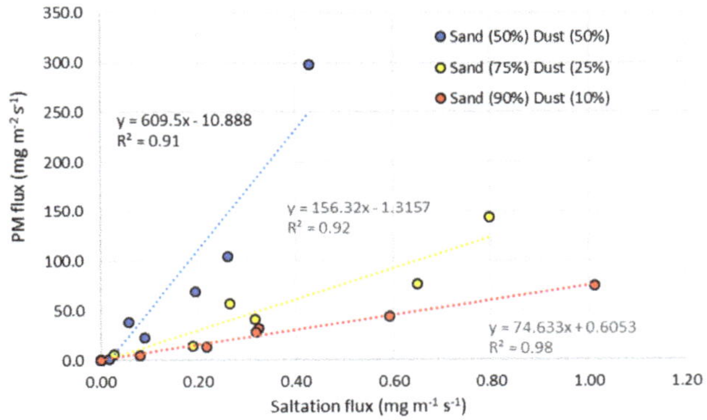

Figure 5. Correlation between saltation flux and PM10 (PM that is less than 10 micrometers in diameter) flux calculated for each soil sample under all the shear velocities applied in the wind tunnel.

3.4. Sandblasting Efficiency

Sandblasting efficiency is the dust flux produced by a unit horizontal of saltation flux (Section 2, Equation (3)). The sandblasting efficiency calculated for the various wind shear velocities is presented in Figure 6. In general, as the amount of dust in the sample increases, the sandblasting efficiency increases as well. The highest values of sandblasting were obtained in the sample of "sand 50%". The values ranged from 0.25 m^{-1} at a shear velocity of 0.36 m s^{-1} to 0.7 m^{-1} at a shear velocity of 0.52 m s^{-1}. Since no dust was recorded from the sample of sand dune in this study, the sandblasting efficiency was not calculated.

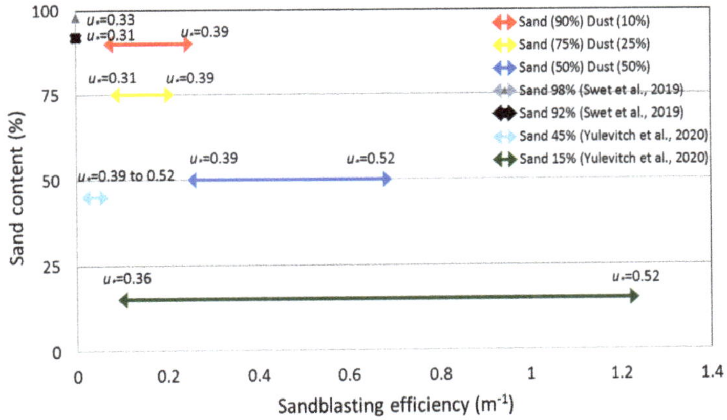

Figure 6. Sandblasting efficiency calculated for soils with different content of sand under a range of wind velocities. In addition, the chart shows the sandblasting efficiency of previously published studies of samples with different sand content [15,16].

4. Discussion

The soil samples tested in this study (Figure 1) represent a range of soil texture in dust sources of sandy and non-sandy soils. The most productive areas for dust are located mainly in arid regions, such as playas (dry lakes) in the Mojave Desert with about 35% sand content [24]. In the Middle-East and North Africa [25–27], the source areas are characterized by a silt-loam (loess) soil with a clay fraction up to 30% [28], including loess soils in the northern Negev that are subjected to intensive agricultural activities [12,29]. Major dust sources in the Chinese and the Mongolian deserts are identified with a range of 0.7–11.9% clay, 1.8–34.1% silt, and 2.3–53.0% sand [30]. Unlike common dust sources, sand dunes have a low content of silt and clay (~2%). However, recently they were considered to be dust sources [15,23] and as such a sample of sand dune was tested in this study.

The results of the dust PM show an increase concentration as the amount of the clay-silt fraction in the sample is higher (Figure 3). Yulevitch et al. [16] have presented a similar trend with highest PM10 flux for a sample with clay-silt content of 85%. Basically, the increase of PM10 concentration under such conditions is related to the higher availability of dust content in the sample. Since the amount of dust particles in the soil samples "sand 90%" and "sand 75%" is relatively low, it can be assumed that the dust emission at velocity > 0.4 m s^{-1} could have been an underestimation for such soils with unlimited dust-particle supply [10,16]. In the sample of "sand 50%;" however, the highest PM10 concentration was recorded in the strongest shear velocity (0.52 m s^{-1}), which indicate on sufficient dust content in the sample for continues emission. the distinct dust threshold for all samples is at 0.24 m s^{-1}, while in sample "sand 75%" a PM concentration was recorder already at 0.19 m s^{-1}.

As for the "sand dune" sample, no dust concentration was recorded at all shear velocities. Though, recent studies have shown the potential of dust generation from sand. Huang et al. [23] showed that under wind shear velocity of 0.3 m s^{-1} the PM10 fluxes in the Oceano sand is smaller than those fluxes from most non-sandy soils (~1 and 150 μg m^{-2} s^{-1}, respectively). Swet et al. [15] recorded PM10 concentration of 0.01 μg m^{-3} from sand dune with <2% clay content under wind shear velocity of 0.3 m s^{-1} in the wind tunnel experiment. They concluded that the dominant dust emission mechanism from sand dunes is clay coatings removal, with a relatively small contribution from re-emission of loose-settled dust. In sands containing higher amounts of dust-sized particles, the relative contribution of the re-emission mechanism increases drastically. It is reasonable to assume that in our study there was a dust emission from the "sand" sample by the clay-coating removal. However, our measurement resolution in this study for samples with <10% dust content were probably not sensitive enough to distinct PM10 concentration values that are significantly above the background value of the experiment (30 μg m^{-3}).

Dust emission is significantly associated with the saltation process. In all the samples of this study, a distinct saltation starts at shear velocity of 0.31 m s^{-1} (Figure 4). A saltation threshold in such velocity was obtained in many studies on sand dunes [8] as well as the trend of increasing sand flux with the shear velocity (Figure 4). The positive linear-correlation between the saltation flux and the PM10 flux was relatively high (R^2 = 0.91–0.99) in all the non-sandy samples (Figure 5). Strong correlations between saltation and dust flux were found also in previews studies on dust emission from non-sandy soils by wind tunnels [17,31,32]. Nonetheless, Sweeney and Mason [33] show that in a Pleistocene loess deposits (Nebraska, USA) the dust may emit without saltation with a lower threshold (u_* = 0.26 to 0.30 m s^{-1}) than the saltation threshold (u_* = 0.33 to 0.44 m s^{-1}).

To further test the relationships between saltation and dust emission, the sandblasting efficiency was examined with comparison to other works (Figure 6). Basically, the sandblasting efficiency increases with the shear velocity as long as there is available dust particle in the soil for emission. The highest values of sandblasting in this study were obtained in the sample of "sand 50%", which contains enough amount of dust for emission. When comparing the sandblasting efficiency values with other works conducted in wind tunnel experiment, it revealed that the sandblasting efficiency increases as a function of the amount of dust in the sample and wind velocity. The highest values were recorded at the sample of "15% sand" (85% dust), 0.10 to 1.3 m^{-1}, under velocities of 0.39 and 0.52 m s^{-1}, respectively [16]. The sandblasting efficiency of the "sand 90%" sample of this study is not presented since no dust was recorded from the sample. However, Swet et al. [15] calculated a sandblasting efficiency of 1.73×10^{-7} m^{-1} from active sand dune at 0.30 m s^{-1}.

Wind tunnels enabled targeted experiments on dust emission processes, in which the empirical data can be used for parametrization of dust models and for validation of model results. Yet, there is little quantitative understanding of how the wide range of particle sizes in realistic dust-emitting soils affects the shear velocity threshold. The dust emission threshold parameterizations used in most models are descriptions of the threshold friction velocity of a dry, bare soil composed of particles of a single grain size. Consequently, dust emissions in models generally increase monotonically with a soil's clay content, and thus do not capture the potentially important mitigating impact of clay particles in raising the threshold friction velocity. Shao and Lu [10] parameterized the threshold as the wind velocity at which the aerodynamic torque on a surface particle exceeds that generated by gravity and the interparticle forces with surrounding sand particles. An improved dust emission scheme incorporates saltation bombardment and aggregates disintegration [34]. The statement of the scheme is that dust emission is proportional to streamwise saltation flux, but the proportionality depends on soil texture. A critical problem in dust research is to estimate size-resolved dust emission rates, which requires reliable dust emission estimates for specific size ranges [35]. Thus, wind tunnel experiments can be used to expand the dust formulation by accounting for the entire particle distribution of the soil (and not just the presence of clay), and accounting for the variability of sand (saltation) particle sizes.

This study has some limitations that are discussed as follow. First, we applied four soil samples, which may not represent the entire range of soil texture. However, these samples are common in many soils that are associated with dust emission processes. Second, the use of a wind tunnel to examine the dust emission may not reflect the natural wind in the field where the velocity and direction are constantly changing. However, the controlled wind velocities by the tunnel makes it possible to examine the emission thresholds of various soil samples. Third, the saltation and dust fluxes calculated in this study by the wind tunnel experiment do not necessarily fit the "real" values measured in the field, but it does allow to a quantitative comparison of sandblasting efficiency.

5. Conclusions

This study examined empirically the dust emission thresholds in loess soils with different content of sand-sized particles. The significance of this study was to provide empirical information on the link between soil particle size distribution and saltation and dust emission thresholds. The results of the wind tunnel experiment show that dust PM10 emission threshold is strongly associated with the saltation threshold. As such, the direct aerodynamic lifting of loose dust particle was not distinct in this study. Under conditions of dust emission: (i) the dust flux increases as the amount of the clay-silt fraction in the soil is higher, (ii) there is a logarithmic increase of dust flux with the of the shear velocity, (iii) the rate of the dust flux under certain saltation is subjected to the available dust amount in the soil during a wind event, and (iv) in soil with non-limited dust supply in the soil, the maximum rate of the dust flux is limited to the capacity of the saltation flux under certain shear velocity. The results of the sandblasting efficiency highlight the significance of the soil disturbance and aggregate disintegration in increasing dust emission from loess soils. The combination of loose sand particles in soils containing distinct amount of dust particles (>10%) is favor for dust emission. The findings of this study can be processed into parametrization in dust emission models. They can also support management strategy of soils in preventing dust emission by keeping the soil aggregation.

Author Contributions: Conceptualization, I.K.; methodology, I.K. and A.R.; formal analysis, A.R.; writing—original draft preparation, A.R.; writing—review and editing, A.R., M.B.-H. and I.K.; funding acquisition, I.K. All authors have read and agreed to the published version of the manuscript.

Funding: This research received no external funding.

Conflicts of Interest: The authors declare no conflict of interest.

References

1. Kok, J.F.; Ridley, D.A.; Zhou, Q.; Miller, R.; Zhao, C.; Heald, C.L.; Ward, D.S.; Albani, S.; Haustein, K. Smaller desert dust cooling effect estimated from analysis of dust size and abundance. *Nat. Geosci.* **2017**, *10*, 274–278. [CrossRef] [PubMed]
2. Katra, I.; Gross, A.; Swet, N.; Tanner, S.; Krasnov, H.; Angert, A. Substantial dust loss of bioavailable phosphorus from agricultural soils. *Sci. Rep.* **2016**, *6*, 24736. [CrossRef] [PubMed]
3. Yitshak-Sade, M.; Novack, V.; Katra, I.; Gorodischer, R.; Tal, A.; Novack, L. Non-anthropogenic dust exposure and asthma medication purchase in children. *Eur. Respir. J.* **2014**, *45*, 652–660. [CrossRef] [PubMed]
4. Krasnov, H.; Katra, I.; Novack, V.; Vodonos, A.; Friger, M.D. Increased indoor PM concentrations controlled by atmospheric dust events and urban factors. *Build. Environ.* **2015**, *87*, 169–176. [CrossRef]
5. Klose, M.; Shao, Y.; Li, X.; Zhang, H.; Ishizuka, M.; Mikami, M.; Leys, J. Further development of a parameterization for convective turbulent dust emission and evaluation based on field observations. *J. Geophys. Res. Atmos.* **2014**, *119*, 10441–10457. [CrossRef]
6. Kok, J.F.; Mahowald, N.; Fratini, G.; Gillies, J.A.; Ishizuka, M.; Leys, J.F.; Mikami, M.; Park, M.-S.; Park, S.-U.; Van Pelt, R.S.; et al. An improved dust emission model—Part 1: Model description and comparison against measurements. *Atmos. Chem. Phys. Discuss.* **2014**, *14*, 13023–13041. [CrossRef]
7. Shao, Y.; Raupach, M.R.; Findlater, P.A. Effect of saltation bombardment on the entrainment of dust by wind. *J. Geophys. Res. Space Phys.* **1993**, *98*, 12719. [CrossRef]
8. Kok, J.F.; Parteli, E.J.; Michaels, T.i.; Karam, D.B. The physics of wind-blown sand and dust. *Rep. Prog. Phys.* **2012**, *75*, 106901. [CrossRef]
9. Bagnold, R.A. *The Physics of Blown Sand and Desert Dunes*; Methuen and Company Limited: London, UK, 1941.
10. Shao, Y.; Lü, H. A simple expression for wind erosion threshold friction velocity. *J. Geophys. Res. Space Phys.* **2000**, *105*, 22437–22443. [CrossRef]
11. Swet, N.; Katra, I. Reduction in soil aggregation in response to dust emission processes. *Geomorphology* **2016**, *268*, 177–183. [CrossRef]
12. Katra, I. Soil Erosion by Wind and Dust Emission in Semi-Arid Soils Due to Agricultural Activities. *Agronomy* **2020**, *10*, 89. [CrossRef]
13. Ben-Hur, M.; Agassi, M. Predicting interrill erodibility factor from measured infiltration rate. *Water Resour. Res.* **1997**, *33*, 2409–2415. [CrossRef]
14. Roskin, J.; Katra, I.; Blumberg, D.G. Particle-size fractionation of eolian sand along the Sinai-Negev erg of Egypt and Israel. *GSA Bull.* **2013**, *126*, 47–65. [CrossRef]
15. Swet, N.; Elperin, T.; Kok, J.F.; Martin, R.L.; Yizhaq, H.; Katra, I. Can active sands generate dust particles by wind-induced processes? *Earth Planet. Sci. Lett.* **2019**, *506*, 371–380. [CrossRef]
16. Yulevitch, G.; Danon, M.; Krasovitov, B.; Fominykh, A.; Swet, N.; Tsesarsky, M.; Katra, I. Evaluation of wind-induced dust-PM emission from unpaved roads varying in silt content by experimental results. *Atmos. Pollut. Res.* **2020**, *11*, 261–268. [CrossRef]
17. Tanner, S.; Katra, I.; Haim, A.; Zaady, E. Short-term soil loss by eolian erosion in response to different rain-fed agricultural practices. *Soil Tillage Res.* **2016**, *155*, 149–156. [CrossRef]
18. Katra, I. Comparison of Diverse Dust Control Products in Wind-Induced Dust Emission from Unpaved Roads. *Appl. Sci.* **2019**, *9*, 5204. [CrossRef]
19. Sharratt, B.; Wendling, L.; Feng, G. Windblown dust affected by tillage intensity during summer fallow. *Aeolian Res.* **2010**, *2*, 129–134. [CrossRef]
20. Singh, P.; Sharratt, B.; Schillinger, W.F. Wind erosion and PM10 emission affected by tillage systems in the world's driest rainfed wheat region. *Soil Tillage Res.* **2012**, *124*, 219–225. [CrossRef]
21. Van Pelt, R.S.; Baddock, M.; Zobeck, T.M.; Schlegel, A.J.; Vigil, M.F.; Acosta-Martinez, V. Field wind tunnel testing of two silt loam soils on the North American Central High Plains. *Aeolian Res.* **2013**, *10*, 53–59. [CrossRef]
22. Asensio, C.; Lozano, F.J.; Gallardo, P.; Giménez-Fernández, A. Soil wind erosion in ecological olive trees in the Tabernas desert (Southeastern Spain): A wind tunnel experiment. *Solid Earth* **2016**, *7*, 1233–1242. [CrossRef]
23. Huang, Y.; Kok, J.F.; Martin, R.L.; Swet, N.; Katra, I.; Gill, T.E.; Reynolds, R.L.; Freire, L.S. Fine dust emissions from active sands at coastal Oceano Dunes, California. *Atmos. Chem. Phys. Discuss.* **2019**, *19*, 2947–2964. [CrossRef]
24. Katra, I.; Lancaster, N. Surface-sediment dynamics in a dust source from spaceborne multispectral thermal infrared data. *Remote Sens. Environ.* **2008**, *112*, 3212–3221. [CrossRef]

25. Ginoux, P.; Prospero, J.M.; Gill, T.E.; Hsu, N.C.; Zhao, M. Global-scale attribution of anthropogenic and natural dust sources and their emission rates based on MODIS Deep Blue aerosol products. *Rev. Geophys.* **2012**, *50*. [CrossRef]
26. Prospero, J.M.; Nicholson, S.E.; Ginoux, P.; Torres, O.; Gill, T.E. Environmental characterization of global sources of atmospheric soil dust identified with the NIMBUS 7 Total Ozone Mapping Spectrometer (TOMS) absorbing aerosol product. *Rev. Geophys.* **2002**, *40*. [CrossRef]
27. Washington, R.; Todd, M.; Middleton, N.J.; Goudie, A. Dust-Storm Source Areas Determined by the Total Ozone Monitoring Spectrometer and Surface Observations. *Ann. Assoc. Am. Geogr.* **2003**, *93*, 297–313. [CrossRef]
28. Gherboudj, I.; Beegum, S.N.; Marticorena, B.; Ghedira, H. Dust emission parameterization scheme over the MENA region: Sensitivity analysis to soil moisture and soil texture. *J. Geophys. Res. Atmos.* **2015**, *120*, 10915–10938. [CrossRef]
29. Shalom, O.; Crouvi, O.; Enzel, Y.; Rosenfeld, D. Locally recycled late Pleistocene loess feeds modern dust storms at the desert margins of the eastern Mediterranean, Israel. *Aeolian Res.* **2020**, *46*, 100612. [CrossRef]
30. Laurent, B.; Marticorena, B.; Bergametti, G.; Mei, F. Modeling mineral dust emissions from Chinese and Mongolian deserts. *Glob. Planet. Chang.* **2006**, *52*, 121–141. [CrossRef]
31. Houser, C.A.; Nickling, W.G. The emission and vertical flux of particulate matter. *Sedimentology* **2001**, *48*, 255–267.
32. Gelbart, G.; Katra, I. Dependence of the dust emission on the aggregate sizes in loess soils. *Appl. Sci.* **2020**, *10*, 5410. [CrossRef]
33. Sweeney, M.; Mason, J.A. Mechanisms of dust emission from Pleistocene loess deposits, Nebraska, USA. *J. Geophys. Res. Earth Surf.* **2013**, *118*, 1460–1471. [CrossRef]
34. Shao, Y. Simplification of a dust emission scheme and comparison with data. *J. Geophys. Res. Space Phys.* **2004**, *109*. [CrossRef]
35. Shao, Y.; Ishizuka, M.; Mikami, M.; Leys, J.F. Parameterization of size-resolved dust emission and validation with measurements. *J. Geophys. Res. Space Phys.* **2011**, *116*. [CrossRef]

MDPI
St. Alban-Anlage 66
4052 Basel
Switzerland
Tel. +41 61 683 77 34
Fax +41 61 302 89 18
www.mdpi.com

Applied Sciences Editorial Office
E-mail: applsci@mdpi.com
www.mdpi.com/journal/applsci

www.ingramcontent.com/pod-product-compliance
Lightning Source LLC
Chambersburg PA
CBHW061510210926
51850CB00008B/1869
* 9 7 8 3 0 3 6 5 6 7 6 4 8 *